Lakes and Water Management

Developments in Hydrobiology 7

Series editor
H. J. Dumont

DR W. JUNK PUBLISHERS   THE HAGUE–BOSTON–LONDON   1982

Lakes and Water Management

Proceedings of the 30 Years Jubilee Symposium of the Finnish Limnological Society, held in Helsinki, Finland, 22–23 September 1980

Edited by
V. Ilmavirta, R. I. Jones and P.-E. Persson

Reprinted from *Hydrobiologia*, vol. 86, no. 1/2 (1982)

DR W. JUNK PUBLISHERS   THE HAGUE–BOSTON–LONDON   1982

Distributors:

for the United States and Canada

Kluwer Boston, Inc.
190 Old Derby Street
Hingham, MA 02043
U.S.A.

for all other countries

Kluwer Academic Publishers Group
Distribution Center
P.O. Box 322
3300 AH Dordrecht
The Netherlands

Library of Congress Cataloging in Publication Data
Main entry under title:

Lakes and water management.

    (Developments in hydrobiology ; 7)
    "Reprinted from Hydrobiologia, vol. 86, no. 1/2
(1982)"
    1. Limnology--Congresses. 2. Water quality
management--Congresses. I. Ilmavirta, V.
II. Jones, R. I. III. Persson, Per-Edvin.
IV. Suomen limnologinen yhdistys. V. Series.
QH96.A3L34 1982   574.5'26322   81-20856
                                    AACR2

ISBN-13: 978-94-009-8005-1   e-ISBN-13: 978-94-009-8003-7
DOI: 10.1007/978-94-009-8003-7

Cover design: Max Velthuijs

Preface

Multiple use of natural waters and watersheds poses many practical problems, some of which are clearly limnological, relating for example to carrying capacities of lakes, deterioration of water quality and fisheries management. It is thus important to consider limnological aspects when making decisions in the field of water management. To further this attitude the Finnish Limnological Society considered 'Lakes and Water Management' to be a suitable theme for the Society's Jubilee Symposium, held in Helsinki on 22–23 September, 1980 at the Viikki Campus of the University of Helsinki, to mark the Society's 30th Anniversary.

The Finnish Limnological Society has arranged national limnological symposia every one or two years. The papers presented at these symposia have been published in the series 'Limnologisymposion' (in Finnish or Swedish, with short summaries in an internationally spoken language). Due to financial difficulties, the series was terminated as of the symposium in 1979. When preparing the 30th Anniversary of the Finnish Limnological Society, the Governing Board of the Society decided to arrange an international symposium on a theme relevant to modern limnology both in Finland and elsewhere. The selected results of the successful meeting on lakes and water management, which was attended by 220 people from 7 countries, are presented in this volume.

In his opening speech, the Chairman of the Finnish Limnological Society, Prof. Reino Ryhänen, emphasized the need for more limnology in practical water management. Similar thoughts were evident in the address of Mr. Reino Uronen, Finnish Deputy Minister of Agriculture and Forestry, who brought the good wishes of the Finnish Government to the Symposium. The opening ceremony of the Symposium was closed by Prof. Emer. Hans Luther, now an Honorary Member of the Society, who reviewed the history of limnology in Finland. The Symposium working sessions then continued for two full days, during which 47 papers were presented and discussed.

It is the hope of the Editors that the present volume contributes towards understanding limnological problems in the management of natural waters. At the same time, it may to some extent be considered a contribution to regional limnology.

Contents

PART ONE: STRUCTURE AND FUNCTION OF LAKE ECOSYSTEMS

1.1. Unpolluted waters

1.2 Eutrophic and polluted waters

VIII

PART TWO: MODELLING

PART THREE: WATER MANAGEMENT

3.1. Water quality control

3.2. Restoration of lakes

3.3. Fisheries

The first half century of limnology in Finland*

Hans Luther

Department of Botany of the University, Unionsgatan 44, SF-00170 Helsingfors 17, Finland

Keywords: Finland, limnology, historical, Järnefelt, rockpools, saproby, lake types, regionality

Abstract

The appearance of the first ecosystem-oriented contribution to hydrobiology is taken as the criterion for the national incipience of limnology. In Finland this is Levander's (1900) classification of archipelago fringe rock pools of shores of the northern Baltic Sea. The first contribution to applied limnology is Häyrén's (1921) saproby classification of the polluted brackish water area off Helsinki, based on plant communities. Freshwater contributions to limnology were scattered and less important in an initial period, from 1898 to the early 1920s. The following 40 years are characterized as the Järnefelt period, during which he published lake descriptions from all parts of Finland, analysed the limnological regions and contributed to aspects of the trophic subdivision of lake types. Other subdivisions were made by Maristo (1941), based on macrophyte vegetation and, for the Åland islands, by P. Palmgren (1936) based on waterfowl populations and by Cedercreutz (1937) based on botanical criteria. Recent limnology has been characterized by teamwork around selected themes, such as energy flow through lake ecosystems, humic materials in freshwater and the composition of brackish water littoral benthos biocoenoses.

Introduction

Compared with other natural sciences, limnology is a young science. When looking for the early beginnings of limnology we have to consider what definition of limnology we should use as our yardstick. The Finnish basic textbook in limnology, 'Vesiemme luonnontalous', written by Heikki Järnefelt (1958a) refers to a definition of 1922, when international collaboration in limnology started. It defines limnology as a science, including all matters regarding inland waters, without any limitation. Such a definition might be applied to a science which has already established its basis but it cannot be used in the search for the early roots of the science in question.

In fact, rather numerous observations on topo-graphical and geophysical hydrology were already available from Finland more than a century ago, but as a rule they contained no biological components so they cannot be considered as representatives of an early limnology. A good review of them has recently been given by Simojoki (1978).

We have to look for a definition which better reflects the basic ideas of limnology as a science of its own. When I began my studies in biology, nearly 50 years ago, it was said that limnology is a science synthesising the analytical results of hydrobiology and non-biological aquatic science. Such a definition arose because the ecosystem concept was not yet established among terrestrial ecologists. In fact, the concept of ecosystems, the relations and interactions between biotic systems and their abiotic environment, was developed in limnology more than half a century before terrestrial ecology was ripe for this stage of ecology, except, to some

---

* Opening address to the symposium.

Hydrobiologia 86, 1–7 (1982). 0018-8158/82/0861–0001/$01.40.

extent, the forest type concept of A. K. Cajander (1909).

As conceived here, limnology does not include such related sciences as, for instance, specialized branches of hydrology, hydrobiology or fish and fishery science.

The real pioneer for this ecosystem-oriented aquatic science was the Swiss scientist F.-A. Forel, who between 1877 and 1901 published successively more complete editions of his monograph on Lac Léman (Lake Geneva). He also adopted the term limnology (Berg 1951). His contemporary was the American Stephen A. Forbes, who in 1887 published his pamphlet 'The lake as a microcosm'. At the beginning of our century they were followed by, among others, the Dane Carl Wesenberg-Lund. In the following decades E. A. Birge and C. Juday in the USA, the German August Thienemann and the Swede Einar Naumann contributed much towards the knowledge and a consolidation of the concepts and definitions of limnology. Thus an international uniformity in research was achieved. This is the international context in which the achievements in Finnish limnology should be judged.

Brackish water limnology

Hydrobiology teaching and research needs equipment which cannot easily be carried around on excursions, except nowadays in mobile laboratories. At our university, then the only one in Finland, hydrobiology was, until 1888, taught only indoors and during university terms. On the initiative of Johan Axel Palmén, Professor of Zoology, and at his own expense, a modest hydrobiological summer laboratory was maintained on Esbo-Lövö (Espoon Lehtisaari) from 1889 to 1899. Palmén himself was not a hydrobiologist but he clearly perceived the need to move hydrobiological teaching and research out from the city laboratories. It was Palmén's assistant, the young Kaarlo Mainio Levander, who was the active leader in research and teaching. In 1900 he published 'Zur Kenntnis des Lebens in den stehenden Klein-gewässern auf den Schäreninseln', which is the first real contribution to limnology in Finland. It deals with more than 50 rock pools in the Esbo-Lövö area, studied over a couple of years, and uses a distinct classification of the rock pools according to the influence of the brackish water of the Baltic Sea

(from surge, splash and spray up to no influence), of rain and melting snow (even of evaporation in dry summer periods), of succession of amphibious and terrestrial mosses and, finally, of peat vegetation. Levander's rock pool classification has remained the basic division of the small basins on coastal rocky outcrops in the Fennoscandian area of Archaean bedrock. His treatment is well balanced in the description of the interactions between abiotic and biotic factors in these, nature's own experimental aquaria. Levander was with this contribution contemporaneous with Wesenberg-Lund as one of the early real limnologists and he is the too often forgotten pioneer in the classicification of small water units.

Palmén could not purchase land at Esbo-Lövö for the establishment of a permanent biological station – fortunately, since this area is now among the most severely polluted in the surroundings of Helsinki/Helsingfors. Instead of this, Palmén purchased a land and water area at Tvärminne on the Hangö peninsula (Palmén in A. Luther, 1957) and in 1902 erected there the Tvärminne zoological station, left in his will to the University of Helsinki, which still maintains the station.

Rock pool research has been continued there (e.g. Järnefelt 1939, 1940b) as well as work on the brackish water of the archipelago and the already classical Pojoviken area of gradually diminished salinity. There the first investigation was made nearly 70 years ago, 1911–1912, as a teamwork of the hydrographic-biological research unit mentioned below (Buch 1914; Granqvist 1914; Witting 1914; Levander 1915).

Levander's other brackish water publications are of faunistic or taxonomic content. As Professor of Zoology 1910–1935 he continued as an active teacher in hydrobiology for generations of biology students. Simultaneously he worked as zoologist at the Hydrographic-Biological Research Unit, administrated by the Societas Scientiarum Fennica. In 1919 the hydrographical part was converted into the governmental Institute of Marine Research, and Levander was appointed director of the remaining Aquatic biology research unit, still under the auspices of Societas Scientiarum Fennica. Levander worked there almost daily until his death in 1943 but the results of his work are mostly hidden in his note books. He did, however, exert a considerable influence upon younger scientists, working in this laboratory (even in the summer at the Tvärminne station).

Ernst Häyrén was botanical assistant at the unit. He had already in 1900 published a short description of the regional subdivision of the archipelago at the south coast of Finland, with the inwardly decreasing marine influence and increasing terrestrial and lacustrine impact. There, and in later editions, especially 1931 and 1948, the regionality of the archipelago waters are described in general terms. Later these features have been studied in more detail in several papers by other authors, especially in the Tvärminne area.

Häyrén's main contribution to Finnish limnology is his treatment of the polluted coastal waters around Helsinki, based on a saproby classification of aquatic plant communities, mainly of the littoral benthos but with some phytoplankton associations added. The treatment is based on the botanical part of the saproby classification of Kolkwitz and Marsson (1902 and later), which was developed for running waters in Germany. Häyrén used the saproby terminology of this German classification: from the katharobe clean water communities, through oligosaprobe and mesosaprobe units to the most polluted polysaprobe units. Häyrén developed an independent classification, based on a consequent subdivision of mainly communities of sessile algae and macrophytes of the brackish water of the shores of this polluted area. The original contribution appeared in 1921, based on field work in the two preceding years. This is the first substantial contribution to applied limnology in Finland and deserved wider attention, but written in Swedish and published in a series of limited distribution it has remained little known outside the Nordic countries. As the plant communities are rather reliable indicators of the stage of pollution, the original investigation has been repeated several times, until quite recently. Changes in taxonomy of algal genera has not hampered the use of the system of Häyrén, as the system is based more on physiognomy of the communities than on species taxonomy. It must be kept in mind that the system was developed for local use: Häyrén himself later on (1944) used it for several other towns on the Finnish south coast, but his comparison between the towns (1945) contains some severe errors. A recent attempt to adapt the saproby system of Häyrén for computerised analysis has not been entirely successful, especially when applied to other archipelago areas and when based on presence/absence of a species and not on the physiognomy of the whole association as in Häyrén's system. This is an admonitory example of misuse of a computer method.

As a part of the same study of the polluted area around Helsinki, Ilmari Välikangas in 1926 published a thesis on the summer season plankton of the harbour area. Further contributions of importance from the Aquatic Research Unit are Sven G. Segerstråle's works on the composition of the animal communities of the littoral *Fucus* belt and mainly of the profundal benthos of clean waters in the archipelago of the Finnish south coast (Pellinge and Tvärminne), from 1928 and 1933 onwards, and reviews of brackish water research and the Baltic Sea by Välikangas (1933) and Segerstråle (1957, 1958, 1959).

Early fresh water limnology

J. A. Palmén, at the turn of the century, decided to move his hydrobiological summer station from Esbo-Lövö to another place on the coast, Tvärminne, assuming that 'one or several inland centres for aquatic research certainly in the future will get established in our country, so rich in lakes', to use his own words (in A. Luther 1957). However, this did not happen until quite recently when a network of biological stations, equipped for work the whole year round, was established in Finland.

Palmén might have had in mind the Field station for fishery investigations (Brofeldt 1920) at Evo, south Häme, established in 1892 in connection with the Institute for higher Forestry Education, one of the two predecessors of the present Faculty of Agriculture and Forestry at our University. An attempt was made to study the limnology of polyhumose (as they were to be later classified) lakes, tarns and brooks of this area, but only two short lake descriptions by Levander appeared (1906a, b) before the academic staff of the Evo Institute in 1908 moved out at the transfer of the academic forestry education to the University of Helsinki. Later on a short review by Valle (1923) appeared.

Other contributions to freshwater limnology were few at this time. K. E. Stenroos (from 1906 Kivirikko) had in 1898 published his doctoral thesis, where the first conscious use in Finland of the term limnology appears, although abiotical data are almost lacking in the work. Stenroos also gives a detailed account of the earlier, mostly faunistical hydrobiological notes from Finland. K. Siitoin described in 1908 the lake Sarajärvi, north of Ladoga, and Arth. Wahlberg published in 1913 a detailed survey of especially the plankton of Littois träsk (Littoistenjärvi) near Turku/Åbo, based on observations over two years. The work of Wahlberg might have raised international attention if it had not been written in Swedish and appeared on the eve of the First World War.

The Järnefelt era

The early era of scattered contributions to freshwater limnology was at the beginning of the 1920s, followed by the Järnefelt era, which comprises 40 years. There were, of course, other scientists working in limnology during this time but the era is clearly dominated by Heikki Järnefelt. He was born in 1891, got his Fil. kand. degree at our University in 1914, and his Ph.D. in 1923. He was from the start of his academic career devoted to hydrobiology and apparently got strong impulses from his main teacher Levander. Like Levander, Järnefelt was more interested in descriptive limnology and hydrobiology than in causality, to be studied by laboratory methods.

Whilst limnology in most countries, less rich than ours in freshwater bodies, has been a problem-oriented science, exercised at field stations in intensively studied lake areas, in Finland during the Järnefelt era it was a descriptive science, which studied lakes all over the country. There was an obvious need to establish regional typology in our vast abundance of lakes, but the attitudes taken by Järnefelt in his publications clearly show his personal preference for descriptive work.

In a series of 20 volumes his 'Zur Limnologie einiger Gewässer Finnlands' appeared between 1925 and 1963. Here Järnefelt gives in 1754 pages descriptions of various length of 120 lakes, scattered all over Finland, from the south coast up to northernmost Lapland. In addition, vol. 16 (1956) treats and classifies 331 lakes on the basis of mostly solitary observations. The purpose was not to write complete lake monographs but instead to get a survey of the regional typology of our lakes. Only

the last volume 20 (1963) provides a synthesis of the thermal and chemical properties of Finnish lakes. This series represents long and tedious work both in the field and in the laboratory analysing the samples. Although superficial in some respects it gives a multitude of information, not least for future studies of these lakes.

The regional distribution of lake types in Finland was treated by Järnefelt several times (e.g. 1929, 1935, 1952a). Eutrophic lakes dominate south of the terminal moraine Salpausselkä, further to the southwest and on the Åland islands, oligotrophic lakes in some areas on and around the terminal moraine as also in a part of South Häme, further on in Lapland. The rest of Finland was regarded by Järnefelt as dystrophic. Later (1958a, b) he considered dystrophy as an additional factor, to be used in the subdivision of both eutrophic and, especially, of oligotrophic lakes. He also discussed the lake types with regard to the bottom fauna (1953a), the trophic indicator value of plankton species (1952b) and clarified and extended the trophic terminology (1953b). He also discussed the developmental history of our lakes (1938) and the vertical distribution of both plankton (1951, 1958c) and the profundal bottom fauna (1955a), tripton problems (1940d) and sedimentation of seston (1955b).

In a number of publications he treated questions of applied limnology, including the influence of the pulp industry (1940a, c, 1949b, 1961) and a powder factory (1952c).

In comparison with the rather numerous lake studies, the problems of running water limnology have been neglected in Finland. Järnefelt published scattered notes on them (e.g. 1949a).

In 1924 Järnefelt was appointed docent in applied limnology at the University of Helsinki and in 1939 he got a personal extraordinary professorship in applied limnology. After the Second World War he had the chance to establish a university department of limnology, which at his time consisted of two overcrowded, small rooms in the Forestry building. At his retirement in 1961, his professorship was changed to a real Chair of Limnology and in 1971 the department got new and enlarged premises.

As said above, other freshwater limnologists were few during this era. Kaarlo Johannes Valle published in 1927–28 'Ökologisch-limnologische

Untersuchungen über die Boden- und Tiefenfauna in einigen Seen nördlich vom Ladoga-See', in two volumes of 400 pages.

Regional reviews of lake types on the Åland islands were published by Pontus Palmgren 1936, based on their bird fauna (oligotrophic lakes were called Colymbus lakes, intermediate ones Podiceps lakes and eutrophic ones Nyroca lakes) and by Carl Cedercreutz 1937, based on plant geography, as compared with the Swedish classifications of Samuelsson and Almquist.

The Linkola school

In the 1930s the botanical lake research school of Kaarlo Linkola arose. This school has published about 20 works which complement Järnefelt's Zur Limnologie series, as Järnefelt almost completely left out macrophytes from his descriptions. Linkola himself published inter alia a trophic classification of the macrophytes (1933) which, however, has some drawbacks. He did not distinguish any eurytrophes and did not have enough knowledge of aquatic macrophytes of areas other than southern and central Finland and thus his classification is of limited value and should be used with the utmost care. Antero Vaarama published in 1938 a vegetation monograph of the great lake Kallavesi in Central Finland, based on a line transect survey and Lauri Maristo published in 1941 a review of the botanical lake types of Finland, based on the macrophyte flora and vegetation physiognomy. Recent attempts to transfer Maristo's extensive data to a form suitable for computer analysis have not been successful as they have not paid enough attention to the fundamental principles Maristo followed.

Current work

It is customary to let historical reviews stop at the time when contemporary scientists began their work. Whilst the late works by Järnefelt and by others, whose activity started early in his era, are included here, it seems to me suitable in other respects to put a limit to my review at 1940, ten years before the Limnological Society of Finland was founded, and thus to leave out the achievements of the generation of still working limnologists. It is neither possible nor desirable to select which persons among us present here should be mentioned, which ones left out.

The Järnefelt era in Finnish limnology has been followed by an era of practical limnological work. More limnologically educated persons than before are working in offices and consultant companies and their routine data tend to get hidden in internal reports. There are, of course, also scientists among us publishing results in an international language.

One feature characterizing the scientific limnology of the last decades is fruitful teamwork around selected themes such as the IBP work: in freshwater on the nature of humus compounds in the water (Lehmusluoto & Ryhänen 1972), in brackish water on the composition of the most important littoral benthos biocoenoses of the northern Baltic Sea (H. Luther et al. 1975). Also the extensive Pääjärvi project on the energy metabolism of a lake deserves mention here (Ruuhijärvi 1974).

However, I have the strong impression that rather more material suitable for scientific publication gets hidden in the bureaucratic machinery and that the capacity, and interest, to think and publish in an international language has a tendency to diminish among our limnologists. With this firebrand I close my opening address and congratulate the Limnological Society of Finland on its successful initiative to arrange this jubilee symposium in English.

References

Berg, K., 1951. The content of limnology demonstrated by F.-A. Forel and August Thienemann on the shore of Lake Geneva. Verh. int. Ver. Limnol. 11: 41–56.

Brofeldt, P., 1920. Evon kalastuskoeasema. Suomen kalatalous 6: 1–141, 20 maps.

Buch, K., 1914. Ueber die Alkalinität, Kohlensäure und Wasserstoffionen-Konzentration in der Pojowiek. Fennia 35: 1–32.

Cajander, A. K., 1909. Ueber Waldtypen. Acta forest. fenn. 1: 1–175 (also: Fennia 28 (2): 1–175).

Cedercreutz, C., 1937. Eine pflanzengeographische Einteilung der Seen Ålands und die regionale Verteilung der verschiedenen Seentypen. Acta Soc. Fauna Flora Fenn. 60: 327–338.

Granqvist, G., 1914. Om de hydrografiska förhållandena i Pojoviken under vintern. Fennia 35 (4): 1–29.

Häyrén, E., 1900. Längs-zonerna i Ekenäs skärgård. Geogr. Fören. i Finl. Tidskrift 12: 222–234.

6

Häyrén, E., 1921. Studier över föroreningens inflytande på strändernas vegetation och flora i Helsingfors hamnområde. Bidrag t. kännedom af Finl. natur och folk 80 (3): 1-128.

Häyrén, E., 1931. Aus den Schären Südfinnlands. Verh. int. Ver. Limnol. 5: 488-507.

Häyrén, E., 1944. Studier över saprob strandvegetation och flora i några kuststäder i södra Finland. Bidrag t. kännedom af Finl. natur och folk 88 (5): 1-120.

Häyrén, E., 1945. Om vattnets flora i några städer vid Finska vikens nordkust. Memoranda Soc. Fauna Flora Fenn. 21: 134-142.

Häyrén, E., 1948. Skärgårdens längszoner. Skärgårdsboken utg. av Nordenskiöld-samf. i Finl. pp. 241-256.

Järnefelt, H., 1925-34. Zur Limnologie einiger Gewässer Finnlands I-XI. Ann. Soc. Vanamo 2, 6, 8, 10, 12.

Järnefelt, H., 1929. Ein kurzer Überblick über die Limnologie Finnlands. Verh. int. Ver. Limnol. 4: 401-407, pl. 1-3.

Järnefelt, H., 1935. Die regionale Verteilung der Gewässertypen in Finnland. Verh. int. Ver. Limnol. 7: 653-656, pl. 14-22.

Järnefelt, H., 1936-63. Zur Limnologie einiger Gewässer Finnlands XII-XX. Ann. Zool. Soc. Vanamo 3, 4, 14, 17, 18, 19, 24, 25.

Järnefelt, H., 1938. Die Entstehungs- und Entwicklungsgeschichte der finnischen Seen. Geol. der Meere und Binnengewässer 2: 199-223.

Järnefelt, H., 1939. Sulla colorazione della vegetazione nei piccoli bacini di acqua sulle isole rocciose della Finlandia. Vol. giubilare Osvaldo Polimanti, 11 pp, 1 pl., 1 map. Perugia.

Järnefelt, H., 1940a. Effect of waste-water from pulpmills on organisms. Suomen Paperi- ja Puutavaralehti 22: 1-7.

Järnefelt, H., 1940b. Beobachtungen über die Hydrologie einiger Schärentümpel. Verh. int. Ver. Limnol. 9: 79-101.

Järnefelt, H., 1940c. Die Einwirkung einer Sulfitfabrik auf den Lievestuoreenjärvi. Verh. int. Ver. Limnol. 9: 180-187.

Järnefelt, H., 1940d. Ein kleiner Beitrag zur Tripton-Frage. Arch. Hydrobiol. 36: 319-329.

Järnefelt, H., 1949a. Der Einfluss der Stromschnellen auf den Sauerstoff- und Kohlensäuregehalt und das pH des Wassers im Flusse Vuoksi. Verh. int. Ver. Limnol. 10: 210-215.

Järnefelt, H., 1949b. Kemiallisen puunjalostusteollisuuden jäteaineet ja maisema. Ref: Die Abfallstoffe der Zellstoffindustrie und die Landschaft. Terra 61: 71-82.

Järnefelt, H., 1951. Beobachtungen über die Vertikalverteilung des Planktons. Verh. int. Ver. Limnol. 11: 213-218.

Järnefelt, H., 1952a. Limnological classification of lakes. Fennia 72: 202-208.

Järnefelt, H., 1952b. Plankton als Indikator der Trophiegruppen der Seen. Ann. Acad. Sci. Fenn. A IV (18): 1-28.

Järnefelt, H., 1952c. Der Vihtajärvi. Ein durch die Abwässer einer Pulverfabrik azidotrophierter See. Hydrobiologia 4: 268-280.

Järnefelt, H., 1953a. Die Seentypen in bodenfaunistischer Hinsicht. Ann. Zool. Soc. Vanamo 15 (6): 1-38.

Järnefelt, H., 1953b. Einige Randbemerkungen zur Seetypennomenklatur. Schweiz. Zeitschr. Hydrol. 15: 198-212.

Järnefelt, H., 1955a. Die vertikale Verteilung der Bodenfauna im Profundal. Memorie Ist. Ital. Idrobiol., Suppl. 8: 165-181.

Järnefelt, H., 1955b. Über die Sedimentation des Sestons. Verh. int. Ver. Limnol. 12: 144-158.

Järnefelt, H., 1958a. Vesiemme luonnontalous. Werner Söderström Oy. Porvoo 325 pp.

Järnefelt, H., 1958b. On the typology of the northern lakes. Verh. int. Ver. Limnol. 13: 228-235.

Järnefelt, H., 1958c. Über die vertikale Tag- und Nachtverteilung des Planktons im Lohjanjärvi. Hydrobiologia 10: 175-197.

Järnefelt, H., 1961. Die Einwirkung der Sulfitablaugen auf das Planktonbild. Verh. int. Ver. Limnol. 14: 1057-1062.

Lehmusluoto, P. O. & Ryhänen, R., 1972. Lake Hakojärvi, a polyhumic lake in Southern Finland. Verh. int. Ver. Limnol. 18: 403-408.

Levander, K. M., 1900. Zur Kenntnis des Lebens in den stehenden Kleingewässern auf den Skäreninseln. Acta Soc. Fauna Flora Fenn. 18 (6): 1-107.

Levander, K. M., 1906a. Beiträge zur Kenntnis des Sees Valkea-Mustajärvi der Fischereiversuchsstation Evois. Acta Soc. Fauna Flora Fenn. 28 (1): 1-28.

Levander, K. M., 1906b. Beiträge zur Kenntnis des Sees Pitkäniemenjärvi der Fischereiversuchsstation Evois. Acta Soc. Fauna Flora Fenn. 29 (3): 1-15.

Levander, K. M., 1915. Zur Kenntnis der Bodenfauna und des Plankton der Pojowiek. Fennia 35 (2): 1-39.

Linkola, K., 1933. Regionale Artenstatistik der Süsswasserflora Finnlands. Ann. Bot. Soc. Vanamo 3 (5): 3-13.

Luther, A., 1957. Tvärminne zoologiska station. Acta Soc. Fauna Flora Fenn. 73: 1-128 (pp. 6-14 by J. A. Palmén, on the establishment at Esbo-Lövö and the transfer to Tvärminne).

Luther, H., Hällfors, G., Lappalainen, A. & Kangas, P., 1975. Littoral benthos of the northern Baltic Sea. I. Introduction. Int. Revue ges. Hydrobiol. 60: 289-296.

Maristo, L., 1941. Die Seetypen Finnlands auf floristischer und vegetationsphysiognomischer Grundlage. Ann. Bot. Soc. Vanamo 15 (5): 1-312.

Palmén, J. A.: see Luther, A.

Palmgren, P., 1936. Über die Vogelfauna der Binnengewässer Ålands. Acta zool. fenn. 17: 1-59.

Ruuhijärvi, R., 1974. A general description of the oligotrophic lake Pääjärvi, southern Finland, and the ecological studies on it. Ann. bot. fenn. 11: 95-104.

Segerstråle, S. G., 1928. Quantitative Studien über den Tierbestand der Fucus-Vegetation in den Schären von Pellinge (an der Südküste Finnlands). Soc. Sci. Fenn. Commentat. Biol. 3 (2): 1-14.

Segerstråle, S. G., 1933. Studien über die Bodentierwelt in südfinnländischen Küstengewässern, I-II. Soc. Sci. Fenn. Commentat. Biol. 4 (8-9): 1-62, 1-79.

Segerstråle, S. G., 1957. Baltic Sea. Geol. Soc. America Memoir 67: 751-800, corrigenda et addenda 1-2.

Segerstråle, S. G., 1958. A quarter century of brackishwater research. Verh. int. Ver. Limnol. 13: 646-671.

Segerstråle, S. G., 1959. Brackishwater classification, a historical survey. Symposium on the classification of brackish waters, Venice 8-14. 4. 1958. Arch. Oceanogr. Limnol. 11 Suppl.: 7-33.

Siitoin, K., 1908. Sarajärven eläimistö. Acta Soc. Fauna Flora Fenn. 29 (10): 1-44.

Simojoki, H., 1978. The History of Geophysics in Finland 1828-1918. The History of Learning and Science in Finland 1828-1918, vol. 5b: 1-157. Soc. Sci. Fenn. Helsinki.

Stenroos, K. E., 1898. Das Thierleben im Nurmijärvi-See. Acta Soc. Fauna Flora Fenn. 17 (1): I-IV, 1-259, pl. 1-4.

Vaarama, A., 1938. Wasservegetationsstudien am Grosssee Kallavesi. Ann. Bot. Soc. Vanamo 13 (1): 1-314.

Välikangas, I., 1926. Planktologische Untersuchungen im Hafengebiet von Helsingfors. I. Acta zool. fenn. 1: 1-298.

Välikangas, I., 1933. Über die Biologie der Ostsee als Brackwassergebiet. Verh. int. Ver. Limnol. 6: 62-112.

Valle, K. J., 1923. Fischwasseruntersuchungen im Staatsrevier Evo. Acta forest. fenn. 25: 1-34.

Valle, K. J., 1927. Ökologisch-limnologische Untersuchungen über die Boden- und Tiefenfauna in einigen Seen nördlich vom Ladoga-See. I. Acta zool. fenn. 2: 1-79.

Valle, K. J., 1928. Ökologisch-limnologische Untersuchungen über die Boden- und Tiefenfauna in einigen Seen nördlich vom Ladoga-See. II. Die Seenbeschreibungen. Acta zool. fenn. 4: 1-231.

Wahlberg, A., 1913. Bidrag till kännedomen om Littois träsk med särskild hänsyn till dess plankton. Acta Soc. Fauna Flora Fenn. 38 (1): 1-201, pl. 1-3.

Witting, R., 1914. Kort översikt af Pojovikens hydrografi. Fennia 35 (1): 1-18.

PART ONE

Structure and Function
of Lake Ecosystems

Dynamics of phytoplankton in Finnish lakes

V. Ilmavirta

Maj and Tor Nessling Foundation, P. Hesperiankatu 3 A 4, SF-00260 Helsinki 26, Finland

Keywords: allochthonous, dynamics, flagellates, phytoplankton, production, succession

Abstract

The studies on lake phytoplankton in Finland are reviewed and the major aspects of the phytoplankton dynamics are discussed. Special attention has been paid to the factors limiting productivity and species succession in different communities. After the early mainly taxonomical and floristic publications on phytoplankton at the end of last century, phytoplankton studies in lakes have proceeded along two different lines: 1) the species composition of communities and taxonomy, and 2) their production ecology or dynamics. Recently, both approaches have been combined, resulting in some profound ecological studies. In many lakes, phosphorus has been shown to be a limiting factor for phytoplankton productivity. However, it has also been shown that the irradiance and water temperature may effectively regulate the seasonal trend of phytoplankton productivity. This is the case especially in polyhumic forest lakes, where allochthonous material seems to play a major role also in primary production ecology.

Introduction

In Finland the density of lakes is extremely high, averaging one lake per 7 km² with the total number of lakes being 55 000. Finnish lakes are peculiar because of the high humic concentration of their water. The brown water colour reflects the high percentage of peatlands and forests on our land surface. Brown water colour, small water volume and shallow water column are the prime factors determining the nature of our lake ecosystems. Dimixy prevails and winter ice-cover is long.

In practical lake management, phytoplankton is the first community disturbed by external loading from surrounding settlements, factories or non point sources. The most rapid and strongest changes in the state of the lake ecosystem are thus seen in the phytoplankton which is consequently the most commonly used indicator of environmental eutrophication and pollution. Thus our understanding of the dynamics of lake ecosystem would perhaps be widest for this community.

In Finland the first attempts to analyse the structure of phytoplankton communities in lakes were made by Levander (1900) and also by Levander & Wuorentaus (1915, 1917) in their studies on over 150 Finnish lakes in the summer of 1913 and 1914. Their published material contains quite detailed lists of phytoplankton species with their relative abundance. This well documented data can be used to study changes in the community structure of phytoplankton in these particular lakes.

Much more extensive material was collected by the first Finnish professor of limnology, Heikki Järnefelt (1952). His studies covered about 300 lakes around Finland, and covered both phytoplanktonic community structure and water chemistry. Based on this material, Järnefelt introduced the first, and still the only, biological typology of Finnish lakes. He applied the classification ideas of Thunmark (1945) and Nygaard (1949) and made Finnish limnology widely known in foreign countries.

Hydrobiologia 86, 11–20 (1982).   0018-8158/82/0861-0011/$02.00.
© Dr W. Junk Publishers, The Hague.

12

In The Water Pollution Control Bureau of The National Board of Agriculture (as from 1.7.1970 The National Board of Waters, Finland) extensive studies of water quality in Finnish lakes have been carried out since 1963, these studies including sampling of phytoplankton. The first assessment of this material was made recently (Heinonen 1980) and these results will be presented elsewhere in this book (Heinonen 1981).

All the studies mentioned above aimed to elucidate the structure of phytoplankton communities in Finnish lakes. However, information on the dynamics of these communities is scarce, since little was known on the productivity of phytoplankton or the factors governing production dynamics. Moreover, the species succession of phytoplankton through different seasons, known to be necessary for ecological monitoring, was poorly known.

The first published attempt to measure the instantaneous production rate of phytoplankton in Finland was by Lehmusluoto (1964). The first annual production rates of phytoplankton were reported by Meriläinen (1967, 1970) for meromictic Lake Valkiajärvi in central Finland. Subsequently measurements of phytoplankton productivity have been more numerous in the 1970s, but few studies have been published in congress languages. The most detailed data including annual production values, are from lakes Päijänne (Granberg 1973),

Pääjärvi (Ilmavirta 1979), Suomunjärvi (Sorsa 1979; Meriläinen, unpubl.) and Hakojärvi (Lehmusluoto 1969).

As shown above, the studies on phytoplankton have proceeded along two different lines: firstly on the structure and taxonomy of phytoplankton communities and the chemistry of their environments, and secondly on the dynamics of their communities. Both these approaches were combined in some recent lake studies, most effectively in the ecological study of the oligotrophic brownwater Lake Pääjärvi, in southern Finland (Ilmavirta 1979; Ruuhijärvi 1974). Detailed information on the role of humic material in lake ecosystems was included in the study of polyhumic oligotrophic lake Hakojärvi (Ryhänen & Lehmusluoto 1972).

In the present paper I will discuss the overall dynamics of phytoplankton in Finnish lakes, and particularly in brown-water lakes. Attention will be focused on the seasonal and diel production dynamics and seasonal species succession of phytoplankton in different types of lakes, but the factors limiting productivity will be briefly discussed.

Primary production

The annual phytoplankton production values, measured in Finnish lakes and known to me, are

Table 1. Phytoplanktonic primary production, total phosphorus concentration and water colour in some Finnish lakes. The latitudes of the lakes are also given. Data: (1) Granberg (1973), Eloranta (1976); (2) Ilmavirta et al. (1974), Heikkola (1980); (3) Ilmavirta et al. (1974); (4) (5) Granberg (1973), Eloranta (1976); (6) Lehmusluoto (1965); (7) Meriläinen (1970); (8) Ilmavirta (1975); (9) Sorsa (1979), Meriläinen (unpubl.); (10) Granberg (1973); (11) Lehmusluoto (1969); (12) Tolonen (1980).

Lake	Prim. prod. $g\ C_{ass}m^{-2}a^{-1}$	Tot. P $\mu g\ l^{-1}$	Colour mg Pt l^{-1}	Latitude N	
1. Jyväsjärvi	124	700	70	62° 15′	
2. Lovojärvi	100	56	80	61° 05′	Eutrophy
3. Ormajärvi	45	24	20	61° 06′	(Rodhe 1969)
4. N-Päijänne (540)	34	25	55	62° 10′	- - - - - - - - - -
4. N-Päijänne (532)	23	36	60	62° 14′	
5. Tuomiojärvi	25	30	50	62° 16′	
6. Hormajärvi	23	20	15	60° 20′	
7. Valkiajärvi	19	?	25	61° 54′	
8. Pääjärvi	15	12	40	61° 04′	Oligotrophy
9. Suomunjärvi	14.5	5	50	63° 07′	
10. S-Päijänne (810)	2.9	15	40	61° 14′	
11. Hakojärvi	2	25	150	61° 15′	
12. Kilpisjärvi	1.5	5	5	69° 00′	

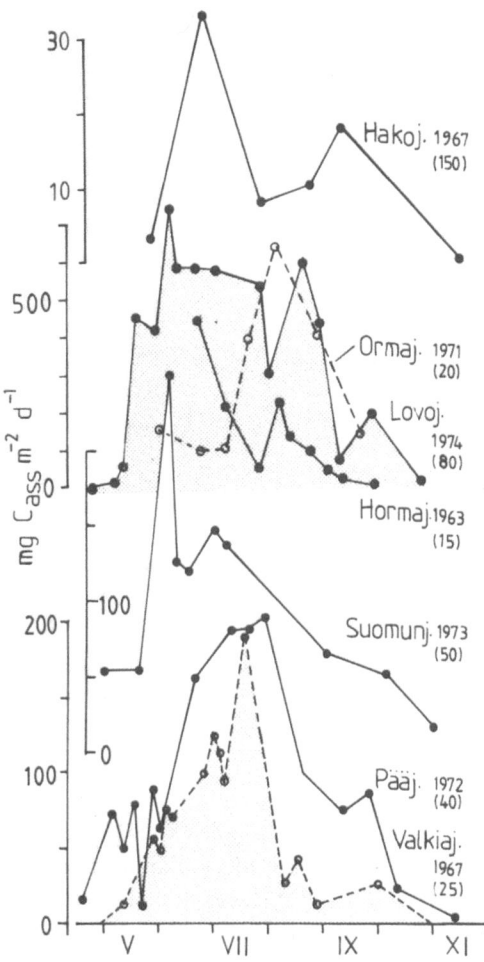

Fig. 1. The seasonal succession of phytoplanktonic productivity in some Finnish lakes. The year of field measurements and the colour of water (in parentheses) are given at the right. (Redrawn after data in Table 1).

The seasonal variations in productivity of phytoplankton in some Finnish lakes are shown in Fig. 1. It is notable that in most humic lakes there is not found the depression of productivity in summer during high stratification, typical of phytoplankton productivity in the Baltic and in many eutrophic lakes in Europe. So phytoplankton do not appear to suffer any major summer depletion of nutrients in these lakes. The poor penetration of light through the water due to brown coloured substances is responsible for the shallow euphotic layer in humic lakes (Fig. 2A). In lakes with effective turbulence the epilimnion is deeper than the euphotic zone and thus phytoplankton seem to have a reserve of nutrient to maintain their production (cf. Lehmusluoto 1971). In the studies at Lake Pääjärvi, Salonen (1981) showed that more than 50% of the suspended organic material was decomposed in the epilimnion, sufficient according to Ilmavirta 1975, 1979, 1981) to provide phytoplankton with nutrients. The bacterial decomposition is strongly supported by allochthonous material running from the drainage area; thus allochthonous input is a major source of energy sustaining phytoplanktonic production in humic lakes. This mechanism has been demonstrated in Lake Pääjärvi and it seems probable that this mechanism also functions in more humic lakes (for Lake Hakojärvi, see Ryhänen 1968; Lehmusluoto 1970; Lehmusluoto & Ryhänen 1972). Instead of external loading it is more relevant to speak of internal nutrient loading, that is of rapid recycling of nutrients in the epilimnion.

The importance of nutrients and energy from allochthonous matter was shown by Ilmavirta (1980) in his studies on 35 small forest lakes. In many of those lakes, where both epilimnion and phytoplankton community indicated strong oligotrophy, the hypolimnion was completely without oxygen. Because of the dimixy of these lakes the oxygen depletion must be caused by high bacterial metabolism of allochthonous matter.

The effect of humic substances in absorbing light is clearly seen in the depth of the euphotic zone of different lakes (Fig. 2); the clearest lakes, Hormajärvi (6) and Kilpisjärvi (12), have the deepest euphotic zones, Lovojärvi (2) the shallowest. In spite of the strong coloration of water, surface inhibition of production is intense in many bottle experiments. Production inhibition is related to the

presented in Table 1. If the trophic classification of Rodhe (1969) can be applied here, most of these lakes seems to be oligotrophic or slightly meso-oligotrophic. Only a few lakes show incident eutrophy. Compared to the published production values for Danish lakes or Mazurian lakes, these values are all low. Most eutrophic lakes in Sweden also have higher values. These low production figures reflect the nutrient poor soil lying on ancient acid bedrock. Clay areas in south and southwest Finland are more nutrient rich but production measurements in these areas are few. Most clearwater lakes in northern and northeastern Finland show ultraoligotrophic production patterns.

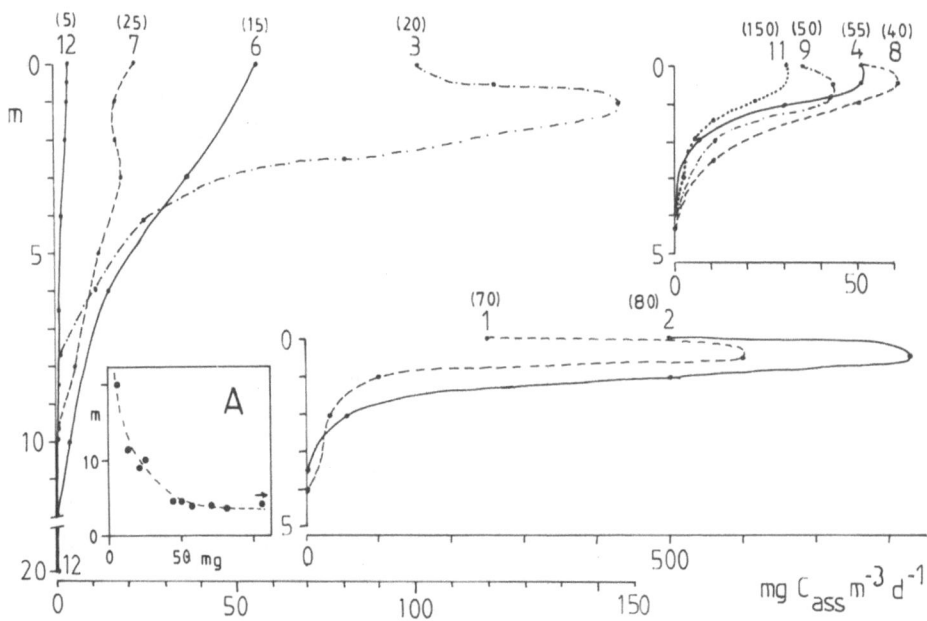

Fig. 2. The depth distribution of phytoplanktonic production in some Finnish lakes in July. Water colour is given in parentheses. The numbers refer to the lakes in Table 1. In Fig. 2A the depth of euphotic layer (metres) is plotted against water colour (mg Pt 1^{-1}). (Redrawn after data in Table 1).

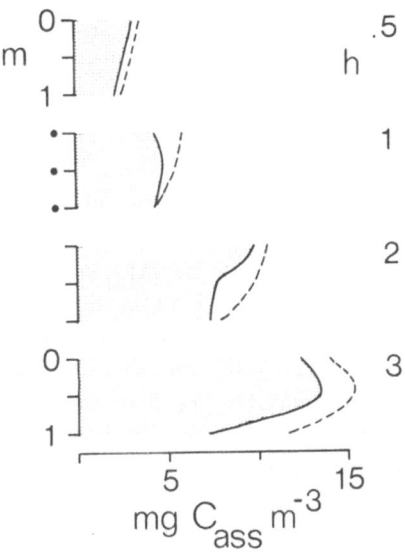

Fig. 3. The difference of phytoplanktonic production in glass (broken line) and acrylic plastic (solid line, shading) bottles during different lengths of exposures (0.5 h–3 h) in Lake Suomunjärvi 5 August 1975. Each experiment bottle of different exposures in three depths contained the same populations at that depth. A notable difference between bottle material is seen already after 0.5 h exposure time. (Data after Ilmavirta (unpubl.)).

lack of turbulence during sample exposure and the prevailing rest period of phytoplankton in the darker layers below the surface, so that prolonged high light intensities cause injury to the photosynthetic mechanism of algae. Inhibition was shown to be higher in acrylic bottles than in ordinary glass bottles during bright days (Ilmavirta 1977), but in dark autumn days acrylic bottles permitted higher production, which is related to the higher transparency of acrylic plastic (see Fig. 3).

To arrive at correct annual production values, day to day variations in productivity must also be known. Ilmavirta (1978) showed that in three humic lakes, phytoplankton production varied over six successive days by a maximum of 37% (± SD %). These variations were highly related to variations in radiation during exposure. Strong variations were also noted in species composition. Thus environmental factors must be known when calculating annual production values.

The diel dynamics of the phytoplankton community seems to be of major importance to the production ecology of phytoplankton. Ilmavirta (1974, 1977) showed that the sum of short 4 h exposures is higher than the production of one 24 h

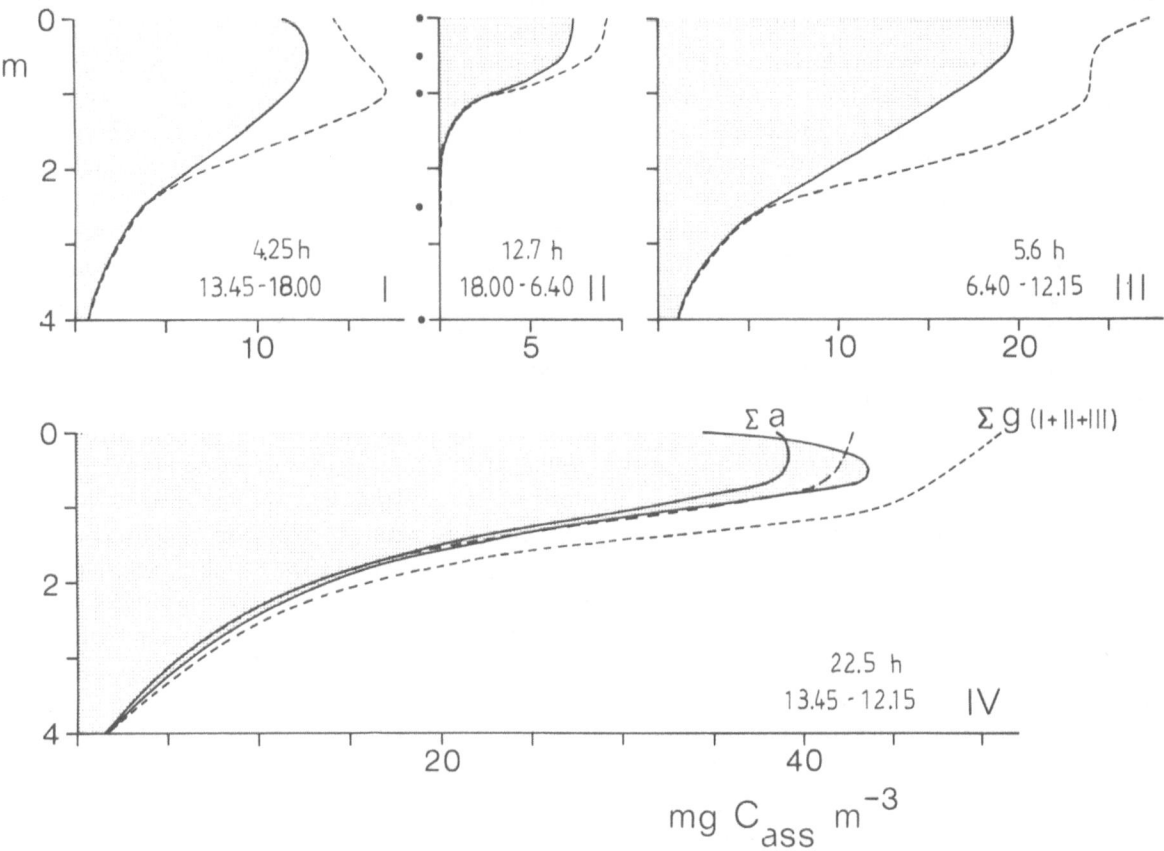

Fig. 4. Diel periodicity of phytoplanktonic productivity in glass (g, broken line) and acrylic plastic (a, solid line, shading) bottles measured during exposures of different length. In Fig. IV, Σa and Σg are the sum curves of short exposures in I–III compared to the long exposures of the same length. Exposure time and the duration of exposure (h) are given below. Measurements were made on 5–6 August 1975 at Lake Suomunjärvi. Because of bright days during exposures productivity in glass bottles was higher. In period II, around night-time, the productive layer was only 2 m. Measuring depths are given with dots in II. Every set of bottles contained a fresh community sampled at the beginning of exposure. (Data after Ilmavirta (unpubl.). For the methods of diel production measurements see Ilmavirta (1974)).

exposure in Lake Pääjärvi. The same phenomenon was found in polyhumic Lake Suomunjärvi (Fig. 4) and in eutrophic polyhumic Lake Lovojärvi (Jones & Ilmavirta 1978). The greatest difference between short and long exposures, 49%, was noted in Pääjärvi, where this phenomenon was strongly correlated with the intensity of radiation during exposures. In Suomunjärvi the difference was 30% and in Lovojärvi 38%. This phenomenon has been thoroughly analysed in Pääjärvi, where it was found that flagellated phytoplankton species showed strong vertical migrations during the day. Biomass was concentrated at the surface in the morning and in the evening during low radiation, but around midday the biomass maximum was at 0.5 m or 1.0 m. This migration rhythm was evident with all flagellated species and the timing of the migrations was slightly different for different species. Such migrations permit phytoplankton to maximize their production without any inhibition of production by too prolonged high light intensity at the surface and the motility of flagellates is an important ecological adaptation of phytoplankton in humic lakes. Vice versa, the ability to move independently also allows flagellated phytoplankton to migrate towards surface during dark days. This migration really seems to be of major importance to the dynamics of many humic lakes, since Ilmavirta (1980) showed that in most very dark forest lakes studied in the Kangasala area in southern Finland the proportion of flagellated species in the total biomass increased with increasing water colour or decreasing Secchi disc transparency.

Factors governing primary production

The model of Vollenweider & Dillon (1974) relates phytoplankton production to the phosphorus load of lakes. According to this model, many Finnish oligotrophic brown-water lakes would be classified as mesotrophic or even eutrophic, as shown in the report of the Nordic OECD eutrophication programme (Ryding 1980). The reason for this is that a great deal of the external allochthonous matter loading the lake is as organic compounds, difficult for water organisms to assimilate rapidly, so only a small part of their nutrients will be directly used in the lake ecosystem.

In many studies phytoplankton productivity has been shown to be nutrient limited, mainly by phosphorus, this conclusion being supported by laboratory bioassay studies. There must then be a firm correlation between phytoplankton annual or seasonal production and phosphorus concentration in such lakes. However, for the material in Fig. 5 this relation is valid only in lakes which have about the same water colour and are in their natural state or loaded only by domestic sewage. Some stations in Lake Päijänne loaded with pulp mill effluents and also Lake Hakojärvi with high humus

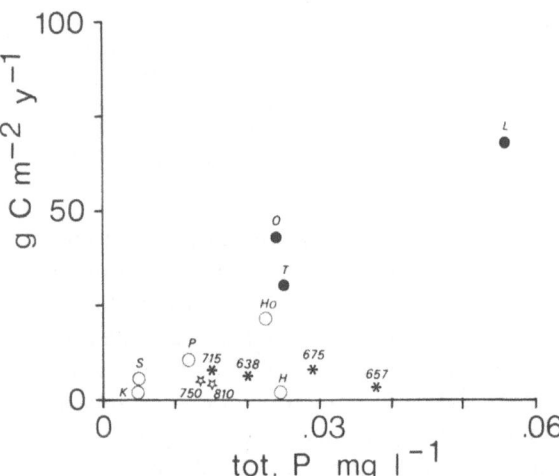

Fig. 5. The relation between phytoplanktonic annual productivity and the total phosphorus concentration of water in some Finnish lakes. Lakes Kilpisjärvi (K), Suomunjärvi (S), Pääjärvi (P), South Päijänne (stars: 750, 810) and Hakojärvi (H) are in their natural stage and oligotrophic. Hormajärvi (Ho), Tuomiojärvi (T), Ormajärvi (O) and Lovojärvi (L) are loaded by domestic sewage or agriculture, but stations in the northern Päijänne (asterisked) with pulp mill effluents. (Drawn after data in Table 1).

concentration (150 mg Pt l^{-1}) do not fit this relation.

Looking at the seasonal production curves of phytoplankton in humic lakes (Fig. 1), one can assume that the seasonal trend of production is mostly regulated by energy parameters, such as water temperature and solar radiation, since all the curves approximate to a parabola in these lakes. In Lake Pääjärvi (cf. Ilmavirta 1981), epipelic production also had the same production patterns, but variations in the nutrient concentrations of the water were negligible (Ilmavirta 1981).

It seems evident that the seasonal production dynamics of phytoplankton is, in fact, strongly regulated by the nutrients released by heterotrophs during the growing season, and heterotrophic metabolism is mainly regulated by water temperature (Salonen 1981). In Lake Pääjärvi, which is the only Finnish lake where this relationship has been studied experimentally, the nutrient supply from heterotrophic activity seems to be high enough to support phytoplankton productivity. Thus the amount of radiant energy for photosynthesis and the water temperature stimulating metabolism are really 'limiting', and the nutrient concentration of water roughly determines the level of productivity, although seasonal succession is related to energy factors. This relation is valid in oligotrophic and mesotrophic brown water lakes, not in all eutrophic ones. In clear water lakes a seasonal production curve with two peaks seems evident. Lakes Hormajärvi and Ormajärvi in Fig. 1 would have spring maxima in productivity, but the available data are too meagre for further discussions.

The rapid seasonal changes in the environmental conditions during the short growing season in Finnish lakes are unfavourable for any particular phytoplankton community and therefore the adaptation of the community must be rapid. This adaptation is achieved through continuous changes in the species composition of phytoplankton and in the pigment concentrations of the algae. As shown in some earlier papers (cf. K. Ilmavirta & Kotimaa 1974; Jones & Ilmavirta 1978; Keskitalo 1977) the seasonal succession of species is very rapid and distinct in a humic lake, although great differences can be found between particular lakes. This continuous seasonal succession of species composition is the second important ecological adaptation of phytoplankton in humic waters. Some of the successional types are descibed in the following section.

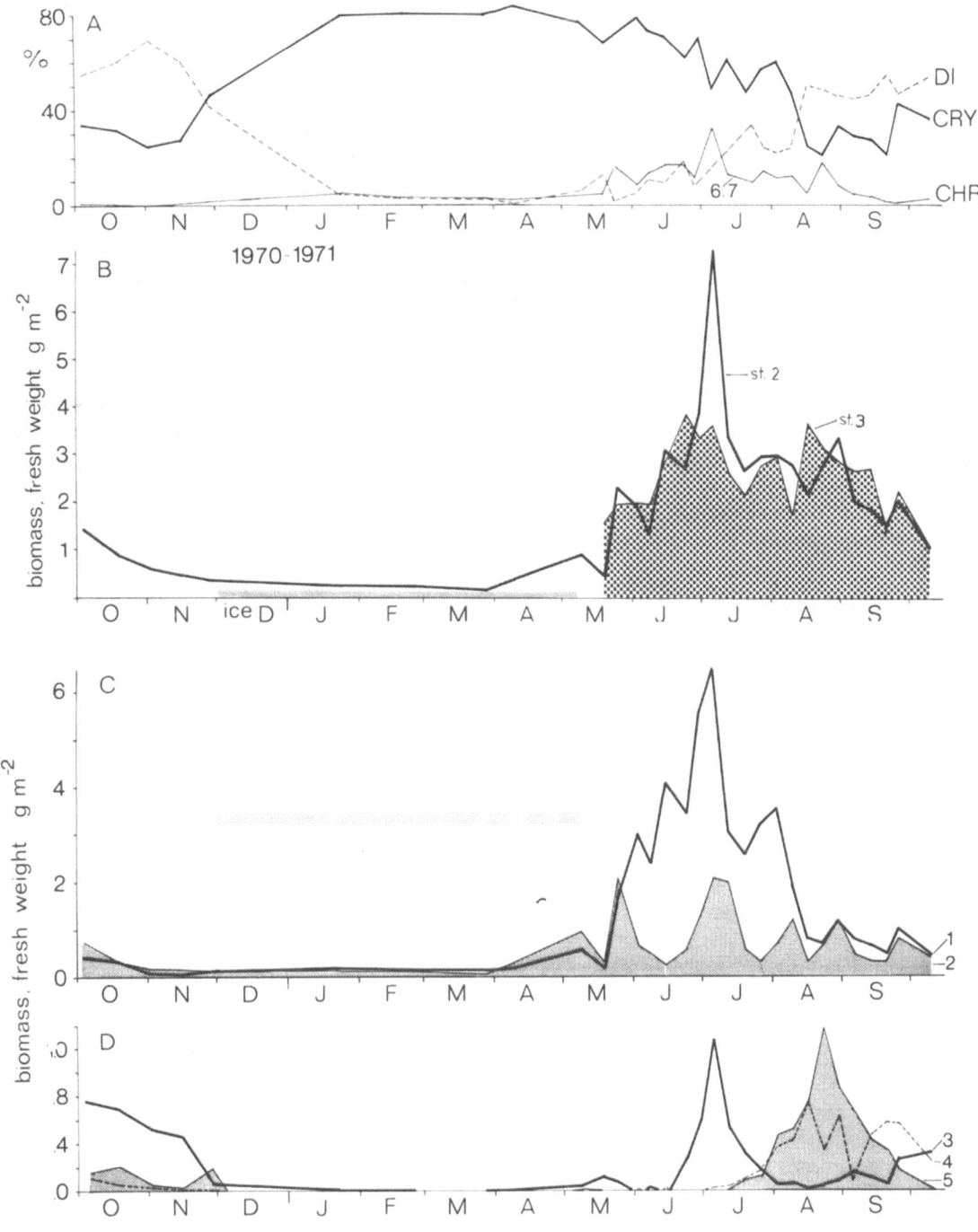

Fig. 6. The seasonal course of the fresh weight biomass of phytoplankton (B) and dominant species (C, D) in the trophogenic layer (0–4 m) of oligotrophic mesohumic lake Pääjärvi (station 2 in Pappilanlahti Bay). A shows the seasonal course of the proportions (% volume) of various groups. DI = Diatomeae, CRY = Cryptophyceae, CHR = Chrysophyta without Diatomeae (other groups negligible). Dominant species: *Cryptomonas erosa* (1), *Chroomonas acuta* (2), *Rhizosolenia longiseta* (3), *Melosira distans* (4), *M. distans* var. *alpigena* (5), *Dinobryon* sp. (6) and *Mallomonas* sp. (7). (Data after K. Ilmavirta & Kotimaa (1974)).

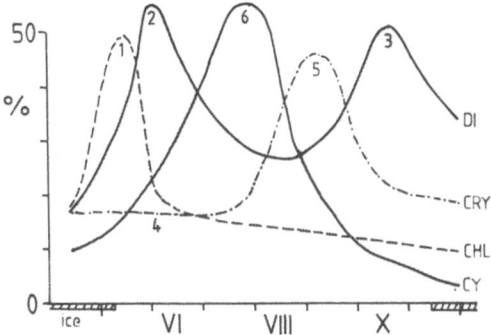

Fig. 7 The seasonal succession of phytoplanktonic species composition in eutrophic brown water lake Lovojärvi expressed as % of total fw biomass. DI = Diatomeae, CRY = Cryptophyta, CHL = Chlorophyta, CY = Cyanophyta (other groups negligible). Dominant species were *Chlamydomonas* spp. (1), *Synedra acus* (2), *Melosira distans* var. *alpigena* (3), *Chroomonas acuta* (4), *Cryptomonas* spp. (4), *Peridinium bipes* (5) and *Anabaeana* spp. (6). (Data after Ilmavirta *et al.* (1974) and Keskitalo (1977). Strongly smoothed).

Seasonal succession of species

Three examples of seasonal species succession can be given here. Lake Pääjärvi is a typical nutrient-poor, mesohumic Finnish lake. Lake Lovojärvi is a eutrophic mesohumic lake, which was enriched by the maceration of flax and hemp some hundred years ago; agriculture and farming is still loading the lake. Lake Hakojärvi is an example of an oligotrophic, very brown-water lake, where light intensities are mostly low.

In Pääjärvi (Fig. 6) species succession is regular and changes in species composition are quite slow. From early spring to the late summer cryptophytes *(Chroomonas acuta, Cryptomonas erosa)* are dominant. Their share of the biomass is mostly over 70%. *Chroomonas* populations show distinct phases of growth with continuous rising and falling of the biomass within three week intervals. After August the proportion of diatoms increases rapidly and they are dominant in late autumn and during winter. *Rhizosolenia longiseta* is dominant during the end of June and the beginning of July, and also in October and November. *Melosira distans* and *Melosira distans* var. *alpigena* have their maxima at the end of August. The most important chrysophytes are *Mallomonas, Kephyrion* and *Dinobryon* species during the whole warm period. Chlorophytes and cyanophytes are negligible.

In Lake Lovojärvi (Ilmavirta *et al.* 1974; Keskitalo 1976), environmental conditions change rapidly during the growing season and phytoplanktonic species succession is also rapid (Fig. 7). Five distinct phytoplankton pulses can be found. In early spring, just when the ice is melting, *Chlamydomonas* had a brief maximum but cryptophytes *(Chroomonas*

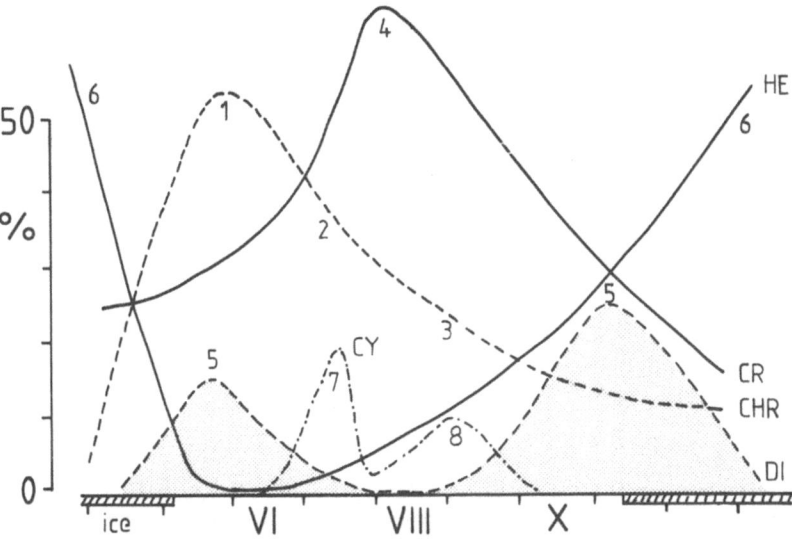

Fig. 8. The seasonal succession of phytoplanktonic species composition in oligotrophic brown water lake Hakojarvi expressed as % of total fw biomass. HE = Heterocontae, CR = Cryptophyta, CHR = Chrysophyceae, DI = Diatomeae. Dominant species were *Mallomonas akrokomos* (1), *M. acaroides* (2), *M. allorgei* (3), *Cryptomonas erosa* and *C. pusilla* (4), *Melosira distans* var. *alpigena* (5), *Botryococcus braunii* (6), *Aphanizomenon flos-aquae* (7) and *Merismopedia tenuissima* (8). (Data after Salovaara (1969), published with the permission of the author. Strongly smoothed).

and *Cryptomonas)* were also important. At the end of May a diatom, *Synedra acus,* had an enormous maximum which lasted three weeks and was followed by a cyanophyte maximum *(Anabaena)* in July during the warmest period. *Peridinium bipes* was dominant during September and diatoms (mostly *Melosira distans)* after that until the end of the ice free period. In cold summers the blue-green maximum did not occur at all.

Lake Hakojärvi (Fig. 8) showed the most extreme and also most complicated species succession (Salovaara 1969). During ice cover, *Botryococcus braunii* comprised over 80% of the total biomass. Straight after the melting of the ice, *Mallomonas akrokomos* (Chrysophyceae) and *Melosira distans* var. *alpigena* (Diatomeae), both typical of oligotrophic brown water lakes, comprised most of the biomass. After the stabilization of thermal stratification, diatoms dropped rapidly and a small blue-green maximum *(Anabaena flos-aquae)* occurred. The most important group in summer was Cryptophyceae *(Cryptomonas erosa* and *Cry. pusilla),* which were responsible for over 70% of the total biomass. A second small maximum of the cyanophyte, *Merismopedia tenuissima,* a typical oligotrophic species, was found. At the end of the autumn turnover, the diatom *Melosira distans* var. *alpigena* had its second maximum. After freezing of the lake, *Botryococcus* was again dominant.

These examples show clearly that the rapid succession of species composition during different seasons is very important for the production dynamics of humic lakes. Rapid changes allow the phytoplankton to maximize its production and its survival under all circumstances and also make this community resistant to variations in environmental conditions. Thus, understanding the species composition of phytoplankton in production studies is essential when interpreting phytoplanktonic production dynamics or when monitoring the stage of environmental pollution. In practice most scientists measure phytoplanktonic biomass using chlorophyll concentration and there is often no information on the species composition of samples or on their changes during the studies. This contributes to the problem in humic lakes where no firm correlation between biomass and production or between biomass and chlorophyll exists, both these correlations being significant in many clear water lakes.

References

Eloranta, P., 1976. Phytoplankton and primary production in situ in the Lakes Jyväsjärvi and North Päijänne in summer 1974. Biol. Res. Rep. Univ. Jyväskylä 2: 51–66.

Granberg, K., 1973. The eutrophication and pollution of Lake Päijänne, Central Finland. Ann. bot. fenn. 10: 267–308.

Heikkola, K., 1980. Phytoplanktonic primary production and factors affecting it in Lake Lovojärvi; Mimeogr., Univ. Helsinki, Dept. Bot. 58 p. (In Finnish).

Heinonen, P., 1980. Quantity and composition of phytoplankton in Finnish inland waters. Nat. Board Waters, Finland. Publ. Water Res. Inst. 37: 1–91.

Heinonen, P., 1981. On the annual variation of phytoplankton biomass in Finnish inland waters. Hydrobiologia (this volume).

Ilmavirta, K. & Kotimaa, A.-L., 1974. Spatial and seasonal variations in phytoplanktonic primary production and biomass in the oligotrophic lake Pääjärvi, southern Finland. Ann. bot. fenn. 11: 112–120.

Ilmavirta, V., 1974. Diel periodicity in the phytoplankton community of the oligotrophic Lake Pääjärvi, southern Finland. I. Phytoplanktonic primary production and related factors. Ann. bot. fenn. 11: 136–177.

Ilmavirta, V., 1975. Dynamics of phytoplanktonic production in the oligotrophic lake Pääjärvi, southern Finland. Ann. bot. fenn. 12: 45–54.

Ilmavirta, V., 1977. Diel periodicity in the phytoplankton community of the oligotrophic lake Pääjärvi, southern Finland. III. The influence of the bottle material on the measurement of production. Ann. bot. fenn. 14: 102–111.

Ilmavirta, V., 1978. Day to day variations in the phytoplankton community in three brown-water lakes in Finland. Verh. int. Verein. Limnol. 20: 881–885.

Ilmavirta, V., 1979. Sources and utilization of energy in Pääjärvi, an oligotrophic, brown-water lake in southern Finland. Arch. Hydrobiol. Beih. Ergebn. Limnol. 13: 212–224.

Ilmavirta, V., 1980. Phytoplankton in 35 Finnish brown-water lakes of different trophic status. Dev. Hydrobiol. 3: 121–130.

Ilmavirta, V., 1981. The ecosystem of the oligotrophic Lake Pääjärvi. 1. Lake basin and primary production. Verh. int. Verein. Limnol. 21: 410–415.

Ilmavirta, V., Ilmavirta, K. & Kotimaa, A.-L., 1974. Phytoplanktonic primary production during the summer stagnation in the eutrophicated lakes Lovojärvi and Ormajärvi, southern Finland. Ann. bot. fenn. 11: 121–132.

Jones, R. I. & Ilmavirta, V., 1978. A diurnal study of the phytoplankton in the eutrophic lake Lovojärvi, Southern Finland. Arch. Hydrobiol. 83: 494–514.

Järnefelt, H., 1952. Plankton als Indicator der Trophiegruppen der Seen. Ann. Acad. Scient. Fennica Ser A IV 18: 1–29.

Keskitalo, J., 1976. Phytoplankton pigment concentrations in the eutrophicated lake Lovojärvi, south Finland. Ann. bot. fenn. 13: 27–34.

Keskitalo, J., 1977. The species composition and biomass of phytoplankton in the eutrophic lake Lovojärvi, southern Finland. Ann. bot. fenn. 14: 71–81.

20

Lehmusluoto, P., 1964. Some notes on the evaluation of primary production by the carbon-14 method in a lake in Southern Finland. Limnologisymposion 1964: 113–116. (In Finnish).

Lehmusluoto, P., 1965. The use of carbon-14 method in estimating net primary production in Lake Hormajärvi in 1963. Mimeogr. 96 p. Univ. Helsinki, Dept. Limnol. (In Finnish).

Lehmusluoto, P., 1969. Notes on the phytoplanktonic primary production of polyhumic Lake Hakojärvi in 1969. Mimeogr. 121 p. Univ. Helsinki, Dept. Limnol. (In Finnish).

Lehmusluoto, P., 1970. Eutrophication in some lakes and coastal areas in Finland, with special reference to polyhumic lakes. In: Murphy, R. S. & Nyquist, D. (eds.) Water Pollution Control in Cold Climates, pp. 48–60. Washington, D.C.

Lehmusluoto, P., 1971. Special features of the phytoplankton primary production in polyhumic lakes. Limnologisymposion 1971: 60–68. (In Finnish).

Lehmusluoto, P. & Ryhänen, R., 1972. Lake Hakojärvi, a polyhumic lake in Southern Finland. Verh. int. Verein. Limnol. 18: 403–408.

Levander, K. M., 1900. Zur Kenntnis der Fauna und Flora finnischer Binnenseen. Acta Soc. pro Fauna Floora Fennica XIX (2): 1–55.

Levander, K. M. & Wuorentaus, Y., 1915. Planktonsammansättningen i Kemi, Uleå och Kumo älf samt Kymmene och Saima system på grund af från juni 1913 till juni 1914 månatligen utförda håfningar (Redogörelse afgifven af arbetsutskottet för undersökning af de finska insjöarnas vatten och plankton, III). Fennia 39(2): 1–36.

Levander, K. M. & Wuorentaus, Y., 1917. Planktonsammansättningen i finska insjöar och floder på grund af håfningar utförda sommaren 1913 (Redogörelse afgifven af arbetsutskottet för undersökning af de finska insjöarnas vatten och plankton, IV). Fennia 40(6): 1–95.

Meriläinen, J., 1967. On the primary production of the meromictic Lake Valkiajärvi, in the Finnish Lake District. Ann. bot. fenn. 7: 29–51.

Meriläinen, J., 1970. On the limnology of the meromictic Lake Valkiajärvi, in the Finnish Lake District. Ann. bot. fenn. 7: 29–51.

Nygaard, G., 1949. Hydrobiological studies on some Danish ponds and lakes. Part II. The quotient hypothesis and some new or little known phytoplankton organisms. Kongl. Danske Vidensk. Sels. Biol. Skr. 12(1): 1–293.

Rodhe, W., 1969. Crystallization of eutrophication concepts in Northern Europe. In: Eutrophication: Causes, Consequences, Correctives, pp. 50–64. Washington, D.C.

Ruuhijärvi, R., 1974. A general description of the oligotrophic Lake Pääjärvi, southern Finland, and the ecological studies on it. Ann. bot. fenn. 11: 95–104.

Ryding, S.-O. (ed.), 1980. Monitoring of inland waters. OECD eutrophication programme, the nordic project. Report from the working group for eutrophication research. Nordforsk Publ. 1980 (2): 1–207.

Ryhänen, R., 1968. Die Bedeutung der Humussubstanzen im Stoffhaushalt der Gewässer Finnlands. Mitt. int. Verein. Limnol. 14: 168–178.

Salonen, K., 1981. The ecosystem of the oligotrophic Lake Pääjärvi 2. Bacterioplankton. Verh. int. Verein. Limnol. 21: 416–420.

Salovaara, E., 1969. Notes on the phytoplankton of the polyhumic Lake Hakojärvi. Mimeogr. Univ. Helsinki, Dept. Limnol. 83 p. (In Finnish).

Sorsa, K., 1979. Primary production of epipelic algae in Lake Suomunjärvi, Finnish North Karelia. Ann. bot. fenn. 16: 351–366.

Thunmark, S., 1945. Zur Sociologie des Süsswasserplanktons. Eine methodologisch-ökologische Studie. Folia limnol. scand. 3: 1–66.

Tolonen, A., 1980. Phytoplanktonic primary production in Lake Kilpisjärvi. Luonnon Tutkija 84: 49–51. (In Finnish).

Vollenweider, R. A. & Dillon, P., 1974. The application of the phosphorus loading concept to eutrophication research. National Research Council Canada No. 13690. 42 p. Burlington.

Dispersion patterns of phytoplankton in lakes

R. I. Jones & R. C. Francis
Department of Human Sciences, Loughborough University of Technology, Loughborough LE11 3TU, England

Keywords: phytoplankton, abundance estimates, dispersion, contagion, micro-patchiness

Abstract

The spatial distribution of phytoplankton is considered in relation to factors which may cause deviations from a random distribution. Such deviations may arise over a wide range of scales and attention is focused on small-scale contagion and its possible significance. Evidence is presented that contagion may occur at scales smaller than have previously been considered. Samples covering a range of volumes from 0.1 ml to 10 ml were taken from a small lake and total numbers of several phytoplankton species were counted in each sample to avoid problems arising from sub-sampling. The counts were used to calculate different indices of dispersion. All indices indicated contagion to occur at the scale being considered. Within the range examined the 'patch size' appeared variable between species. The possible derivation and relevance of such 'micro-patchiness' is discussed.

Introduction

Since the earliest investigations of phytoplankton in lakes it has been recognized that the abundance of any species within a lake is markedly variable. Indeed, the search for explanations to account for such variability has provided the chief stimulus to research in phytoplankton ecology. In particular, the way in which the community structure of phytoplankton can alter dramatically over short periods of time (often providing a strong contrast with the situation in terrestrial plant communities) has provoked many investigations employing divers approaches. These range from elegant studies of the induction and control of growth in a particular species in a selected lake (e.g. Lund 1950, 1954) to more general discussions of the nature of seasonal succession in phytoplankton communities (e.g. Round 1971).

A second aspect of the variation in phytoplankton abundance which has attracted considerable interest is the vertical heterogeneity often evident down the water column. Many relevant environmental conditions show strong vertical gradients within lakes, especially during periods of stratification, and it has long been a belief that study of the distribution of phytoplankton species in relation to such gradients would enhance understanding of their ecological requirements. Studies of this kind have often established strong links between spatial heterogeneity in the vertical plane and shorter-term temporal variability, through the evidence that many motile species of algae undertake distinct diurnal vertical migrations (e.g. Berman & Rodhe 1971; Ilmavirta 1974; Happey-Wood 1976; Heaney & Talling 1980). Certain non-motile species are also able to regulate their buoyancy and may show highly variable vertical distribution patterns (Reynolds 1972).

However, in addition to this temporal and vertical spatial variation, it has become clear that there are other sources of variation affecting estimates of phytoplankton abundance which are poorly understood. Figure 1 summarizes the different modes and

Hydrobiologia 86, 21–28 (1982). 0018-8158/82/0861-0021/$01.60.

22

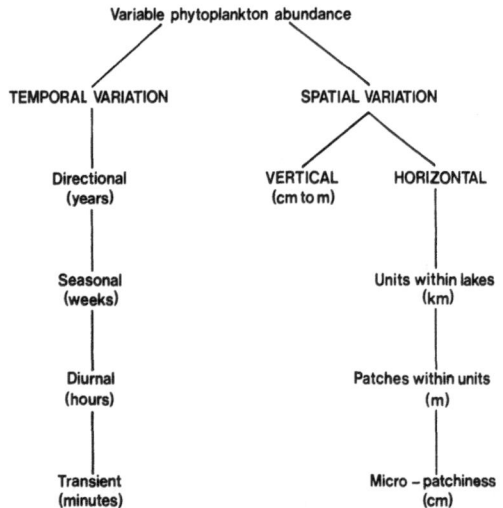

Fig. 1. Sources of variation which may influence estimates of phytoplankton abundance in lakes.

scales of operation within modes which may contribute to variation in measured abundance within lakes.

At the upper end of the scale of temporal variation, it is to be expected that increased understanding of long term trends will result from the steady accumulation of routine measurements now being made at many limnological stations (e.g. Lund 1979). Although such 'first hand' records may at present be exceptional, there is a wealth of 'second hand' information available in the sedimentary record of lakes which can be of immense value in putting current observations into perspective (e.g. Elner & Happey-Wood 1980). At the opposite end of the scale of temporal variation, we know of no work which considers the possible importance of what we have termed transient variations, that is the changes in distribution which may occur within a time scale measured in minutes. It is known, however, that physiological processes may show marked changes within such time periods (Harris & Piccinin 1977; Heaney & Talling 1980).

It is with the second component of spatial variation that the remainder of this paper will be primarily concerned, namely heterogenity in the horizontal plane. The assumption of a homogeneous distribution of phytoplankton was once commonplace, but attempts to produce annual production budgets for whole lakes have demonstrated the need to take greater account of spatial

distribution patterns. Indeed, Ganf (1975) was forced to concede that the original IBP aim of estimating organic production at each trophic level in tropical Lake George 'has not been possible for the phytoplankton, although a rough estimate of daily photosynthetic productivity may be given for a single mid-lake station'; this conclusion was based partly on the difficulty of integrating spatial heterogeneity in phytoplankton distribution. Although vertical heterogeneity may be insufficient to negate general models for describing production under unit surface area (Ganf & Horne 1975), or in extreme cases may necessitate less generalized, more empirical solutions (Fee 1973), knowledge of horizontal heterogeneities is so limited that reliable whole lake production estimates must be based on (prohibitively?) large numbers of sampling stations. Moreover, until more information is available on dispersion patterns of phytoplankton in different kinds of lakes it will remain difficult to optimize a sampling programme in such a way as to obtain adequate precision of results for minimal sampling effort (see Harjula 1979).

Horizontal heterogeneity may occur in lakes over a wide range of scales. Some authors have stressed the existence of particular scales of discontinuity (Platt *et al.* 1970; Sandusky & Horne 1978) while others have considered the range of spatial scales to be a continuum (Haury *et al.* 1977). In practice it may prove difficult to generalize, with any one lake producing a range of discontinuity scales dependant upon its particular morphometry and other internal and external factors affecting mixing of water. In Figure 1 we have distinguished three functional scales in the horizontal plane.

Firstly, there may be recognized within lakes units which show consistently different properties and may effectively be considered as separate water bodies. Such units may be frequent in large lakes, particularly when thorough mixing is impeded by the presence of islands or other morphometric irregularities, although the boundaries between units may be indistinct (Harjula 1979). However, under some circumstances the boundaries between units may be quite abrupt. For example in Lough Neagh, Northern Ireland, a large lake which generally shows a remarkably uniform distribution of phytoplankton (Gibson *et al.* 1971), Kinnego Bay at the southern end of the Lough maintains a phytoplankton standing crop three to four times higher than that in the bulk of Lough Neagh (Jewson 1976; Jones 1977). Generally the distinction between the bay and the rest of the Lough is an abrupt one (Fig. 2), although occasional periods of admixing do occur (Jones, 1977).

The second functional scale recognized in Figure 1 is that of patches within the units discussed above. Such patches may be only a crude localization of individuals as a result of environmental forces, such as the accumulation of buoyant blue-green algae at the downwind end of a lake (George & Edwards 1976). However, there is an increasing realization that more subtle contagious distributions may exist in the open water of lakes, perhaps of a kind that would justify use of the term 'pattern'. The study of small scale pattern is well advanced in terrestrial plant ecology (see recent overview by Greig-Smith 1979), where the relative positional stability of individuals (at least over the period required to take a series of samples!) is highly amenable to the investigation of spatial pattern in series of adjacent samples either from transects or from grids of

quadrats. A recent attempt to apply the same techniques to phytoplankton (Richards & Happey-Wood 1979) suggested surface distributions of algae were contagious with patterns of horizontal distribution which differed between species in dimensions and intensity. However, it does seem questionable whether the sampling procedure used warranted the statistical treatment of results. The pattern analysis technique depends on organisms being sampled at fixed spatial intervals; in the aquatic environment where phytoplankton move freely through the sampling medium under the influence of turbulent flow, a transect line with samples at fixed intervals may effectively generate results no different from those obtained from an equivalent number of random samples. This difficulty might be obviated if samples were taken simultaneously, perhaps by a pneumatically-operated multiple sampler of the sort described by Heaney (1974). Interestingly, if the pattern of water turbulence was known, the same effect might be achieved by sampling one fixed spatial point at appropriate time intervals.

In spite of this criticism, there is no doubt that contagion at a scale of several metres is a widespread phenomenon amongst lake phytoplankton. Contagion may be particularly marked for strongly swimming algae and has been frequently recorded for populations of dinoflagellates (Heaney 1976; Pollinger & Berman 1975). George & Heaney (1978) found frequent contagion in the distribution of phytoplankton in the epilimnion of Esthwaite Water when either *Ceratium* or blue-green algae were the dominant types, but no evidence for departure from randomness when other algae were dominant. These results were based on samples from stations at intervals of some 100 m, so would not have detected contagion on a smaller scale. Richards & Happey-Wood (1979), however, presented evidence of contagious distributions for a number of non-motile species with a scale of pattern at about 24 m and some suggestion of contagion at a scale of 2–4 m.

Against this background of apparent departures from random distribution of phytoplankton at successively smaller scales, we decided to investigate the possibility of micro-scale contagion using small sample volumes. Samples were taken from Charnwood Water, a disused gravel pit near Loughborough, Leicestershire. This lake (area 1186

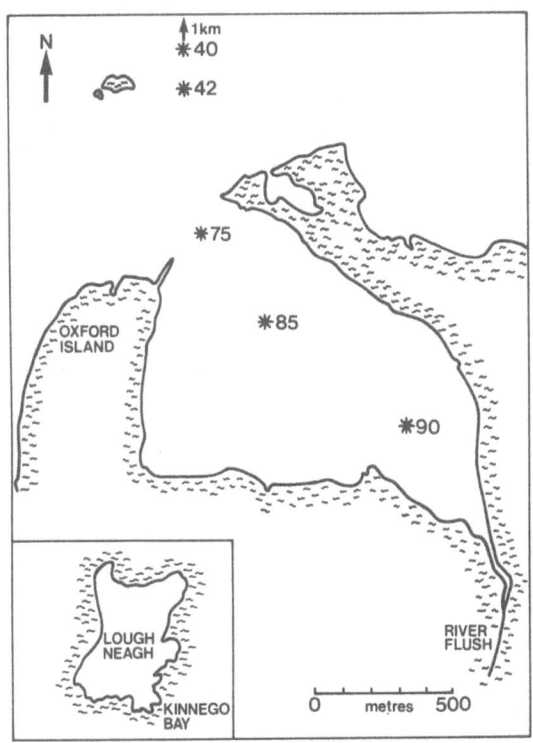

Fig. 2. Values of chlorophyll *a* (μg l^{-1}) along a transect from Lough Neagh into Kinnego Bay on 10 September 1973, showing the abruptness of the transition.

m², maximum depth 8 m) does not stratify, although vertical discontinuities in temperature and oxygen content may be present for periods of a few days during the summer months. Chlorophyll *a* values in the lake are generally around 2 μg l⁻¹, although values up to 40 μg l⁻¹, have been recorded and considerable variation between sets of samples taken with a 5 l Friedinger sampler is common (N. Billington, personal communication).

Methods

Sampling was carried out from a stationary small boat by dipping syringes to a depth of 5 cm and withdrawing the desired volume of water. Twenty samples of each of five sample sizes (10 ml, 5 ml, 1 ml, 0.5 ml and 0.1 ml) were taken in random order. These samples were transferred to small vials and preserved with Lugol's iodine solution. Four common and easily distinguishable species were selected for counting: *Chroomonas* sp., *Cryptomonas* sp., *Euglena* sp. and *Melosira italica* (E) Kg. Algae were counted using a Zeiss IM35 inverted microscope and a magnification of × 400, following

sedimentation for a minimum of 12 hours in perspex settling chambers. Each entire sample was sedimented and the whole of the bottom of the chamber was examined. This procedure was adopted because it was felt that any errors arising from sub-sampling (either prior to settling or in selecting fields of view to count) would at best make extremely complicated and at worst invalidate the proposed statistical treatment of the counts. The final counts of each species are therefore as close as it was possible to get to the true numbers in each sample. Thus our counting procedure was very much more rigorous than that frequently employed when a count of 100 organisms gives an accuracy of ± 20% (Lund *et al.* 1958).

Results

The results from the counts are summarized in Table 1 and it is evident that there was considerable variation between samples. The calculated mean density of cells also varied between sample sizes, although without any consistent pattern. This suggests that, with the sample volumes being used,

Table 1. The range, mean and variance of counts for the four species investigated, with the mean cells ml⁻¹ for each sample size.

Species	Sample size (ml)	Range	$\bar{x}$	s^2	mean cells ml⁻¹
Chroomonas sp.	10	2 798–4 566	3 763	289 665	376
	5	990–3 494	2 068	373 426	414
	1	191– 513	342	10 674	342
	0.5	70– 274	168	5 556	336
	0.1	13– 106	44	697	440
Euglena sp.	10	34– 190	93	2 054	9.3
	5	12– 54	38	96	7.6
	1	3– 32	13	69	13.0
	0.5	0– 25	8.6	44	17.2
	0.1	0– 16	5.8	22	11.6
Cryptomonas sp.	10	2– 20	9.1	19.7	0.9
	5	0– 11	4.5	9.1	0.9
	1	0– 3	1.2	1.11	1.2
	0.5	0– 3	0.6	0.57	1.2
	0.1	0– 2	0.4	0.33	3.5
Melosira italica	10	4– 26	10.7	30.7	1.1
	5	0– 14	6.3	16.2	1.3
	1	0– 2	1.9	0.58	1.9
	0.5	0– 3	0.8	0.62	1.6
	0.1	0– 11	0.1	0.09	1.0

twenty samples would not be sufficient to give a close approximation to the true mean density. It should be noted that the values given in Table 1 are arithmetic means; confidence limits have not been calculated since the variance is not independent of the mean (see below) and consequently a suitable transformation of the data would have to be made to normalize the data before confidence limits could be applied.

If the algae were located randomly in the water their dispersion would be correctly described by the Poisson distribution, in which case the variance is equal to the mean and $s^2/\bar{x} = 1$. Departures of this ratio from unity can be tested for significance using χ^2 (Elliott 1977). Table 2 shows that all the species examined gave significant departure from random dispersion, with values of $s^2/\bar{x} > 1$, indicating a contagious distribution. *Chroomonas* and *Euglena* showed significant contagion (p < 0.001) at all sample volumes, but *Cryptomonas* and *Melosira* showed a sudden change from apparently random to significantly contagious dispersion between sample sizes 1 ml and 5 ml. The effect of quadrat size on the detection of non-randomness has been extensively considered in terrestrial plant ecology (see Kershaw 1973) and a sudden change in the significance of the variance: mean ratio of the kind described above occurs when the sample size coincides with the size of the aggregation (Kershaw 1973). This suggests that *Cryptomonas* and *Melosira* were contagiously distributed with discontinuities at a scale of about 5 ml..

In many ecological situations a suitable description of a contagious distribution is provided by the negative binomial distribution. However, the present data for phytoplankton did not show an adequate fit to the negative binomial distribution (Statistic T – Elliot 1977). George (1974) found the distribution of zooplankton in a Welsh reservoir to be usually contagious but to seldom agree with the negative binomial. Cassie (1962) derived an alternative distribution which he termed Poisson-lognormal, to describe the dispersion of plankton. This distribution has greater skewness than the negative binomial. However, values of Statistic T calculated from the present data were always negative and generally strongly so, which indicates less skewness than the negative binomial (Elliott 1977). A number of possible models might then be more suitable, such as the Polya-Aeppli distribution. Cassie (1962) rejected this distribution for plankton since 'it derives from circumstances in which a habitat is initially colonized at random by parent organisms all arriving simultaneously and later producing clusters of offspring . . . In the case of plankton, initial colonization is not instantaneous but continuous.' However, if one considers a water body with phytoplankton randomly distributed but precluded from growing by unfavourable environmental conditions, then a change in those conditions to permit commencement of growth would produce the same effect as simultaneous colonization. So under some circumstances the Polya distribution and similar types might be as suitable for plankton as they have been shown to be for certain other organisms (Barnes & Stanbury 1951).

While providing a test for non-randomness, the $s^2/\bar{x}$ ratio is a poor measure of the intensity of contagion since it is strongly influenced by the number of organisms in the samples (Elliott 1977). Many alternative indices of dispersion have been suggested and have been critically reviewed by Elliott (1977). Elliott discussed a number of attributes which should be possessed by the ideal index of dispersion and concluded that no one index fulfils all the necessary criteria. However, the

Table 2. Variance: mean ratio ($s^2/\bar{x}$) at each sample size for the four species investigated with significance of departure from random expectation ($s^2/\bar{x} = 1$) (***$p < 0.001$, **$p < 0.01$, *$p < 0.05$).

	Sample size (ml)				
	10	5	1	0.5	0.1
Chroomonas sp.	538***	181***	31.1***	33.1***	16.0***
Euglena sp.	22.0***	2.54***	5.19***	5.14***	3.86***
Cryptomonas sp.	2.16**	2.01**	0.92	0.93	0.94
Melosira italica	2.88***	2.58***	0.55	0.80	0.90

parameter, b, of Taylor's power law and Morisita's index are considered useful and may be applied to the present data.

Taylor (1961) showed that the distribution of individuals in a population is such that the variance is not independent of the mean and that their relationship obeys a power law. This law holds in a continuous series of distributions from regular through random to contagious and is of the form $s^2 = a \bar{x}^b$, where a and b are constants, a being largely a sampling factor and b an index of aggregation characteristic of the species. As an index of dispersion b has the advantage of being independent of n, $\bar{x}$ and Σx, and is therefore of value in interspecific comparisons. However, calculation of b does require several estimates of s^2 and $\bar{x}$, and strictly b cannot be obtained from simple regression techniques since the independent variate ($\bar{x}$) is not error free (see Southwood 1978); in practice this latter criticism is generally ignored. Figure 3 shows plots of s^2 against $\bar{x}$ on log scales for the four species of algae we counted. All slopes indicate departure from random expectation with the calculated regressions showing very good fits to the data. Thus contagion is strongly indicated for all species, but for *Melosira* and *Cryptomonas* the regression line crosses the Poisson line (slope $b = 1$) showing a change from contagious to random distribution at lower values of $\bar{x}$. This supports the evidence from the ratio $s^2/\bar{x}$ (Table 2) that these two species have a 'clump size' of about 5 ml.

Morisita's index of dispersion (Elliott 1977) is essentially similar to Lloyd's measure of patchiness (Lloyd 1967), but has the advantage of a significance test for departure from randomness. The index was calculated as:

$$I\delta = n \; \frac{\Sigma(x^2) - \Sigma x}{(\Sigma x)^2 - \Sigma x}$$

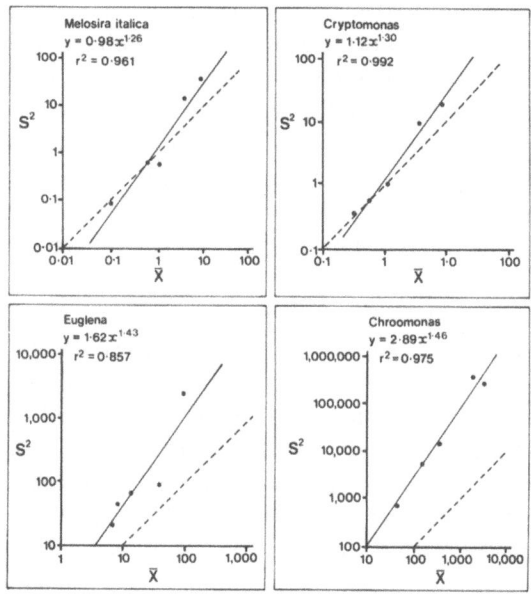

Fig. 3. Plots of variance (S^2) against mean (x) on $\log_{10}$ scales for the four algae counted. The line of random expectation (---) and the fitted power law equations are also given.

and departures from randomness were judged significant when:

$$I\delta \, (\Sigma x - 1) + n - \Sigma x$$

was outside the appropriate significance levels of χ^2 for $n - 1$ degrees of freedom (Elliott 1977). Values of this index and indications of departure from randomness are shown in Table 3. The index is independent of sample mean ($\bar{x}$) and total numbers in the sample (Σx) but is a strong function of the number of sample units (n) (Elliott 1977). In the present study where the number of sample units was kept constant, values of the index may be compared between sample sizes and between species.

Deviations of the index values from that expected for a random distribution are small, but in some instances highly significant (Table 3). Both

Table 3. Values of Morisita's Index at each sample size for the four species investigated with level of significance (*** $p < 0.001$, ** $p < 0.01$, * $p < 0.05$).

	Sample size (ml)				
	10	5	1	0.5	0.1
Chroomonas sp.	1.02***	1.09***	1.09***	1.19***	1.35***
Euglena sp.	1.23***	1.04***	1.32***	1.57***	1.51***
Cryptomonas sp.	1.14**	1.24**	0.98	1.04	0.95
Melosira italica	1.18***	1.26***	0.60	0.79	0.00

Cryptomonas and *Melosira* only show significant contagion for the 5 ml and 10 ml sample sizes, confirming the pattern indicated by other indices. *Chroomonas* and *Euglena* show significant contagion ($p < 0.001$) at all sample volumes, although the value of the index for these two species is higher at the smallest sample size (0.1 ml and 0.5 ml) suggesting a smaller scale of patchiness than for the other two species.

Discussion

Data of the sort presented here raise three particular questions to be considered: firstly how real are the patterns described; secondly how does the recognition of such dispersion patterns affect the study of phytoplankton, especially sampling procedures; and thirdly how meaningful are these patterns in the ecology of the organisms involved?

Sufficient published reports exist for there to be no doubt that random dispersion of phytoplankton is the exception rather than the rule, with contagious dispersion occurring over a range of scales. Evidence for contagion at the scale demonstrated by the present data has not been published previously, although Haury *et al.* (1977) discuss micro-scale pattern of the same order of magnitude. It is difficult to comment on the reality of such contagion when so few data are available. We have obtained significant departures from random distribution with four species at a range of densities and, having taken precautions to overcome problems such as that posed by subsampling, we feel that the contagion we have shown is real and not a methodological artefact. However, the intensity of contagion (i.e. the contrast between 'patches' and background) is generally low; this is important to consideration of the ecological meaning of the dispersion patterns.

Large scale departures from a random distribution clearly have important consequences for the estimation of phytoplankton abundance within a water body. Micro-scale contagion of the kind we have demonstrated will not affect the sampling programme, since the dimensions of pattern are well below those of the samplers in common use. However, since algal counts are generally made on small subsamples taken from a larger sampler, care must be taken to ensure that subsamples are truly representative. Moreover, with the increasing interest in the use of close-interval samplers to investigate vertical distribution of phytoplankton (Heaney 1974; Blaker 1979) more attention will have to be paid to the interdependence of the horizontal and vertical planes of patchiness (George & Heaney 1978).

The ecological importance of micro-scale patchiness must remain obscure, in particular when nothing is known of the stability of such distributions. It is difficult to see how this information could be obtained, but without some idea of rates of development, breakdown and duration, interpretation of patterns remains speculation. Different species appear to show different scales of pattern and no correlation was found between species abundance through the samples, i.e. 'patches' for different species did not coincide spatially. These observations do not appear consistent with patterns resulting from physical processes such as small scale water turbulence, and we therefore feel that microdistribution patterns are more probably of biological origin. Patchiness could result from differential growth or grazing, as is known to occur at much larger scales. Depending on the duration of distribution patterns, such processes could produce a locally heterogeneous environment for algal cells, both biologically and chemically, not reflected by the relatively coarse environmental measurements available to the limnologist. Although such suggestions are only speculation, it does seem clear that more consideration needs to be given to the conditions experienced by an algal cell if we are to improve understanding of phytoplankton growth and production processes.

References

Barnes, H. & Stanbury, F. A., 1951. A statistical study of plant distribution during the colonization and early development of the vegetation on china clay residues. J. Ecol. 39: 171–181.

Berman, T. & Rodhe, W., 1971. Distribution and migration of Peridinium in Lake Kinneret. Mitt. int. Ver. Limnol. 19: 266–276.

Blakar, I. A., 1979. A close interval water sampler with minimal disturbance properties. Limnol. Oceanogr. 24: 983–988.

Cassie, R. M., 1962. Frequency distribution models in the ecology of plankton and other organisms. J. Anim. Ecol. 31: 65–92.

Elliott, J. M., 1977. Some Methods for the Statistical Analysis of Samples of Benthic Invertebrates (2nd ed.). Scient. Publs. Freshwat. Biol. Ass. 25, 160 pp.

Elner, J. K. & Happey-Wood, C. M., 1980. The history of two linked but contrasting lakes in North Wales from a study of pollen, diatoms and chemistry in sediment cores. J. Ecol. 68: 95–121.

Fee, E. J., 1963. A numerical model for determining integral primary production and its application to Lake Michigan. J. Fish. Res. Bd Canada 30: 1447–1468.

Ganf, G. G., 1975. Photosynthetic production and irradiance – photosynthesis relationships of the phytoplankton from a shallow equatorial lake (Lake George, Uganda). Oecologia (Berl.) 18: 165–183.

Ganf, G. G. & Horne, A. J., 1975. Diurnal stratification, photosynthesis and nitrogen-fixation in a shallow, equatorial lake (Lake George, Uganda). Freshwat. Biol. 5: 13–39.

George, D. G., 1974. Dispersion patterns in the zooplankton populations of a eutrophic reservoir. J. Anim. Ecol. 43: 537–551.

George, D. G. & Edwards, R. W., 1976. The effect of wind on the distribution of chlorophyll a and crustacean plankton in a shallow eutrophic reservoir. J. appl. Ecol. 13: 667–690.

George, D. G. & Heaney, S. I., 1978. Factors influencing the spatial distribution of phytoplankton in a small productive lake. J. Ecol. 66: 133–155.

Gibson, C. E., Wood, R. B., Dickson, E. L. & Jewson, D. H., 1971. The succession of phytoplankton in L. Neagh 1968–70. Mitt. int. Ver. Limnol. 19: 146–160.

Greig-Smith, P., 1979. Pattern in vegetation. J. Ecol. 67: 755–779.

Happey-Wood, C. M., 1976. Vertical migration patterns in phytoplankton of mixed composition. Br. phycol. J. 11: 355–369.

Harjula, H., 1979. Analysis of errors in estimating phytoplankton primary productivity and chlorophyll a with special reference to Lake Päijänne. Ann. bot. fenn. 16: 307–337.

Harris, G. P. & Piccinin, B. B., 1977. Photosynthesis by natural phytoplankton populations. Arch. Hydrobiol. 80: 405–457.

Haury, L. R., McGowan, J. A. & Wiebe, P. H., 1977. Patterns and processes in the time-space scales of plankton distributions. In: Steele, J. H. (ed.) Spatial Pattern in Plankton Communities, pp. 277–327. Plenum, New York.

Heaney, S. I., 1974. A pneumatically-operated water sampler for close intervals of depth. Freshwat. Biol. 4: 103–106.

Heaney, S. I., 1976. Temporal and spatial distribution of the dinoflagellate Ceratium hirundinella O. F. Muller within a small productive lake. Freshwat. Biol. 6: 531–542.

Heaney, S. I. & Talling, J. F., 1980. Dynamic aspects of dinoflagellate distribution patterns in a small productive lake. J. Ecol. 68: 75–94.

Ilmavirta, V., 1974. Diel periodicity in the phytoplankton community of the oligotrophic lake Pääjärvi, southern Finland. I. Phytoplanktonic primary production and related factors. Ann. bot. fenn. 11: 136–177.

Jewson, D. H., 1976. The interaction of components controlling net phytoplankton photosynthesis in a well-mixed lake (Lough Neagh, Northern Ireland). Freshwat. Biol. 6: 551–576.

Jones, R. I., 1977. Factors controlling phytoplankton production and succession in a highly eutrophic lake (Kinnego Bay, Lough Neagh). I. The phytoplankton community and its environment. J. Ecol. 65: 547–559.

Kershaw, K. A., 1973. Quantitative and Dynamic Plant Ecology (2nd ed.). Edward Arnold, London.

Lloyd, M., 1967. 'Mean crowding'. J. Anim. Ecol. 36: 1–30.

Lund, J. W. G., 1950. Studies on Asterionella formosa Hass. II. Nutrient depletion and the spring maximum. J. Ecol. 38: 1–35.

Lund, J. W. G., 1954. The seasonal cycle of the plankton diatom, Melosira italica (Ehr.) Kutz. subsp. subarctic O. Müll. J. Ecol. 42: 151–179.

Lund, J. W. G., 1979. Changes in the phytoplankton of an English lake, 1945–1977. Hydrobiol. J. 14; 6–21.

Lund, J. W. G., Kipling, C. & Le Cren, E. D., 1958. The inverted microscope method of estimating algal numbers and the statistical basis of estimations by counting. Hydrobiologia 11: 143–170.

Platt, T., Dickie, L. M. & Trites, R. W., 1970. Spatial heterogeneity of phytoplankton in a near-shore environment. J. Fish. Res. Bd Can. 27: 1453–1473.

Pollinger, U. & Berman, T., 1975. Temporal and spatial distribution of the dinoflagellate blooms in Lake Kinneret, Israel. Verh. int. Ver. Limnol. 19: 1370–1382.

Reynolds, C. S., 1972. Growth, gas vacuolation and buoyancy in a natural population of a blue-green alga. Freshwat. Biol. 2: 87–106.

Richards, M. C. & Happey-Wood, C. M., 1979. The application of pattern analysis to freshwater phytoplankton communities. Limnol. Oceanogr. 24: 950–956.

Round, F. E., 1971. The growth and succession of algal populations in freshwaters. Mitt. int. Ver. Limnol. 19: 70–99.

Sandusky, J. C. & Horne, A. J., 1978. A pattern analysis of Clear Lake phytoplankton. Limnol. Oceanogr. 23: 636–648.

Southwood, T. R. E., 1978. Ecological Methods (2nd ed.). Chapman & Hall, London.

Taylor, L. R., 1961. Aggregation, variance and the mean. Nature (Lond.) 189: 732–735.

On the annual variation of phytoplankton biomass in Finnish inland waters

Pertti Heinonen

Water Research Institute, National Board of Waters, Finland P.O. Box 250, SF-00101 Helsinki, Finland

Keywords: phytoplankton, eutrophication, biological monitoring

Abstract

Annual variations in phytoplankton biomass in 63 lakes in Southern and Central Finland are discussed. Biomass is rather small during winter (January–April), usually <0.05 mg l^{-1} (fresh weight) and there are no differences between oligotrophic and eutrophic lakes. In early spring and in autumn biomass varies widely, depending mainly on water temperature. Phytoplankton biomass is smaller in July than in June and August in oligotrophic lakes (biomass <0.20 mg l^{-1} fresh weight) and mesotrophic (biomass 1.0–2.5 mg l^{-1}) lakes, but greater in eutrophic (biomass 2.5–10.0 mg l^{-1}) and hypereutrophic (biomass >10.0 mg l^{-1}) lakes. The standard deviation of phytoplankton biomass in Finnish inland waters is usually smallest in July, which facilitates the comparison of phytoplankton between different kinds of lakes.

Introduction

Several growth factors, of which the most important are temperature, light intensity and the availability of nutrients, particularly phosphorus (cf. e.g. Granberg 1973; Ilmavirta 1974; Heinonen 1980), affect phytoplankton biomass in Finnish inland waters. Due to the low nutrient content of the soil, most Finnish lakes are in an oligotrophic state. Eutrophication is observed only in waters contaminated by effluents or which for some other reason have abnormally high nutrient levels.

In this research the variations in phytoplankton biomass in different types of water bodies were investigated. The division into oligotrophic, mesotrophic, eutrophic and hypereutrophic lakes was made on the basis of July phytoplankton biomass levels (Heinonen 1980). In particular, phytoplankton biomass in the summer months is considered and an attempt is made to estimate the correct timing of sampling in view of the fact that research funds are very often limited.

Material and methods

The phytoplankton samples (15 lakes) from the winter season (January–April) are from the years 1967–1969 (Lepistö & Kerminen 1970) from different parts of Finland. The summer samples (June–August) are from 1971 (Lepistö *et al.* 1979) and represent waters in Southern and Central Finland (63 lakes). The samples were profile samples, in most cases extending from the surface to a depth of 2 m, and were preserved with formalin. Phytoplankton biomass was assayed by using the Utermöhl technique (Utermöhl 1958). The species volume information required for calculation of biomass was obtained from the mean values for each algal species recorded by Naulapää (1972).

Results and discussion

The surface water temperature of Finnish lakes in winter is 0.1–0.4 °C and the intensity of illumination is very low due to the covering of ice and snow.

Hydrobiologia 86, 29–31 (1982).   0018-8158/82/0861–0029/$00.60.

Table 1. Variations in phytoplankton biomass in the winter season of years 1967–1969 in different types of lakes in Finland.

Type of lake	Phytoplankton biomass in winter (January–April) mg l^{-1} (fresh weight)					
	$\bar{x}$	s	V%	min	max	n
Oligotrophic	0.02	0.014	61	0.01	0.06	25
Mesotrophic	0.04	0.011	28	0.02	0.05	7
Eutrophic	0.03	0.015	55	0.01	0.05	13
Hypereutrophic	0.02	0.015	86	0.01	0.04	4

For these reasons the production and biomass of phytoplankton was very low (Table 1).

The winter biomass levels varied between 0.01 and 0.06 mg l^{-1}, with no significant differences between the different lake types. Most of the species observed were diatoms, the most dominant species being *Asterionella formosa* Hass and *Melosira italica* (E.) Kg. The level of phytoplankton was independent of nutrient concentrations, since growth was inhibited by low illumination and temperature. Examination of indicator species could possibly yield differences on which to base the classification of lakes, but this was not carried out in this study because of the scarcity of material.

The major emphasis in the monitoring of the annual cycle of phytoplankton was centred on the three summer months when for example the recreational use of water bodies is at its maximum. The summer levels of phytoplankton varied considerably between the different lake types (Table 2).

The range in biomass levels for the whole summer season was in oligotrophic lakes 0.07–2.75 mg l^{-1}, in mesotrophic lakes 0.17–5.41 mg l^{-1}, and in eutrophic and hypereutrophic lakes 0.74–13.17 and 7.05–26.91 mg l^{-1}, respectively. In oligotrophic and

mesotrophic lakes the biomass maximum was in June and the minimum in July. With increasing eutrophication the maximum values were observed in July and the minimum in June (Fig. 1). A similar shift in the biomass maximum for eutrophic lakes in Denmark has been described by Mathiesen (1971).

For the purposes of classification of lakes it is important to be able to distinguish reliably between

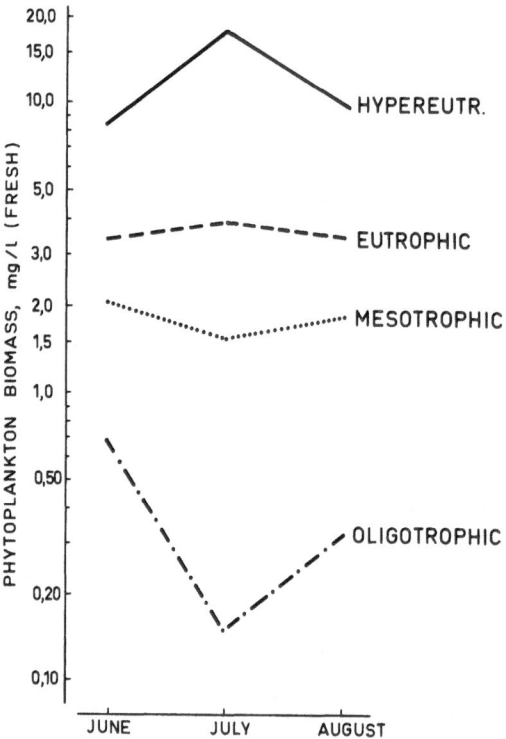

Fig. 1. Variations of phytoplankton fresh weight biomass in summer 1971 in different types of lakes in Southern and Central Finland (n = 63).

Table 2. Variations in phytoplankton biomass in the summer months of year 1971 in different types of lakes in Southern and Central Finland.

Type of lake	Phytoplankton biomass, mg l^{-1} (fresh weight)									
	n	June			July			August		
		$\bar{x}$	s	V%	$\bar{x}$	s	V%	$\bar{x}$	s	V%
Oligotrophic	25	0.68	0.599	88	0.15	0.041	28	0.32	0.162	51
Mesotrophic	23	2.02	1.410	70	1.56	0.433	28	1.79	1.435	80
Eutrophic	12	3.39	3.329	98	3.93	1.519	39	3.46	2.561	74
Hypereutrophic	3	8.46	1.522	18	17.46	8.473	49	9.77	1.325	14

adjacent groups. On the basis of the June samples only the difference between eutrophic and hypereutrophic lakes was statistically significant. In August the oligotrophic and mesotrophic groups were distinguishable with high significance and the eutrophic and hypereutrophic groups showed significant difference. In July the differences between the oligotrophic and mesotrophic and also between the mesotrophic and eutrophic groups were highly significant. These results suggest that in those cases in which it is necessary to rely on infrequent sampling, the monitoring of phytoplankton biomass should be conducted in July. In this way the differences would be observable with maximum reliability.

References

Granberg, K., 1973. Näkökohtia vesibiokenoosin talvitutki-muksista: Kasviplankton. Summary: Some aspects on the phytoplankton investigations of the polluted lakes during the wintertime. Limnologisymposion 1970: 20-27.

Heinonen, P., 1980. Quantity and composition of phytoplankton in Finnish inland waters. National Board of Waters, Finland. Publ. Wat. Res. Inst. 37: 1-91.

Ilmavirta, V., 1974. Diel periodicity in the phytoplankton community of the oligotrophic lake Pääjärvi, southern Finland. I. Phytoplanktonic primary production and related factors. Ann. bot. fenn. 11: 136-177.

Lepistö, L. & Kerminen, S., 1970. Kuukausittainen plankton-tutkimus 38 Suomen järvestä vv. 1967-1969. Maatalous-hallitus. Vesiensuojelutoimiston tiedonantoja 61: 1-195.

Lepistö, L., Kokkonen, P. & Puumala, R., 1979. Kasviplank-tonin määristä ja koostumuksesta Vuoksen, Kymijoen ja Kokemäenjoen vesistöalueilla kesällä 1971. National Board of Waters, Finland. Report 172: 1-250.

Mathiesen, H., 1971. Summer maxima of algae and eutrophication. Mitt. int. Verein. Limnol. 19: 161-181.

Naulapää, A., 1972. Eräiden Suomessa esiintyvien planktereiden tilavuuksia. Summary: Mean volumes of some Plankton Organisms found in Finland. National Board of Waters, Finland. Report 40: 1-47.

Utermöhl, H., 1958. Zur Vervollkommung der quantitativen Phytoplanktonmethodik. Mitt. int. Verein. Limnol. 9: 1-38.

Dynamics of phytoplankton in the brackish-water inlet Pojoviken, southern coast of Finland

Åke Niemi

Tvärminne Zoological Station, SF-10850 Tvärminne, Finland

Keywords: Baltic Sea, brackish-water firth, salinity stratification, water exchange, phytoplankton production, regulating factors, critical salinity range

Abstract

The water exchange between the brackish-water firth Pojoviken and the Baltic Sea is restricted by a shallow sill (6 m). An outflowing, oligohaline surface layer is isolated from the nutrient-rich mesohaline deep water by a pycnocline at a depth of 6–10 m. During the ice-free period phytoplankton production is chiefly regulated by the river discharge regime. Contrary to the situation in the outer archipelago and the sea zone, in Pojoviken phytoplankton production continues until late autumn, because the stable salinity stratification prevents the phytoplankton from sinking below the critical depth for production. The phytoplankton composition seems to be regulated chiefly by salinity. The salinity interval 2–2.5‰ is apparently the critical range where brackish-water phytoplankton changes to an assemblage composed of typical freshwater species.

General

Pojoviken (Pohjanpitäjänlahti) is a stratified sea inlet east of Hangö (Hanko) peninsula (Fig. 1). A sill at 6 m causes stagnation in summer, which leads to decreased oxygen concentrations below the pycnocline. In winter inflowing saline water of high density renews the deep water (Figs. 2, 3, 5). Pojoviken is a classic research area. The hydrography has been studied since the beginning of this century (Granqvist 1914; Buch 1914; Witting 1914; Halme 1944; Launiainen 1972; Niemi 1973, 1977, 1978; Virta 1977), as has the phytoplankton (Levander 1915; Halme & Mölder 1958; Niemi *et al.* 1970; Niemi 1973). The water exchange is typical of inlets with a shallow sill in the Northern Baltic Sea.

Phytoplankton production, succession and regulating factors

The phytoplankton production is regulated by light and nutrients. In an estuarine firth like Pojoviken, the light available for the phytoplankton community is dependent on the water stratification, i.e. the location of the pycnocline, and on the transparency of the water. In winter, when the snow-covered ice allows insufficient light to reach the surface water, there is no marked phytoplankton production (Niemi 1973, 1978).

The amounts of nutrients available during the growing period, inorganic phosphorus and nitrogen and reactive silicon, are greatly dependent on the discharge of the Svartån, which again shows marked seasonal and year-to-year fluctuations connected with the meteorological conditions (Niemi 1977, 1978).

After the break-up of the ice, usually in late April, the surface layer contains abundant nutrients mobilized during the winter (Fig. 3). The period after the ice breaks up (April–June) is the time of greatest stream discharge, and the surface layer is turbid due to suspended clay particles. This leads to a poor light climate, which depresses primary

Hydrobiologia 86, 33–39 (1982). 0018-8158/82/0861-0033/$01.40.
© Dr W. Junk Publishers, The Hague.

34

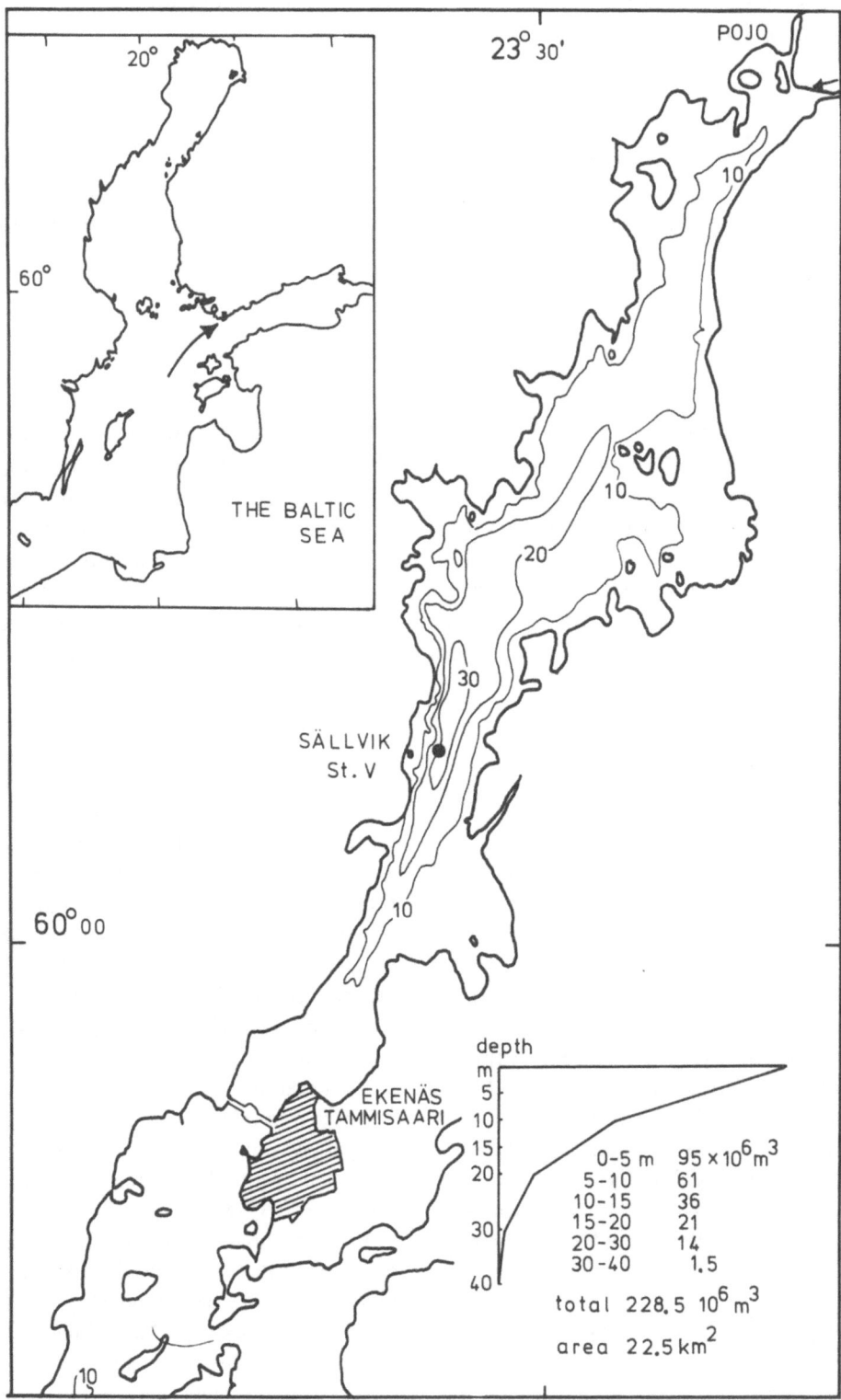

Fig. 1. Map of the study area and depth profile of Pojoviken, and volume between different depth contours (according to Skult 1976).

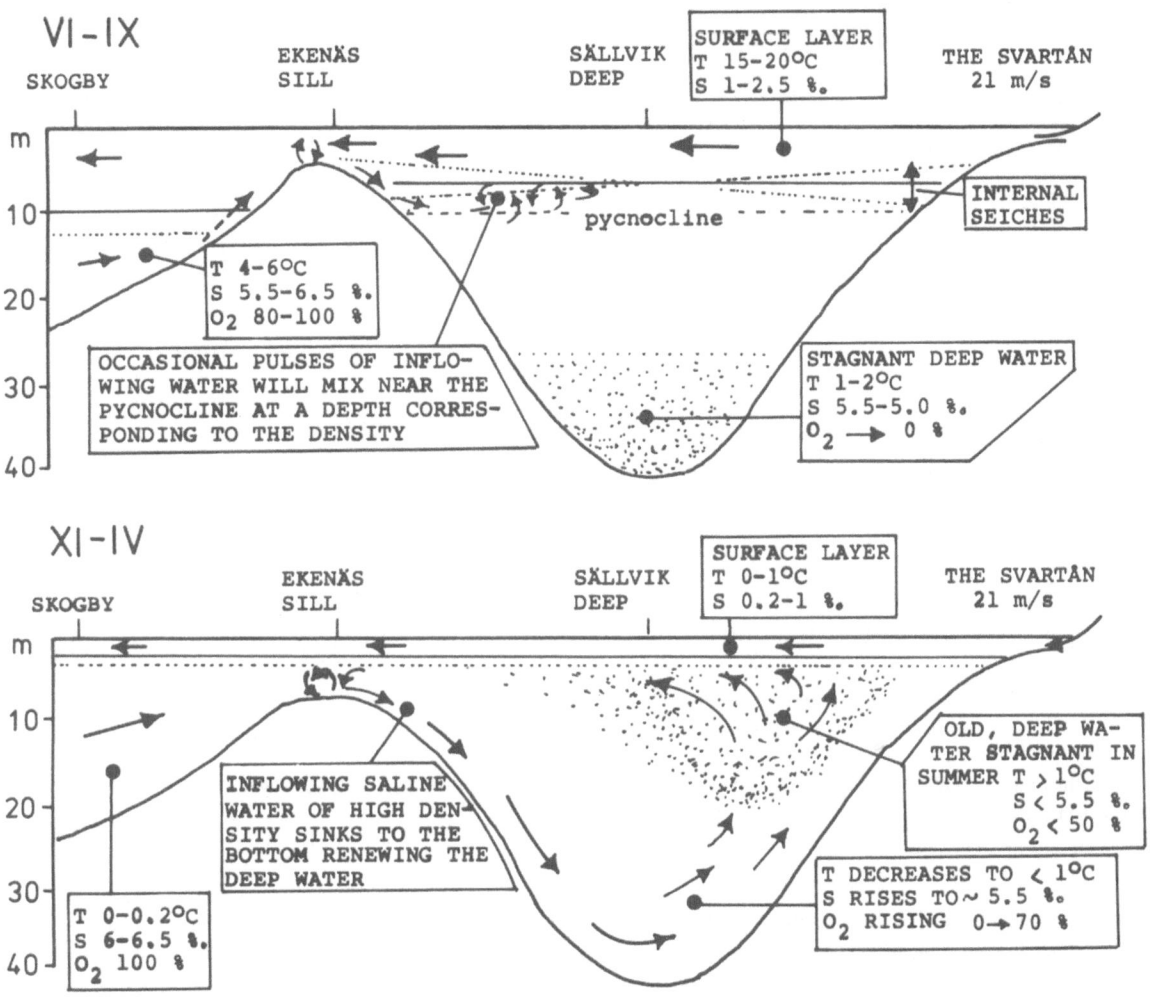

Fig. 2. Water exchange and stratification in Pojoviken in summer (June–September, above), and in winter (November–April, below).

production (cf. van den Hoek *et al.* 1979). As the turbidity decreases in May, phytoplankton production gradually increases, generally reaching a maximum in mid June (0.5–1.0 g C m^{-2} day^{-1}). The unfavourable light conditions in spring are probably the reason why the production maximum is later in Pojoviken than in the outer archipelago and sea zone, where the vernal production peak occurs in early May (Niemi 1972, 1975). During the phytoplankton maximum in June, which is made up chiefly by diatoms, the nutrients PO$_4$-P, NO$_3$-N and SiO$_4$-Si decrease, so that by late June their concentrations are low. The more rapid depletion of PO$_4$-P than of inorganic nitrogen (Fig. 3) indicates that phosphorus is an important factor regu-

lating phytoplankton production in June. In late June the SiO$_4$-Si concentration reaches its lowest annual level.

In July follows a low-production stage, provided that the river discharge is small, as it usually is. Owing to the sinking of biogenic material from the productive surface layer, the concentrations of total phosphorus and nitrogen decrease during summer in this layer (Niemi 1978), and the concentrations of inorganic phosphorus and nitrogen become very low. SiO$_4$-Si increases, however, because the amounts of Si-consuming diatoms are small. The primary producers chiefly consist of nanoplankton: chrysophyceans, cryptophyceans and chlorophyceans. During this stage the primary production is

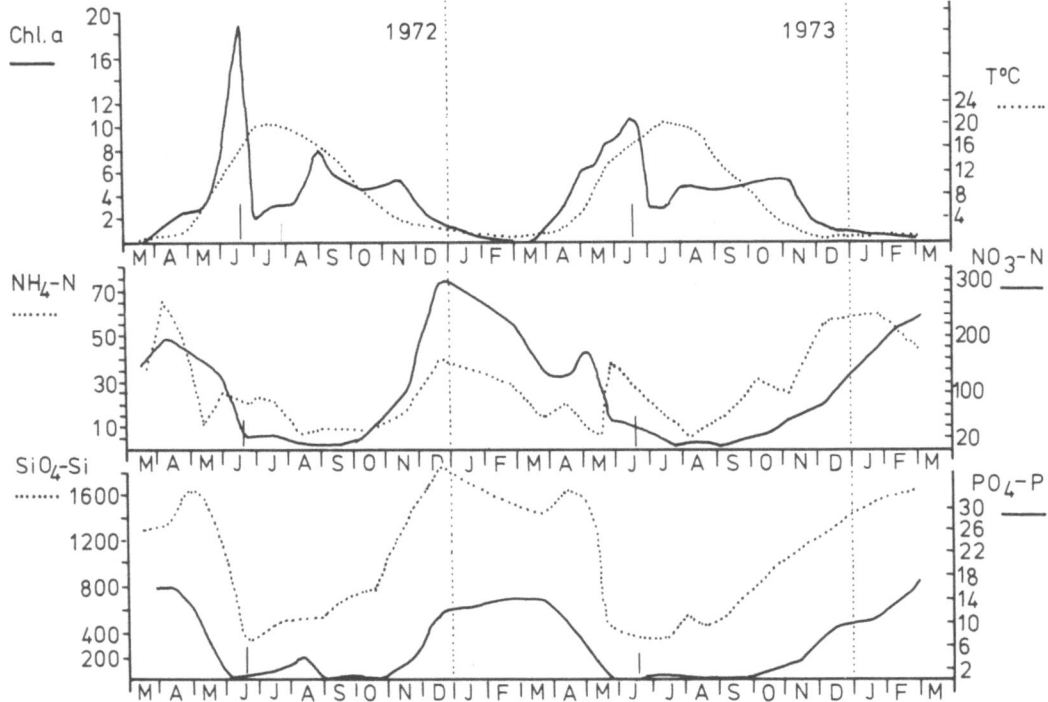

Fig. 3. Seasonal fluctuation of chlorophyll a, NO_3-N, NH_4-N, PO_4-P, SiO_4-Si (μg l^{-1}) and temperature (°C), mean values of the surface layer (0–6 m) at St. Sällvik.

ca. 0.2 g C m^{-2} day^{-1}, and it seems to be regulated by a complex of factors, chiefly the mobilization rate of nutrients (PO_4-P and NH_4-N), temperature, and grazing.

A typical feature in such a strongly stratified inlet is the concurrence of the mixed surface layer and the productive layer. Although the Secchi depth may be 5–5.5 m (the 1-%-light depth accordingly ca. 10–11 m), the primary production extends only to a depth of 6–7 m, i.e. only to the pycnocline (Fig. 4). This is due to the fact that phytoplankton in and below the lower part of the pycnocline will sink and circulate to depths where light is insufficient for net production. Thus the pycnocline makes the productive layer much shallower than the transparency of the water would allow (Niemi 1978).

The phytoplankton succession in Pojoviken includes a late summer or autumnal diatom maximum. Its timing is dependent on an increase of the discharge of nutrient-rich water from the Svartån, possibly also on mixing of surface and subsurface water by internal seiches in the pycnocline. The

phytoplankton production is usually already increasing in August. In September diatoms predominate in the plankton. From late September measurable amounts of inorganic phosphorus and nitrogen appear in the surface layer, and light evidently becomes a limiting factor for production. However, the phytoplankton production is still considerable in November, despite the weak irradiance.

In autumn the hydrographic conditions are more favourable for primary production in an estuarine firth such as Pojoviken than in the outer archipelago and the open Baltic Sea (Fig. 5). From October on, the increasing cyclonic activity over the Baltic Sea area causes deep mixing of the water, reaching down to the bottom or to the permanent halocline at a depth of 60–70 m, i.e. below the critical depth for phytoplankton production (according to Sverdrup 1953). This hinders phytoplankton production in the sea area at this time (Niemi 1975). Only during periods of exceptionally fair weather does an autumnal diatom peak occur in the open sea area (Niemi & Ray 1977; Hällfors & Niemi

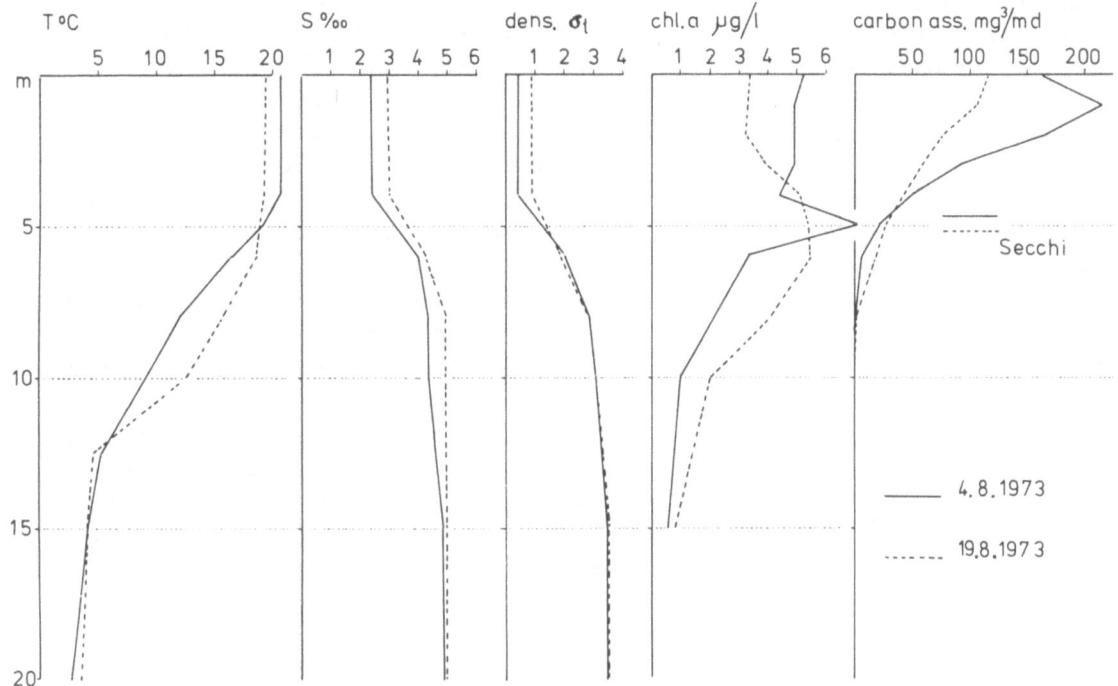

Fig. 4. Primary production, nutrients and hydrographic data on 4 and 19 August in 1973 at Sällvik.

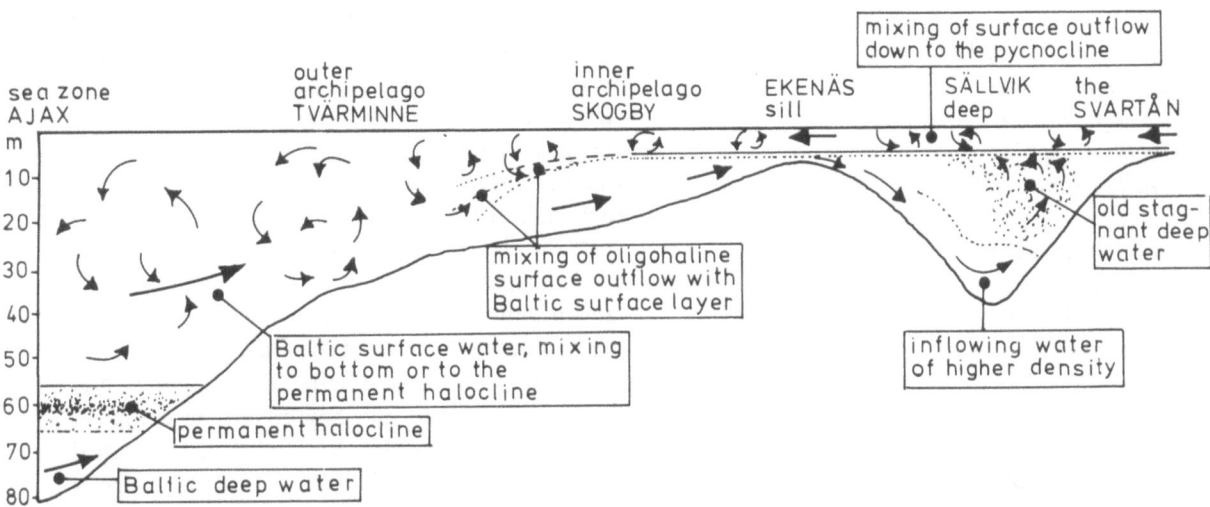

Fig. 5. Water exchange and stratification in the Tvärminne-Pojoviken area in October–November, when the autumnal turnover has taken place in the sea area.

1981). In Pojoviken, on the other hand, stratification prevents mixing below 5–6 m, which is favourable for phytoplankton growth during weak light conditions. As the algae remain in the euphotic layer the whole time, they can use the short daylight efficiently even in November. The increasing discharge of the Svartån in late autumn makes the oligohaline surface layer still shallower, thus partly compensating for the decreasing light intensity. Only the development of ice-cover terminates the growing period.

Thus in an inlet like Pojoviken the superficial permanent pycnocline enhances phytoplankton production in autumn, whereas it must be considered unfavourable in summer, when it makes the productive layer shallower than the light penetration would permit.

Composition of phytoplankton

Only a few remarks on the phytoplankton composition will be made here. After the break-up of the ice *Chlamydomonas* spp. are abundant, occasionally colouring the water green, a feature which is typical of eutrophicated archipelago waters (Välikangas 1925; Häyrén 1929; Melvasalo 1971; Niemi 1972).

The species composition of the diatom maximum in June seems to be regulated mainly by the salinity of the surface water. If the salinity is below 2‰, the phytoplankton consists of freshwater diatoms, which do not occur in water of higher salinity:

Cyclotella comta (Ehr.) Kütz.
C. meneghiniana Kütz.
Melosira spp.
Stephanodiscus spp.
Asterionella formosa Hassal

Fragilaria construens v.
  subsalina Hust.
F. crotonensis Kitton
Synedra acus Kütz.
S. ulna (Nitzsch) Ehr.
Tabellaria fenestrata (Lyngbye) Kütz.

(cf. Halme & Mölder 1958; Niemi 1973).

If the salinity reaches *ca.* 2.5‰, *Diatoma elongatum* (Lyngbye) C.A. Ag. becomes dominant and the above-mentioned species decrease in abundance. If the salinity rises to over 3‰, as in 1980, marine phytoplankton species appear in Pojoviken: *Gonyaulax catenata* (Lev.) Kofoid, *Protoperidinium bipes* (Paulsen) Balech, *P. granii* (Ost.) Balech, *Chaetoceros wighamii* Brightw., *Skeletonema costatum* (Grev.) Cleve. Thus the critical salinity range in Pojoviken seems to be 2–3‰; at these values the freshwater element changes to a brackish-water or marine element. The same salinity range was critical for the brackish-water and freshwater elements of the benthic macrophytes in Pojoviken (Luther 1951).

Little information exists about the influence of salinity on the phytoplankton composition during the summer minimum stage. Generally speaking, an increase in salinity depresses the freshwater chrysophyceans (*Dinobryon bavaricum* Imhof, *Mallomonas* spp.), Chlorococcales and Desmidiales, but favours species occurring in coastal waters, such as *Uroglena americana* Calkins and *Dinobryon divergens* Imhof.

The autumnal diatom bloom consists chiefly of freshwater diatoms owing to the low surface salinity, which is caused by the increased discharge of the Svartån.

Concluding remarks

The phytoplankton production and composition in a strongly stratified estuarine firth like Pojoviken have a regular seasonal pattern. High turbidity in the water several weeks after the break-up of the ice leads to poor light conditions, and restricts the development of a vigorous phytoplankton bloom. After the maximum in June, nutrients are scarce and production is greatly influenced by the freshwater discharge to the inlet. Stratification gives rise to a shallow oligohaline surface layer, which prevents the phytoplankton from circulating below the critical depth. Thus production may continue into late autumn, when it has already come to an end in the outer archipelago and the sea zone.

In estuarine inlets salinity is one of the chief factors determining the phytoplankton composition (cf. also Florin 1957; Willén 1962). The critical salinity seems to be 2–3‰, in which range a freshwater element changes to an assemblage of brackish-water and marine species.

References

Buch, K., 1914. Ueber die Alkalinität, Kohlensäure und Wasserstoffionen-Konzentration in der Pojowiek. Fennia 35 (3): 1-32.

Granqvist, G., 1914. Om de hydrografiska förhållandena i Pojoviken under vintern. Fennia 35 (4): 1-29.

Florin, M.-B., 1957. Plankton of fresh and brackish waters in the Södertälje area. Acta phytogeogr. suec. 37: 1-144.

Hällfors, G. & Niemi, Å., 1981. Vegetation and primary production. In: Voipio, A. (ed.) The Baltic Sea, p. 220-238. Amsterdam.

Halme, E., 1944. Planktologische Untersuchungen in der Pojo-Bucht und angrenzenden Gewässern. I. Milieu und Gesamtplankton. Ann. Zool. Soc. fenn. Vanamo 10 (2): 1-180.

Halme, E. & Mölder, K., 1958. Planktologische Untersuchungen in der Pojo-Bucht und angrenzenden Gewässern. III. Phytoplankton. Ann. Bot. Soc. fenn. Vanamo 30 (3): 1-71.

Häyrén, E., 1929. Zwei Notizen über das Meereseis und die Algen. Memoranda Soc. Fauna Flora Fennica 5: 134-140.

van den Hoek, C., Admiraal, W., Colijn, F. & de Jonge, V. N., 1979. The role of algae and seagrasses in the ecosystem of the Wadden Sea: a review. In: Wolff, W. J. (ed.) Flora and Vegetation of the Wadden Sea, Report 3: 3-118. Rotterdam.

Launiainen, J., 1972. Pohjanpitäjänlahden hydrografiasta 1971. Manuscript. Dept. of Geophys., Univ. Helsinki.

Levander, K. M., 1915. Zur Kenntnis der Bodenfauna und des Planktons der Pojowiek. Fennia 35 (2): 1-39.

Luther, H., 1951. Verbreitung und Ökologie der höheren Wasserpflanzen im Brackwasser der Ekenäs-Gegend in Südfinnland. I. Acta Bot. fenn. 49: 1-231.

Melvasalo, T., 1971. Havaintoja Helsingin ja Espoon merialueiden kasviplanktonlajistosta ja -biomassoista vuosina 1966-1970. (Summary: Observations on phytoplankton species and biomass in the sea area of Helsinki and Espoo in 1966-1970). Rep. Wat. Cons. Lab., Helsinki City Engineer's Office 3 (10): 1-97.

Niemi, Å., 1972. Observations on phytoplankton in eutrophied and non-eutrophied archipelago waters of the southern coast of Finland. Memoranda Soc. Fauna Flora Fennica 48: 63-74.

Niemi, Å., 1973. Ecology of phytoplankton in the Tvärminne area, SW coast of Finland. I. Dynamics of hydrography, nutrients, chlorophyll *a* and phytoplankton. Acta bot. fenn. 100: 1-68.

Niemi, Å., 1975. Ecology of phytoplankton in the Tvärminne area, SW coast of Finland. II. Primary production and environmental conditions in the archipelago and the sea zone. Acta bot. fenn. 105: 1-73.

Niemi, Å., 1978. Ecology of phytoplankton in the Tvärminne area, SW coast of Finland. III. Environmental conditions and primary production in Pojoviken in the 1970s. Acta bot. fenn. 106: 1-28.

Niemi, Å., 1977. Hydrography and oxygen fluctuations in Pojoviken, southern coast of Finland, 1972-1975. Meri (Helsinki) 4: 23-35.

Niemi, Å. & Ray, I.-L., 1977. Phytoplankton production in Finnish coastal waters. Report 2: Phytoplankton biomass and composition in 1973. Meri (Helsinki) 4: 6-22.

Niemi, Å., Skuja, H. & Willén, T., 1970. Phytoplankton from the Pojoviken-Tvärminne area, S. coast of Finland. Memoranda Soc. Fauna Flora Fennica 46: 14-28.

Skult, P., 1976. Gemensam vattendragskontroll för Svartåns nedre lopp, Fiskarsån, Pojoviken samt Ekenäs-Tvärminne havsområde. Sammanställning av resultaten år 1975. Manuscript Tvärminne Zool. Stat., Univ. Helsinki.

Sverdrup, H. U., 1953. On conditions for the vernal blooming of phytoplankton. J. Cons. int. Explor. Mer 18: 287-295.

Välikangas, I., 1925. (Erään Chlamydomonas-lajin vihreäksi värjäämää vettä Kaisaniemen lahdesta.) Meddel. Soc. Fauna Flora fenn. 48: 170, 274.

Virta, J., 1977. Estimating the water and salt budgets of a stratified estuary. Nordic Hydrology 8: 11-32.

Willén, T., 1962. The Utål Lake Chain, Central Sweden and its phytoplankton. Oikos, Suppl. 5: 1-156.

Witting, R., 1914. Kort öfversikt af Pojovikens hydrografi. Fennia 35 (1): 1-18.

Carbon, phosphorus and nitrogen budgets of the littoral Equisetum belt in an oligotrophic lake

Jouko Sarvala[1], Timo Kairesalo[2], Irma Koskimies[3], Anja Lehtovaara[3], Jukka Ruuhijärvi[3] & Inkeri Vähä-Piikkiö[3]

[1] Tvärminne Zoological Station, University of Helsinki, 10850 Tvärminne, Finland
[2] Department of Limnology, University of Helsinki, Viikki, E-building, 00710 Helsinki 71, Finland
[3] Lammi Biological Station, University of Helsinki, 16900 Lammi, Finland

Keywords: oligotrophic lakes, Equisetum, carbon, phosphorus, nitrogen

Abstract

Stores and flows of carbon, phosphorus and nitrogen in a littoral Equisetum stand were studied in 1978–1980 in the oligotrophic, mesohumic lake Pääjärvi, southern Finland. The major carbon and nutrient stores were sediment and Equisetum. The seasonal cycle of the macrophyte vegetation had a profound influence on the whole littoral ecosystem. In spring, when only dead remains of Equisetum were present above ground, there were few differences in nutrient, chlorophyll a and zooplankton concentrations between the littoral and the open lake; phytoplankton and epiphytes were the major producers.

In early June, when new shoots of Equisetum reached the water surface, water exchange between the littoral and the open lake started to diminish, and the characteristic features of a closed macrophyte zone gradually developed: by August the P, Chl a and zooplankton concentrations in the littoral were 5–10 times those in the open lake. From late June until autumn Equisetum was overwhelmingly dominant both in biomass and in production.

The measured total primary production and respiration values indicated a high rate of internal cycling of carbon and nutrients. The daily P requirements of plant growth exceeded the total P stored in the water by a factor of 2–4, and also exceeded the release of nutrients in excretion. High N:P ratios in the water (total 10–64, inorganic 18–171) suggested that P was probably always the limiting nutrient.

The P content of the annual production of Equisetum in Pääjärvi was 2.3% of the mean annual P load, and 5.3% of the mean total P storage in the water volume of the lake.

Introduction

Littoral macrophyte stands are the most productive areas in lakes. Even in oligotrophic lakes the patches of emergent vegetation, growing particularly in the neighbourhood of stream inlets, usually have a clearly eutrophic character, and differ distinctly from the open lake. The role of macrophytic communities in the production and nutrient budgets of lakes has received increasing attention during recent years (e.g. Pieczyńska 1976; Wetzel 1979; see reviews by Hutchinson 1975 and Wetzel 1975). Knowledge of the ecology of littoral communities may be useful for lake management. The macrophytic community as a whole seems to form an effective 'filter', which retains nutrients and organic matter coming from the land (e.g. Tilton & Kadlec 1979), and favours their accumulation in the littoral sediment, which then in turn supports the luxuriant growth of emergent macrophytes. During summer, nutrients coming from land and derived from sediment are bound and stored in organic matter, but in the autumnal decay of the vegetation, they are largely released into the open lake. Harvesting macrophytes from the lake before their decay could thus be one way to diminish the nutrient loading of lakes (e.g. Björk 1972).

The ecosystem of the mesohumic, oligotrophic

Hydrobiologia 86, 41–53 (1982). 0018-8158/82/0861-0041/$02.60.

lake Pääjärvi in southern Finland has been the subject of intensive ecological studies since the early 1970s, and the primary and secondary production as well as the energy budget of the whole lake are documented (see e.g. Ilmavirta 1979, 1981; Salonen 1981b; Sarvala et al. 1981). However, the ecology of the littoral helophyte zone was poorly known, so in 1978–1980, a study of the carbon, phosphorus and nitrogen economy of the littoral Equisetum stands in Pääjärvi was undertaken. The aims of this investigation were 1) to work out the distribution of carbon and nutrients in the littoral subsystem, 2) to elucidate the internal dynamics of the system, and 3) to evaluate the importance of the littoral subsystem to the nutrient economy of the whole lake, particularly from the management viewpoint. The present paper is the first attempt towards a synthesis of our results and is based mainly on the results from 1979. The present paper was compiled by the author J. S., after discussion with the other authors.

Study area

Pääjärvi is a brown-water, nutrient-poor lake at an altitude of 103 m in southern Finland (ca. 61°04′N, 25°08′E). Its area is 13.4 km², maximum length 10 km, shoreline development 2.5, maximum depth 87 m, mean depth 14.4 m and volume 206 million m³. The lake is dimictic, and there is no oxygen deficit in the hypolimnion in any season (see Ruuhijärvi 1974; Ilmavirta 1979, 1981). Macroscopic vegetation extends to a depth of 1–1.5 m, and the lower limit of microscopic algae lies at 4–5 m. Helophyte vegetation covers only 2.4% of the total lake area (Ilmavirta 1979, 1981; Kansanen & Niemi 1974). The present study was concentrated in a dense and relatively homogenous Equisetum fluviatile L. stand in the western bay (Pappilanlahti Bay) of the lake. In this area, Carex-species dominate in the eulittoral meadows, while Equisetum extends as an almost monospecific stand over a distance of about 40–50 m from the mean summer shoreline to a depth of about 1 m. In the shallowest part of the Equisetum zone some loosely floating plants (Lemna spp., Ricciocarpus natans) occur, and beyond the Equisetum there is a sparse growth of Potamogeton and isoëtids (Fig. 1).

Sampling methods

To allow repeated sampling without disturbance, a permanent pier system was built from the shoreline through to the outer edge of the emergent vegetation. This lattice-like pier system was provided with removable boards so that sampling was possible practically everywhere within a grid of 10 × 40 m.

Different zones could be distinguished according to the height and abundance of macrophytes, but also according to the other components of the community (Fig. 1). Stratified sampling was applied, the strata being defined parallel to the shoreline, and, in most cases, all samples from a stratum were bulked to reduce labour.

Water temperature was recorded continuously at five points, and occasional checks were made in connexion with weekly sampling. Surface irradiance was also monitored continuously using a Kipp & Zonen solarimeter.

For carbon, nutrient and pigment analyses as well as for phytoplankton studies water samples were collected weekly with a 15.2 cm² acrylic plastic tube, which was lowered close to the bottom, closed at the upper end, lifted so that the lower end could be manually closed, and then emptied into a collecting vessel. Usually at least 8–10 water columns were bulked to form one sample, so that more than 3 l of water was collected in the shallowest area, and usually about 8 l in the deeper parts.

Phytoplankton biomass was estimated from chlorophyll a, using provisionally a C/Chl a ratio of 35 (Golterman & Kouwe 1980). Nutrient levels in phytoplankton were estimated assuming that levels obtained in 1980 for particles going through a 25 μm mesh were representative of phytoplankton.

Macrophyte abundance and shoot length were followed through the season. The carbon, phosphorus and nitrogen levels of Equisetum are based on replicate determinations on only two samples from the time of the maximum development of the stems, data on the seasonal variation in nutrient levels not yet being complete.

Seasonal succession of epiphytic communities was followed from May to November. Along two transects, underwater parts of the Equisetum stems were carefully cut and collected into acrylic plastic tubes, and the epiphytes were later removed with a

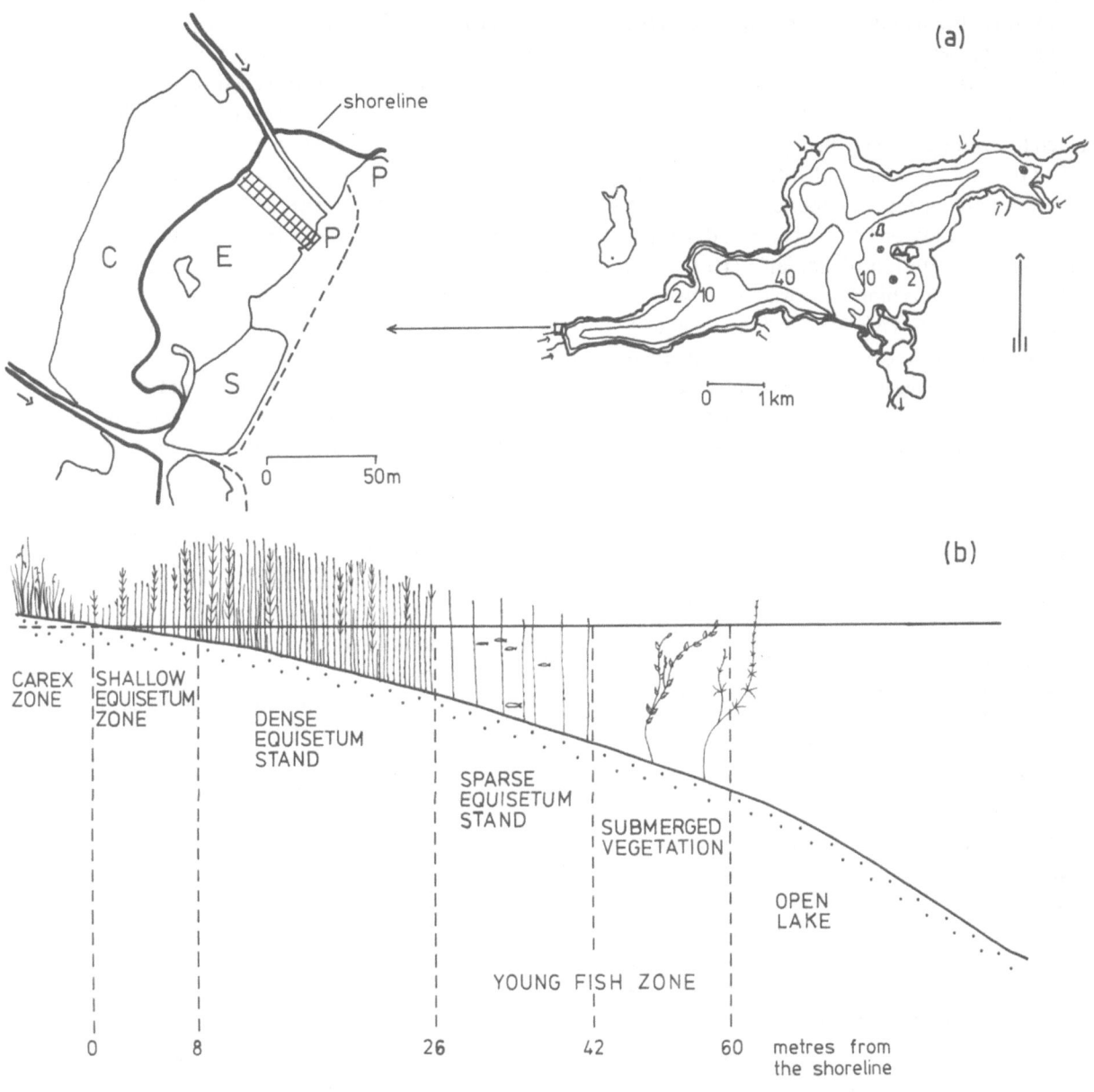

Fig. 1. (a) The location of the study area and (b) a schematic representation of the littoral vegetation there. The present paper deals only with results from the dense *Equisetum* stand. The location of the permanent pier grid in the western bay of the lake Pääjärvi is shown in the upper maps. C = *Carex* zone, S = *Sparganium* vegetation, E = *Equisetum*, P = *Potamogeton*. The broken line shows the outer border of isoetid vegetation.

plastic brush. Biomass parameters were organic carbon, phosphorus and nitrogen content and chlorophyll *a* concentration.

In 1979, zooplankton biomass was estimated weekly by direct determination of the total carbon content of the whole zooplankton community, using methods developed by Salonen (1979 and

unpublished) and during the study. Four sampling strata were established. A number of vertical cores of water (usually 10–15) were taken with a 15.2 cm² tube from the bottom to the surface, and these were pooled to give one sample, which was then strained through a 25 μm nylon mesh. Nutrient levels in zooplankton were determined on five occasions, twice in 1979 and three times in 1980; the nutrient/carbon ratios obtained in 1980 were applied to the 1979 carbon values.

Zoobenthos was sampled in this study only once, in August 1979. Two transects with a 50 cm² corer attached to a rod were taken through the vegetation belt. For the other seasons zoobenthos values were roughly estimated using the experience obtained in a simultaneous study on the zoobenthos dynamics in the littoral vegetation of Pääjärvi (Hiisivuori, unpubl.). Zoobenthos biomasses were expressed as ash-free dry weights (AFDW), using previously established AFDW/length and carbon/length relationships for the various taxonomic units (carbon percentage in AFDW was assumed to be 50%). Phosphorus and nitrogen contents of zoobenthos were estimated from the carbon contents, assuming similar ratios between carbon and nutrients as in zooplankton.

Weekly quantitative samples of young fish were collected with 1 m² buoyant nets (Bagenal 1974; Hewitt 1979) in May–September. Collection was made both inside and outside helophyte stands in the depth zone of 0.3–2.0 m. AFDW/length regressions were calculated for the different species, and 50% of AFDW was assumed to be carbon (this was confirmed in 1978 and 1980). Phosphorus and nitrogen levels in the 0+ and 1+ age groups of different species were determined in July and September 1980, and the mean phosphorus/AFDW and nitrogen/AFDW ratios from this material were used to estimate the nutrient contents of young fish in 1979.

Twenty to forty centimetre deep sediment cores for chemical analyses were taken with a 15.2 cm² tube from different depth zones during the open water period. Underground parts of *Equisetum* were separated from the sediment before nutrient analyses.

Chemical analyses

All forms of carbon were determined with the carbon analyzer and methods developed by Salonen (1979, 1981a). PO_4-phosphorus and NO_2- and NO_3-nitrogen were determined with an AKEA automatic chemical analysis system, using for phosphate the molybdate–ascorbic acid method introduced by Murphy & Riley (1962; see also Grasshoff 1976), and for nitrite and nitrate the method of Wood *et al.* (1967), reducing nitrate to nitrite in cadmium–copper columns. NH_4^+ was determined with an ammonium electrode (Orion). Total nitrogen and total phosphorus were determined with the AKEA system after boiling the samples in an autoclave for 30 min with peroxodisulphate in alkaline solution (Koroleff 1979). In sediments, total phosphorus was determined after baking for 6 h at 500 °C and soaking for 2 h at 60 °C in 0.2 M HCl solution (cf. Solórzano & Sharp 1980). For part of the samples, total carbon and nitrogen were determined with a Hewlett-Packard CHN-analyzer (Department of Limnology, University of Helsinki). For comparison, some samples were analyzed for total phosphorus and nitrogen in Viljavuuspalvelu Oy, and in the Water Research Laboratory of the National Board of Waters.

In 1979, chlorophyll *a* was extracted with methanol (Golterman *et al.* 1978; Riemann 1978).

Process measurements

Primary production of phytoplankton was measured with the routine ^{14}C method (Steemann-Nielsen 1952), measuring bottles lying sideways on acrylic plastic racks (cf. Ilmavirta *et al.* 1977). Total primary production and respiration were measured in large acrylic plastic chambers (volume 160 l, height 1 m). During measurements, the air and water inside the chamber was continuously circulated by two battery pumps. Primary production and respiration were measured from the total change of CO_2 contents inside the chambers; during respiration measurements the transparent chambers were covered with white boxes. Samples of both air and water were taken with syringes using a needle inserted through the silicon stoppers of the chambers. CO_2 was determined using an infrared gas analyzer (Salonen 1981a). Incubation periods ranged from 15 min to 1 h.

The share of sediment respiration was estimated from a temperature regression established in 1978 (Bergström & Sarvala unpubl.), using closed sediment cores.

Results

The distribution of organic carbon, nitrogen and

phosphorus in the main components of the littoral ecosystem in 1979 was estimated for four periods representing different phases in the seasonal development of the macrophyte vegetation: spring (17–28 May), early summer (11–21 June), late summer (11–23 August) and early autumn (3–17 September). Basic data from 1979 were complemented with information on the carbon to nutrient ratios in 1980.

The carbon to nutrient ratios (by weight) in the particle fraction passing a 25 μm mesh in summer 1980 were as follows:

	C:P	C:N	N:P
June 1980	73.0	9.1	8.0
July 1980	41.7	5.6	7.5

In the open lake the phosphorus levels of this fraction seemed to be clearly lower (C:P 135.1–158.7). The carbon to chlorophyll *a* ratio of this particle fraction was high, 137–167, suggesting that considerable amounts of detritus or animals were included. There was no difference in this ratio between the *Equisetum* stand and the open lake; microscopical analyses confirmed the presence of abundant detritus particles. Therefore, a C/Chl *a* ratio of 35 was adopted from Golterman & Kouwe (1980).

The chlorophyll *a* concentrations in the studied zone (see tabulation below) were higher (Fig. 2), but the resulting biomass per surface area was lower than values obtained from the open lake:

	Chl *a* μg l⁻¹	Chl *a* mg m⁻² (mean)
May	0.8– 3.1	1.0
June	0.5–13.2	2.0
August	13.0–22.7	8.3
September	5.0–19.9	2.6

In August, 1979, *Equisetum* stems from the central part of the stand contained on average 12.0% ash (range 11.7–12.5%), 1.56% N (n = 2) and 0.24% P (n = 6) per dry weight, which gives 6.5 as the N:P ratio. According to direct carbon determinations the organic carbon content was 43.4% of AFDW. The dry weight of above-ground *Equisetum* at any sampling time was estimated from the number of stems per m² and the height of the stems.

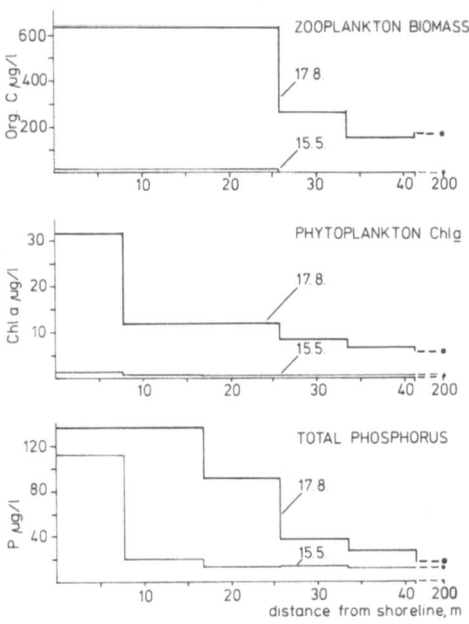

Fig. 2. The total phosphorus and chlorophyll *a* concentrations and zooplankton biomass in May and August 1980 along a transect from the shoreline to the open water at a depth of 5 m in the Pappilanlahti bay. Helophyte vegetation extends to a distance of about 40 m from the shoreline.

The mean dry weight of 1 cm of *Equisetum* stem in the zone studied was 13.8 mg (range 13.0–14.7) in early August 1979 and early September 1980.

The carbon level of epiphyton varied from 8.8 to 28.6% of dry weight. The ratio of carbon to phosphorus varied seasonally, being highest during the time of the maximum biomass in June:

	C:P mean	Range
25 May	79.4	73.0– 84.7
14 June	169.5	156.3–196.1
13 August	85.5	75.8– 99.0
17 September	59.9	53.8– 69.9

In the calculations the mean C:P ratios of epiphyton given above were used for each period. No seasonal trends were observed in the C:N ratio, and thus the overall mean value of 7.4 was used for the whole summer (range 6.5–8.5).

For zooplankton, the mean C:P and C:N ratios obtained in June, July and September, 1980, were used:

46

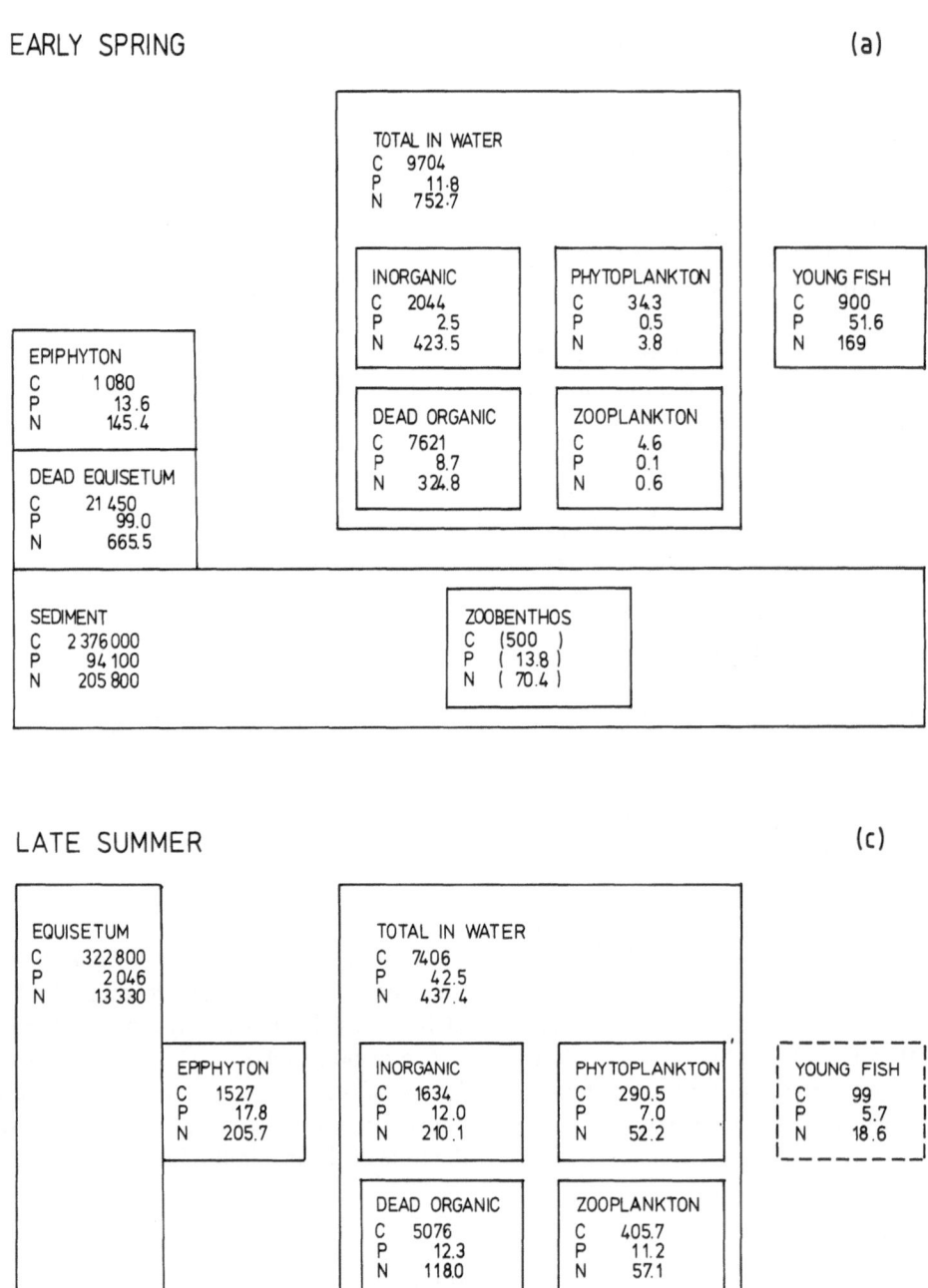

Fig. 3. (a)–(d) Block diagrams showing the distribution of carbon, nitrogen and phosphorus in the main components of the littoral ecosystem. All figures are mg m^{-2}. Figures in parentheses denote estimates not based on actual measurements. Broken lines of the young fish block after June indicate that the fishes do not move within the studied zone, although they partly rely on its production.

EARLY SUMMER   (b)

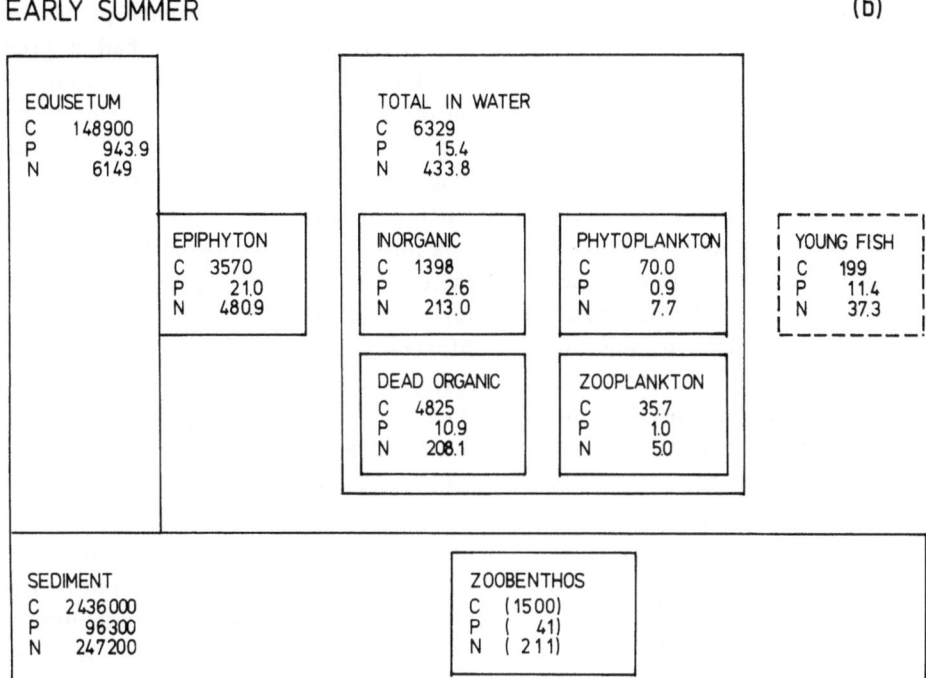

AUTUMN   (d)

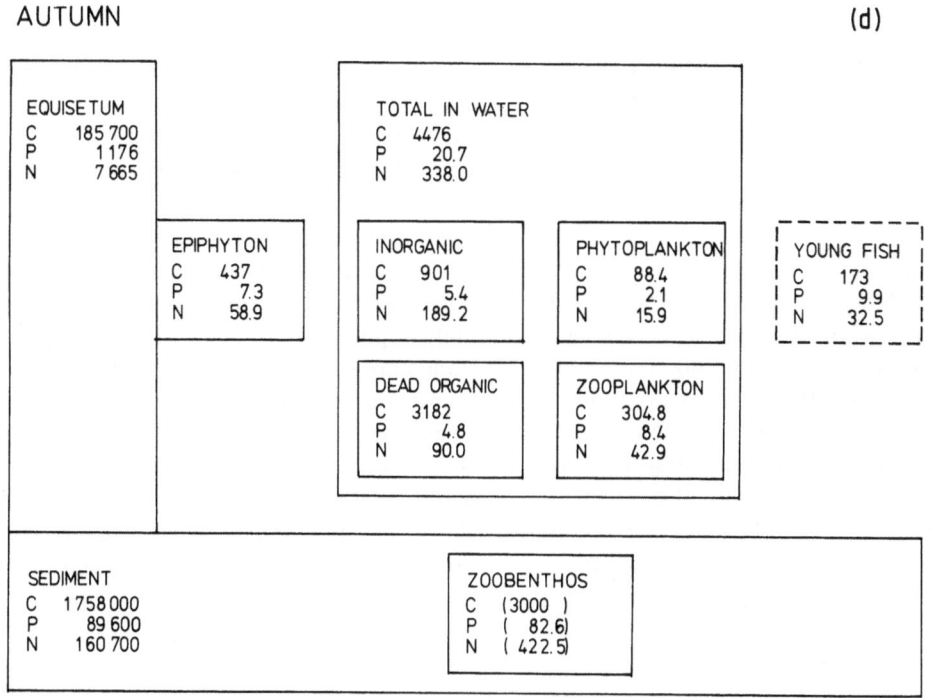

	Mean	Range
C:P	36.3	27.4-49.0
C:N	7.1	5.7- 9.4
N:P	5.1	4.4- 5.8

In young fish the nitrogen and phosphorus levels varied between the age groups and species in a consistent way. The phosphorus level increased, while the nitrogen level decreased, with increasing ash content. Much of the phosphorus was thus probably bound in the skeleton of the fishes, whereas nitrogen was associated mainly with proteins. Roach had the highest nutrient levels, and perch the lowest ones. For the present purpose overall means for all species and age groups were calculated:

	Mean	Range
N % of dry weight	9.39	7.08-10.17
P % of dry weight	2.87	2.26- 3.06
N:P	3.28	2.7 - 3.9
Ash %	12.0	9.8 -14.8

The nitrogen contents of the sediment varied from 0.014 to 0.087% of dry weight during May–September 1979, and phosphorus contents from 0.025 to 0.043%. On a dry weight basis, the nitrogen level in July–September was clearly lower than that in May–June. Phosphorus levels per dry weight remained similar through the summer. N:P ratio varied from 1.41 to 2.62, being highest in spring and early summer. The relatively low levels of nitrogen and their decrease during summer might be due to denitrification: during periods of intensive production, oxygen concentrations at night fell close to zero in the inner parts of the *Equisetum* stand. Loss of ignition varied from 1.4 to 2.8% of dry weight, and carbon level from 0.34 to 1.04% of dry weight; both were highest in spring and early summer, decreased until late summer, and increased again in autumn.

The distribution of organic carbon, nitrogen and phosphorus in the main components of the littoral ecosystem in May, June, August and September 1979 is shown in Figs. 3a–3d. Through all seasons, the major carbon and nutrient store in the littoral zone was the sediment. The second major store and by far the most important living component of the littoral ecosystem was the macrophyte *Equisetum*. The macrophyte vegetation had a pronounced seasonal cycle which exerted a profound influence on the whole littoral ecosystem. In spring (May, Fig. 3a) only dead remains of *Equisetum* stems from the previous year were present above ground, partly as stumps projecting from the bottom, partly floating as a dense layer near the shoreline, creating there a narrow zone where temperature rose very rapidly and nutrient concentrations became high. Most of the *Equisetum* zone thus remained open in spring, and water exchange between the littoral and the open lake was unhampered. This was reflected in the nutrient concentrations, which were similar both in the littoral zone and in the open lake (Fig. 2). Most of the carbon and nitrogen storage of the littoral outside the sediment was in dead organic matter, but the largest phosphorus store was, somewhat unexpectedly, the community of young fish, then moving and feeding all over the *Equisetum* zone. According to biomass, the dominant producers in spring were the epiphytes, for which the remnants of *Equisetum* formed a favourable substratum. Phytoplankton biomass was low, and zooplankton scarce.

New shoots of *Equisetum* appeared in late May, and by the middle of June the shoots had reached the water surface in the central part of the stand. From this time onwards the water exchange between the littoral and the open lake was progressively diminished, and the characteristic features of the closed macrophyte zone soon developed: high nutrient concentrations, abundant phytoplankton and zooplankton. In 1979, the hydrological conditions were somewhat exceptional. Water level was high in spring and remained high until late June. Then the water level rose again in July as a consequence of heavy rains, and the normal summer level was not attained again until early August. Thus the macrophyte zone 'closed' later than usually, and was diluted again in July, so that the maximum development of the community was postponed until late August.

In June (Fig. 3b), epiphytes had their peak biomass, and their nutrient contents alone exceeded the stores in water. By then, however, *Equisetum* had taken the place of the dominant producer, and its carbon content was 27-fold, phosphorus content 61-fold and nitrogen content 14-fold that of the total storage in water. Young fish were excluded from the inner parts of the *Equisetum* zone, and phytoplankton and zooplankton biomasses had started to increase.

In August (Fig. 3c), *Equisetum* had its peak height and abundance, and its carbon and nutrient stores were almost hundredfold those of the other living components of the system, and were 30–50 times higher than the total carbon and nutrient contents of water. The biomasses of phytoplankton, zooplankton and zoobenthos were maximal, but epiphytes had already decreased. The total storage of phosphorus had increased with the increasing plankton biomass, but also the inorganic phosphorus storage had increased. In August, the phosphate concentration in the littoral might be tenfold that in the open lake (Fig. 2). There were no changes in total and inorganic nitrogen from early summer. In general, most of the nutrients outside the sediment were then bound into living biomass.

In September (Fig. 3d), the macrophyte vegetation had started to degenerate, and all the living components of the system were declining. Total and inorganic nutrients in water had also decreased, although this was largely due to the decreasing volume of the littoral. The general features of the littoral system were in September still the same as in

August: nutrients were bound into biomass, and the share of dead organic matter in the nutrient pool was at its lowest.

Table 1 shows the development of the N:P ratio in water, both in the total and inorganic fractions. These ratios were very high in early summer and decreased towards autumn, but were always so high that there is no doubt that phosphorus was always the limiting nutrient in this system (cf. Forsberg *et al.* 1978; Niemi 1979; Ryding 1980). The dynamic nature of the littoral ecosystem is evident from Table 2. The daily gross primary production and total respiration values were measured as carbon. The daily nutrient requirements of photosynthesis were calculated according to the mean elementary composition of the dominant producers at each time (phytoplankton and epiphytes in May, epiphytes and *Equisetum* in June, and *Equisetum* in August and September; Kairesalo, unpubl.). The nutrient release in connexion with respiration/excretion was calculated from the mean composition of the dominant living components at each time: phytoplankton, epiphytes and zoobenthos in May (fish were ignored, since a large part of their phosphorus is not available for recycling), epiphytes, *Equisetum* and zoobenthos in June, and *Equisetum* in August and September. The daily requirements of plant growth exceeded the total phosphorus storage of water by a factor of 2–4, and the inorganic storage by a factor of 8–25. These figures suggest that diel variation in the nutrient concentrations should be detectable in the littoral zone, and our studies on the diel dynamics have, indeed, confirmed this (Kairesalo *et al.*, unpubl.). In May, when phytoplankton and epiphytes were the major producers, the amount of phosphorus

Table 1. N:P ratio by weight of total and inorganic fractions in water, in the littoral *Equisetum* zone in the oligotrophic lake Pääjärvi in 1979.

	17–28 May	11–21 June	11–23 August	3–17 September
Total nutrients	64.1	28.2	10.3	16.4
Inorganic nutrients	171.1	81.9	17.5	35.2

Table 2. Some indices of carbon, nitrogen and phosphorus metabolism in the littoral *Equisetum* belt in the oligotrophic lake Pääjärvi. All figures are mg m^{-2} for stores and mg m^{-2}day^{-1} for processes. T = temperature of water, C = organic carbon, N = nitrogen, P = phosphorus.

	17–28 May T = 8.4 °C			11–21 June T = 13.7 °C			11–23 August T = 18.9 °C			3–17 September T = 11.6 °C		
	C	N	P	C	N	P	C	N	P	C	N	P
Total storage in water	9 704	753	12	6 329	434	15	7 406	437	42	4 476	338	21
Inorganic storage in water	2 044	424	2	1 398	213	3	1 634	210	12	901	189	5
Gross primary production	3 271	382	41	9 493	880	64	15 440	631	97	8 074	330	51
Equisetum length growth	–	–	–	5 025	207	32	–	–	–	–	–	–
Total respiration/excretion	2 201	276	35	6 902	653	57	9 900	405	62	6 881	273	42
Benthic respiration	234	33	6	424	60	12	660	93	18	341	48	9

released in excretion almost equalled the requirements of primary production, but later, when *Equisetum* had attained its dominant position, the plant growth requirements substantially exceeded the amount of phosphorus released by excretion.

In June, the linear growth of *Equisetum* was a major part of the measured gross primary production. The very rapid growth of the young shoots of *Equisetum* up to the surface of the water happened simultaneously and at the same rate in all parts of the system, and presumably depended on carbon and nutrients stored in the underground parts during the previous year. *Equisetum* was probably also responsible for the major part of the total production and respiration later in summer. The contribution of sediments to the total respiration was highest in spring, when the decomposition of dead *Equisetum* was intense, but even then it was only about 10% of the total.

Discussion

The average carbon to nutrient ratios obtained in the littoral for the particle fraction going through a 25 μm mesh corresponded to atomic ratios of C:N:P of 147:17:1. Although these figures differed from the ratios generally accepted for marine phytoplankton (atomic C:N:P 106:16:1; see e.g. Corner & Davies 1971), the C:P ratio was fairly close to that given by Golterman (1975) for freshwater phytoplankton. The large proportion of detritus in this particle fraction weakened the value of the present analyses, and also made the measured carbon to chlorophyll *a* ratios meaningless. Although widely varying C/Chl *a* ratios have been published (see e.g. de Jonge 1980 for a review), the average ratio of 35 given by Golterman & Kouwe (1980) and used here may be reasonable as a first approximation. In the present study, a mean C/Chl *a* ratio of about 30 was obtained for epiphyton (Kairesalo, unpubl.). Most of the C/Chl *a* ratios found by Melvasalo (1980) for Baltic brackish-water phytoplankton varied from 35 to 44, and de Jonge (1980) gave average annual values of 40–61 for mobile benthic diatoms.

The nutrient levels found for *Equisetum* were clearly lower than those of phytoplankton. They were also in the lower end of the range published for macrophytes in general (see review by Hutchinson

1975), but were very similar to the levels reported for *Equisetum fluviatile* by Bernatowicz (1969; 1.44% N and 0.33% P) in the Polish lake Warniak. The mean ash content of *Equisetum* found here (12.0%) was the same as given for emergent macrophytes by Straskraba (1968), and only slightly lower than that given by Bernatowicz (1969) for *Equisetum* in Poland (13.2%). The carbon to nutrient ratios of epiphyton were intermediate between those of phytoplankton and macrophytes.

The mean carbon to nutrient ratios obtained for zooplankton corresponded to atomic ratios of C:N:P of 94:11:1, which differed considerably from the generally accepted ratios for marine zooplankton (atomic C:N:P of 117:23:1; see Corner & Davies 1971). The levels of phosphorus observed in the littoral zooplankton were thus higher than in most earlier studies, but in good accordance with phosphorus levels given for zooplankton by Gutel'makher (1977) and Vijverberg & Frank (1976). C:N ratios were slightly higher than usually given for marine zooplankton (e.g. Razouls 1977; Roman 1980). Nitrogen levels were also lower than found by Vijverberg & Frank (1976) for freshwater zooplankton, but agreed well with those obtained by Salonen et al. (1976) for zooplankton in Pääjärvi and neighbouring lakes. Most probably the nutrient levels in zooplankton are dependent on the nutrient concentrations in their food although taxonomic differences may also exist: the littoral zooplankton is dominated by cladocerans, whereas the values for marine zooplankton refer to copepod-dominated communities.

In the present study, the carbon to nutrient ratios of zooplankton were applied to zoobenthos as well. This will not lead to serious errors, since the nutrient levels given for zoobenthos seem to correspond closely to those of zooplankton (e.g. Iyengar et al. 1963: C:N by weight 4.4–6.3, N:P 8–13, P 0.6–1.3% of dry weight).

The nutrient contents of the littoral sediment, calculated in terms of dry weight, were lower than values found for the profundal sediments in Pääjärvi (Simola, unpubl.; National Board of Waters, Finland), but this was mainly due to the high mineral content of the littoral sediment. The littoral C:N ratios of the sediment (around 15) were slightly higher and the N:P ratios slightly lower than in the profundal.

The daily phosphorus requirements of primary

production were very high compared to the nutrient storage in the water, and also exceeded the release of nutrients in excretion. From late June until autumn the major producer was *Equisetum*. Like other macrophytes, *Equisetum* probably takes most of its nutrients from the sediment (see e.g. Carignan & Kalff 1979; Barko & Smart 1980). Allowing for this fact, and the very rough nature of these calculations, the fixation of phosphorus in primary production and the release of phosphorus in excretion may approximately balance each other in the water for most part of the summer. Since the storage of inorganic phosphorus increased during summer, phosphorus was in fact released slightly in excess of plant requirements. Extra dissolved nutrients were certainly also transported to the open lake, where concentrations were much lower.

The nutrient stores in the sediment were enormous in comparison with the daily needs or the maximum biomass of *Equisetum*. However, the sediment stores are at least partly composed of refractory compounds which are not readily usable. According to recent studies (Carignan & Kalff 1979), only a small mobile fraction of phosphorus in the sediments would be available for the macrophytes. Therefore, the growth of *Equisetum* may ultimately be limited by the availability of sediment nutrients.

The measured gross primary production and respiration values indicate a very high rate of internal cycling of carbon and nutrients within the littoral ecosystem. These values may well be overestimated because of the continuous stirring during the measurements. On the other hand, equivalent cycling rates have been measured elsewhere. The role of zooplankton in phosphorus cycling has been stressed previously (e.g. Scavia 1979). Zooplankton within the *Equisetum* zone was dominated by cladocerans. Daily phosphorus release by cladoceran zooplankton may be of the order of 35–60% of their total body phosphorus, and for rotifers values 3–4 times higher are expected (see Lehman 1980 and references therein). Thus, in August, zooplankton in the littoral should have released 4–7 mg P $m^{-2}day^{-1}$, which is not far below the daily requirements of phytoplankton production. Likewise, phosphorus release rates of 5–10 mg P $m^{-2}day^{-1}$ by lake benthos have been measured at summer temperatures (Kamp-Nielsen 1975; Gallepp 1979; Granéli 1979). To these estimates must be added remineralization by bacteria. In the littoral, however, macrophytes may be the major source of released nutrients (cf. Barko & Smart 1980).

In 1970–1976 the mean phosphorus concentration in Pääjärvi was 9.4 mg P m^{-3}, which means a total storage of 1 939 kg P in the whole lake (calculated from the measurements by the National Board of Waters, published in Ryding 1980). During the same period, the average annual phosphorus load of the lake was 4 411 kg P. Using the production of organic matter by *Equisetum* in Pääjärvi in 1972–1973 (Kansanen *et al.* 1974) and the mean phosphorus content of *Equisetum* from the present study, the annual contribution of the decomposing *Equisetum* stands to the phosphorus pool of the lake could be at maximum 102 kg P, or 2.3% of the annual phosphorus load and 5.3% of the average total phosphorus storage in the lake. Thus the littoral macrophyte vegetation has a relatively minor role in the nutrient economy of the whole lake Pääjärvi. The *Equisetum* stands cover only 0.88% of the whole lake area. Earlier studies emphasizing the importance of the littoral vegetation have dealt with shallower lakes having a much larger coverage of macrophytes (see Wetzel 1975; Barko & Smart 1980). However, even in deep oligotrophic lakes like Pääjärvi, macrophyte vegetation may be a locally important source of nutrients for the pelagial. On the other hand, the littoral community also effectively fixes extra nutrients coming from land (Kairesalo *et al.*, unpubl.; Tilton & Kadlec 1979); it is not known to what extent these nutrients become permanently buried in the littoral sediment.

In conclusion, the littoral system is a dynamic one. There are two predominant stores: sediments and macrophytes. The development of the other components seems to be dependent on the water level relative to the height of the macrophytes, which influences the exchange of water and nutrients between the littoral and the open lake. Within the littoral, the exchange of nutrients between water and sediments, between macrophytes and sediments, and between macrophytes and water are crucial, and need to be investigated further.

Acknowledgements

Our thanks are due to the University of Helsinki

52

for providing us with good working facilities at Lammi Biological Station. Some nutrient analyses were made in the Department of Limnology, University of Helsinki, and in the Water Research Laboratory, National Board of Waters, Finland. This study was made possible by the financial aid from the Maj and Tor Nessling Foundation, which is gratefully acknowledged.

References

Bagenal, T. B., 1974. A buoyant net designed to catch freshwater frish larvae quantitatively. Freshwat. Biol. 4: 107–109.

Barko, J. W. & Smart, R. M., 1980. Mobilization of sediment phosphorus by submersed freshwater macrophytes. Freshwat. Biol. 10: 229–238.

Bernatowicz, S., 1969. Macrophytes in the lake Warniak and their chemical composition. Ekol. polsk. A, 17: 447–467.

Björk, S., 1972. Swedish lake restoration program gets results. Ambio 1: 153–165.

Carignan, R. & Kalff, J., 1979. Quantification of the sediment phosphorus available to aquatic macrophytes. J. Fish. Res. Bd Can. 36: 1002–1005.

Corner, E. D. S. & Davies, A. G., 1971. Plankton as a factor in the nitrogen cycles in the sea. Adv. mar. Biol. 9: 101–204.

Forsberg, C., Ryding, S.-O., Claesson, A. & Forsberg, Å., 1978. Water chemical analyses and/or algal assay? Sewage effluent and polluted lake water studies. Mitt. int. Ver. Limnol. 21: 352–363.

Gallepp, G. W., 1979. Chironomid influence on phosphorus release in sediment–water microcosms. Ecology 60: 547–556.

Golterman, H. L., 1975. Physiological Limnology. An Approach to the Physiology of the Lake Ecosystems. Amsterdam: Elsevier Scientific.

Golterman, H. L., Clymo, R. S. & Ohnstad, M. A. M. (eds.), 1978. Methods for Physical and Chemical Analysis of Fresh Waters. IBP Handbook No. 8. Oxford: Blackwell Scientific.

Golterman, H. L. & Kouwe, F. A., 1980. Chemical budgets and nutrient pathways. In LeCren E. D. & Lowe-McConnell R. H., (eds.) The Functioning of Freshwater Ecosystems, pp. 85–140. Cambridge: Cambridge University Press.

Granéli, W., 1979. The influence of Chironomus plumosus larvae on the exchange of dissolved substances between sediment and water. Hydrobiologia 66: 149–159.

Grasshoff, K., 1976. Procedures for the automatic determination of seawater constituents. In: K. Grasshoff (ed.), Methods of Seawater Analysis, pp. 276–289. Weinheim: Verlag Chemie.

Gutel'makher, B. L., 1977. Quantitative evaluation of the role of zooplankton in the phosphorus cycle in waterbodies (in Russian, with English summary.). Zh. Obsch. Biol. 38: 914–922.

Hewitt, D. P., 1979. Tests to confirm quantitative sampling of young fish by the Bagenal buoyant net. Freshwat. Biol. 9: 339–341.

Hutchinson, G. E., 1975. A Treatise on Limnology. 3: Limnological Botany. New York: J. Wiley.

Ilmavirta, V., 1979. Sources and utilization of energy in Pääjärvi, an oligotrophic, brown-water lake in southern Finland. Ergebn. Limnol. 13: 212–224.

Ilmavirta, V., 1981. The ecosystem of the oligotrophic lake Pääjärvi. 1. Lake basin and primary production. Verh. int. Ver. Limnol. 21: 410–415.

Ilmavirta, V., Jones, R. I. & Kairesalo, T., 1977. The structure and photosynthetic activity of pelagial and littoral plankton communities in Lake Pääjärvi, southern Finland. Ann. bot. fenn. 14: 7–16.

Iyengar, V. K. S., Davies, D. M. & Kleerekoper, H., 1963. Some relationships between Chironomidae and their substrate in nine freshwater lakes of southern Ontario, Canada. Arch. Hydrobiol. 59: 289–310.

Jonge, V. N. de, 1980. Fluctuations in the organic carbon to chlorophyll a ratios for estuarine benthic diatom populations. Mar. Ecol. Prog. Ser. 2: 345–353.

Kamp-Nielsen, L., 1975. A kinetic approach to the aerobic sediment-water exchange of phosphorus in Lake Esrom. Ecol. Modelling 1: 153–160.

Kansanen, A. & Niemi, R., 1974. On the production ecology of isoetids, especially Isoëtes lacustris and Lobelia dortmanna, in Lake Pääjärvi, southern Finland. Ann. bot. fenn. 11: 178–187.

Kansanen, A., Niemi, R. & Överlund, K., 1974. Pääjärven makrofyytit. (Macrophytes [of Pääjärvi].) Luonnon Tutkija 78 (4–5): 111–118 (in Finnish, with English summary).

Koroleff, F., 1979. Methods for the chemical analysis of seawater. Meri 7: 1–60 (in Finnish).

Lehman, J. T., 1980. Release and cycling of nutrients between planktonic algae and herbivores. Limnol. Oceanogr. 25: 620–632.

Melvasalo, T., 1980. Phytoplankton biomass in relation to chlorophyll a in the northern Baltic Sea. Nat. Board Waters of Finland, Publ. of Water Res. Inst. (manuscript).

Murphy, J. & Riley, J. P., 1962. A modified single-solution method for the determination of phosphate in natural waters. Analyt. Chim. Acta 27: 31–36.

Niemi, Å., 1979. Blue-green algal blooms and N:P ratio in the Baltic Sea. Acta bot. fenn. 110: 57–61.

Pieczyńska, E. (ed.), 1976. Selected problems of lake littoral ecology. Warszawa: Univ. Warsaw, Inst. Zool., Dept. Hydrobiol.

Razouls, S., 1977. Analyse pondérale, élémentaire et calorimétrique des stades juvéniles de Copépodes pélagiques au cours d'une année. J. exp. mar. Biol. Ecol. 26: 265–273.

Riemann, B., 1978. Carotenoid interference in the spectrophotometric determination of chlorophyll degradation products from natural populations of phytoplankton. Limnol. Oceanogr. 23: 1059–1066.

Roman, M. R., 1980. Tidal resuspension in Buzzards Bay, Massachusetts. III. Seasonal cycles of nitrogen and carbon: nitrogen ratios in the seston and zooplankton. Estuar. Coastal. Mar. Sci. 11: 9–16.

Ruuhijärvi, R., 1974. A general description of the oligotrophic lake Pääjärvi, southern Finland, and the ecological studies on it. Ann. bot. fenn. 11: 95–104.

Ryding, S.-O., 1980. Monitoring of inland waters. OECD Eutrophication Programme. The Nordic Project.-Nordforsk, Secretariat of Environmental Sciences, Publ. 1980 (2): 1–207.

Salonen, K., 1979. A versatile method for the rapid and accurate determination of carbon by high temperature combustion. Limnol. Oceanogr. 24: 177–183.

Salonen, K., 1981a. Rapid and precise determination of total inorganic and gaseous organic carbon in water. Water Res. 15: 403–406.

Salonen, K., 1981b. The ecosystem of the oligotrophic lake Pääjärvi. 2. Bacterioplankton. Verh. int. Ver. Limnol. 21: 416–421.

Salonen, K., Sarvala, J., Hakala, I. & Viljanen, M.-L., 1976. The relation of energy and organic carbon in aquatic invertebrates. Limnol. Oceanogr. 21: 724–730.

Sarvala, J., Ilmavirta, V., Paasivirta, L. & Salonen, K., 1981. The ecosystem of the oligotrophic lake Pääjärvi. 3. Secondary production and an ecological energy budget of the lake. Verh. int. Ver. Limnol. 21: 422–427.

Scavia, D., 1979. Examination of phosphorus cycling and control of phytoplankton dynamics in Lake Ontario with an ecological model. J. Fish. Res. Bd Can. 36: 1336–1346.

Solórzano, L. & Sharp, J. H., 1980. Determination of total dissolved phosphorus and particulate phosphorus in natural waters. Limnol. Oceanogr. 25: 754–758.

Steemann Nielsen, E., 1952. The use of radioactive carbon (C[14]) for measuring organic production in the sea. J. Cons. int. Explor. Mer. 18: 117–140.

Straskraba, M., 1968. Der Anteil der höheren Pflanzen an der Produktion der stehenden Gewässer. Mitt. int. Ver. Limnol. 14: 212–230.

Tilton, D. L. & Kadlec, R. H., 1979. The utilization of a freshwater wetland for nutrient removal from secondarily treated waste water effluent. J. Envir. Qual. 8(3): 328–334.

Vijverberg, J. & Frank, Th. H., 1976. The chemical composition and energy contents of copepods and cladocerans in relation to their size. Freshwat. Biol. 6: 333–345.

Wetzel, R. G., 1975. Limnology. Philadelphia: W. B. Saunders.

Wetzel, R. G., 1979. The role of the littoral zone and detritus in lake metabolism. Ergebn. Limnol. 13: 145–161.

Wood, E. D., Armstrong, F. A. J. & Richards, F. A., 1967. Determination of nitrate in sea water by cadmium-copper reduction to nitrite. J. mar. biol. Ass. U.K. 47: 23–31.

The diversity, biomass and production of zooplankton in Lake Inarijärvi

Pirkko Selin & Lasse Hakkari
Hydrobiological Research Centre, University of Jyväskylä, Seminaarinkatu 15, SF-40100 Jyväskylä 10, Finland

Keywords: arctic, zooplankton, production, fish food, regulation

Abstract

About 650 zooplankton samples were collected from Lake Inarijärvi in 1977–1979 from the littoral and pelagial zones of the lake. One hundred and twenty-three zooplankton taxa were found and most of them can be considered euplanktonic.

The most important species were *Holopedium gibberum, Daphnia cristata, Cyclops* spp. and *Eudiaptomus* spp. Mean pelagial zooplankton biomass was 0.29 g m^{-3} in the 0–5 m depth zone, 0.17 g m^{-3} in 5–10 m and 0.11 g m^{-3} in 10–20 m.

The zooplankton biomass at a sandy shore was about 0.09 g m^{-3}, at a stony shore 0.05 g m^{-3} and at a vegetated shore 0.76 g m^{-3}. About 70% of the whole zooplankton production consisted of crustaceans.

The sum of herbivore and carnivore zooplankton production in the pelagial area during the summer was 210–330 kg ha^{-1} × 3 months.

Introduction

Lake Inarijärvi is one of the largest lakes in Finland, situated north of the polar circle. Lake Inarijärvi differs from other big Finnish lakes in its extended ice cover (7 months), with a growing season of only about 3 months. L. Inarijärvi is relatively deep and clear (mean Secchi disc transparency 6.5 m) and its total phosphorus and total nitrogen contents are low (4–10 μg P l^{-1} and 150–260 μg N l^{-1}).

L. Inarijärvi is the central lake of the River Paatsjoki basin and is regulated by a power plant in the USSR. The water area of the lake is 1085 km^2 and the maximum depth 95 m. The water leaving L. Inarijärvi flows via the River Paatsjoki to the Arctic Ocean.

This paper outlines the zooplankton composition, biomass and production in L. Inarijärvi in 1977–1979. The results of the zooplankton studies are to be used for assessing the role of zooplankton as food for fish.

Material and methods

Sampling and analysis

Zooplankton was collected from pelagial and littoral stations. The pelagial stations were Munuaissaari, Roiro, Väylä and Varttassaari (only in 1977), and the littoral stations were near the station Varttasaari in sandy, stony and vegetated shores (Fig. 1).

The pelagial material was collected with a Sormunen tube sampler from different depth zones so that three vertical samples were composed of plankton animals from 0 to 5 m, 5 to 10 m and 10 to 20 m (e.g. Hakkari 1978). The material from littoral zones was taken so that one zooplankton sample consisted of the animals in 13 l taken by a Ruttner sampler from a water column of 0.2–1.0 m. Some parallel samples were also collected and the whole collection comprised 459 pelagial samples and 183 littoral samples. All the samples were taken during the ice free season, June–September.

The samples were filtered through a 50 μm mesh

Hydrobiologia 86, 55–59 (1982). 0018-8158/82/0861-0055/$01.00.

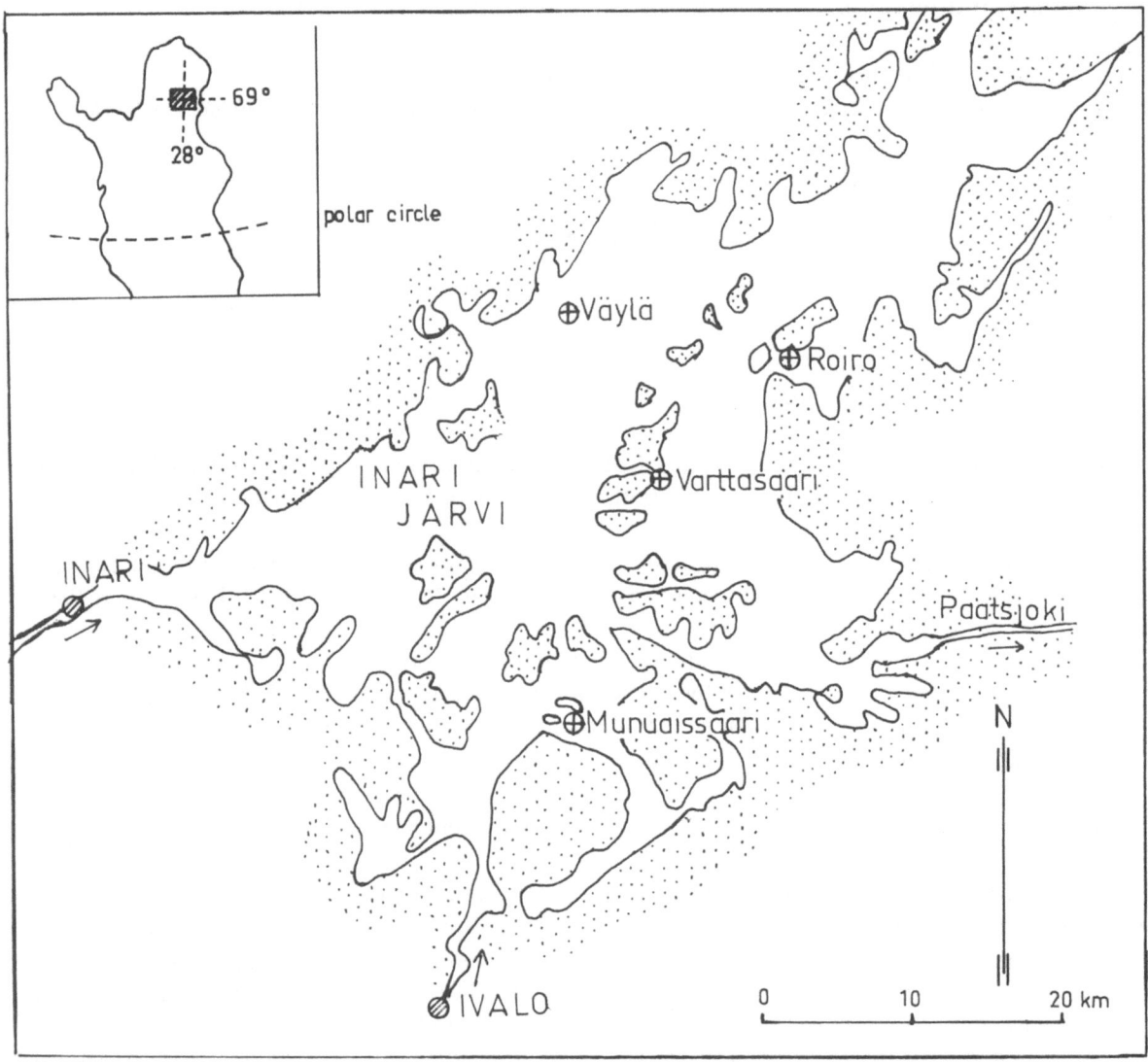

Fig. 1. The sampling stations in Lake Inarijärvi.

and all the material was preserved with 4% formalin. Examination was carried out using an inverted microscope (e.g. Hakkari 1978). Smaller organisms may pass through the 50 μm mesh and therefore the numbers of small protozoans are not quantitative. When possible, the developmental stages of cladocerans and copepods were determined.

Biomass and production

The zooplankton biomass of the large animals was partly estimated by calculating the volumes of the geometric configurations of the organisms or from the literature. The specific gravity was taken as 1.0 (Hutchinson 1967). The mean biomass is certain depth zones was calculated by weighting the biomass of the vertical samples with the volumes of the corresponding water layer.

The production of the most common rotifers was estimated by using the formula of Galkovskaya (Winberg 1971). The production of individual-rich cladocerans and copepods was estimated in most cases by the graphic method (e.g. Winberg 1971;

Edmondson & Winberg 1971). The functions of time, the duration of development and the individual growth increment were obtained partly from empirical data and partly from the literature (Bottrell *et al.* 1976; Hakkari 1978). The production of some large and rarely found cladocerans were assessed using the formula of Pechen (1965).

The diversity of the zooplankton fauna in L. Inarijärvi was assessed by the Shannon and Wiener index (Hutchinson 1967).

Results

Zooplankton composition

Altogether 123 zooplankton taxa (genera, species and subspecies) were identified in 1977–1979, and most of them can be considered euplanktonic. Altogether 50 rotifer species, 16 cladoceran species and 11 copepods were found from the pelagial stations, and 48 rotifers, 22 cladocerans and 6 copepods from the littoral stations. The most important species were *Holopedium gibberum* Zaddach, *Daphnia cristata* Sars, *Cyclops* spp. and *Eudiaptomus* spp. In other respects the composition of zooplankton fauna in L. Inarijärvi resembles the fauna of some other Arctic lakes (e.g.

Axelson 1961; Langeland 1972). The number of rotifer taxa exceeded those of other groups, and in spite of their small volume they form a considerable part of the whole zooplankton population because of their great numbers.

The zooplankton population in 1977–1979 in the pelagial area was nearly the same in all stations but the vegetated shore differed from other littoral areas because of the rich zooplankton population.

The diversity indices compared to other Finnish lakes were as follows:

	Diversity-index
L. Inarijärvi	
pelagial station	4.8
sandy shore	5.0
stony shore	4.9
vegetation shore	5.5
Central Finland:	
Kuusvesi (Granberg & Hakkari 1977)	3.6
L. Päijänne (Hakkari 1978)	3.5

Biomass and production

The mean zooplankton biomasses in L. Inarijärvi in different vertical zones are presented in Table 1. The differences between pelagial stations are not statistically significant (ANOVAR) whereas the

Table 1. The mean zooplankton biomass (g m^{-3}) in different vertical depth zones of the pelagial and in littoral stations in Lake Inarijärvi 1977–1979.

	June	July	August	September	$\bar{x}$
Pelagial					
Munuaissaari					
0–5 m	0.05	0.57	0.34	0.15	0.28
5–10 m	0.03	0.27	0.18	0.20	0.17
10–20 m	0.02	0.09	0.16	0.09	0.09
Roiro					
0–5 m	0.06	0.38	0.53	0.15	0.28
5–10 m	0.05	0.25	0.37	0.12	0.19
10–20 m	0.02	0.09	0.09	0.10	0.08
Väylä					
0–5 m	0.30	0.20	0.59	0.19	0.32
5–10 m	0.05	0.15	0.20	0.14	0.14
10–20 m	0.02	0.06	0.05	0.45	0.15
Littoral					
0.2–1.0 m					
Sandy shore	0.02	0.11	0.14	0.09	0.09
Stony shore	0.06	0.04	0.06	0.04	0.05
Vegetation shore	0.24	1.06	1.11	0.62	0.76

biomass at the vegetated shore was significantly higher than at the other littoral stations.

The biomass in L. Inarijärvi is about 30–50% of those reported in L. Päijänne and Konnevesi, Central Finland (e.g. Hakkari 1978), and 20% of the mean biomass in Lake Pääjärvi, Southern Finland (Latja 1974). The mean biomass of Crustacea in the pelagial area of L. Inarijärvi was about 55% of that found in L. Krivoe and 10% in L. Krugloe, subarctic area (Alimov *et al.* 1972). A similar biomass was found in humic mountain lakes with low nutrient content (Langeland 1972).

The zooplankton production (fresh weight) during the growing season (June–September) in the pelagial area of L. Inarijärvi was as follows:

Site	Production of herbivores kg ha^{-1}	Production of predators kg ha^{-1}	Total kg ha^{-1}
Munuaissaari	281	51	332
Roiro	153	54	207
Väylä	177	51	228

The production in winter can only be about 5% of that during summertime (Mednikov 1962: ref. Winberg 1971).

The fresh weight values of zooplankton can be transformed into energy units assuming that 1 g wet weight is equal to 2.1 kJ (Hillbricht-Ilkowska 1977). The mean total zooplanktonic production in the pelagial area of L. Inarijärvi was 43–70 kJ m^{-2}. The production during the vegetation period in L. Krivoe was about 82 kJ m^{-2} and in L. Krugloe 54 kJ m^{-2} (Alimov *et al.* 1972). In oligotrophic parts of L. Päijänne the total zooplankton production was about 114 kJ m^{-2} (Hakkari 1978). The annual zooplankton production in oligotrophic L. 239 (Schindler 1972) was about 146 kJ m^{-2}, and in Char Lake, Canada, 6.7 kJ m^{-2} (Rigler & MacCallum 1974).

Most of the zooplankton production in L. Inarijärvi was from crustaceans (*ca.* 75%).

If zooplankton predators use four times their own production of herbivorous production (Hillbricht-Ilkowska 1977) we can estimate the part of the energy leaving for the next trophic level, in many cases for fish. The mean energy content remaining from zooplankton for the next trophic

level during the growing season in the pelagial area would then be about 76 kg ha^{-1} or 16 kJ m^{-2}.

Discussion

In spite of the water level regulation the zooplankton composition in L. Inarijärvi resembles that of other oligotrophic lakes in Finland. The amount of large crustaceans was relatively great. As the results show, the Shannon-diversity of the zooplankton in L. Inarijärvi was quite high. So neither diversity nor number of species showed any special extremes in spite of the northern position of the lake, whereas in the Arctic Char Lake, Canada (74°42′N), the zooplankton fauna consisted only of *Limnocalanus macrurus, Keratella cochlearis* and some ciliates (Rigler & MacCallum 1974).

The mean chlorophyll-*a* content in July 1977 and 1979 was about 2 mg Chl-*a* m^{-3} (National Board of Waters, Finland; unpublished results). This shows that the primary production per volume during the growing season may be very low. However, the water of L. Inarijärvi is very clear (transparency 6.5 m). So the euphotic zone is thick and primary production per unit surface area may be relatively high. In spite of the lower biomass, the zooplankton production in L. Inarijärvi was similar to other oligotrophic, Arctic lakes.

The length of the growing season in L. Inarijärvi (*ca.* 90 days) is about one month shorter than that in the lake district of Finland. The mean summer water temperature was *ca.* 2–3 °C lower than in L. Päijänne, but clear differences were found in the development times of the planktonic crustaceans in these lakes.

Lakes Krivoe and Krugloe are situated on the coast of the White Sea almost on the polar circle, and zooplankton production in these small lakes was similar to that in L. Inarijärvi. Char Lake is situated at almost the same latitude but the ice-free time is only some 30 days and the water temperature is *ca.* 3–5 °C. The difference between the mean summer temperatures in L. Inarijärvi and Char Lake is 8–10 °C, and the zooplankton production in Char Lake was only 13% of that in L. Inarijärvi.

Such comparisons show how zooplankton production is affected by length of the growing season, water temperature and light conditions.

Acknowledgements

This study was financed by The National Board of Waters, Finland. We are much obliged for this support.

References

Alimov, A. F., Boullion, V. V., Finogenova, N. P., Ivanova, M. B., Kuzmitskaya, N. K., Nikulina, V. N., Ozeretskovskaya, N. G. & Zharowa, T. V., 1972. Biological productivity of Lake Krivoe and Krugloe. In: Productivity Problems of Freshwaters, pp. 39–56. PWN, Warsaw.

Axelson, J., 1961. Zooplankton and impoundment of two lakes in Northern Sweden (Ransaren och Kultsjön). Inst. Freshw. Res. Rep. 42: 84–168.

Bottrell, H. H., Duncan, A., Gliwicz, Z. M., Grygierek, E., Herzig, A., Hillbricht-Ilkowska, A., Kurasawa, A., Larsson, P. & Weglenska, T., 1976. A review of some problems in zooplankton production studies. Norw. J. Zool. 24: 419–456.

Edmondson, W. T. & Winberg, G. G. (eds.), 1971. A Manual on Methods for the Assessment on Secondary Productivity in Fresh Waters. IBP-Handbook 17. 358 pp. Oxford and Edinburgh.

Granberg, K. & Hakkari, L., 1977. The study of zooplankton and bottom fauna in the watercouse of Äänekoski-Vaajakoski, Central Finland, in 1976. (English summary). Hydrobiologian tutkimuskeskus, tiedonantoja 88: 1–112

Hakkari, L., 1978. On the productivity and ecology of zooplankton and its role as food for fish in some lakes in Central Finland. Biol. Res. Rep. Univ. Jyväskylä 4: 1–87.

Hillbricht-Ilkowska, A., 1977. Trophic relations and energy flow in pelagic plankton. Pol. ecol. Stud. 3: 3–98.

Hutchinson, G. E., 1967. A Treatise on Limnology 1. 1015 pp. New York and London.

Langeland, A., 1972. A comparison of the zooplankton communities in seven mountain lakes near Lillehammer, Norway (1896 and 1971). Norw. J. Zool. 20: 213–226.

Latja, R., 1974. Pääjärven eläinplankton. Luonnon Tutkija 78: 153–156.

Pechen, G. A., 1965. Produktsiya vetvistousykh rakoobraznykh ozernogo zooplankton. (Summary: Production of cladocerous crustaceans of lacustrine zooplankton). Gidrobiol. Zh. 1: 19–26.

Rigler, F. H. & MacCallum, M. E., 1974. Production of zooplankton in Char Lake. J. Fish. Res. Bd Can. 31: 637–646.

Schindler, D. W., 1972. Plankton production in Canadian shield lakes. In: Productivity Problems of Freshwaters, pp. 311–331. PWN, Warsaw.

Winberg, G. G. (ed.), 1971: Methods for the Estimation of Production of Aquatic Animals. 175 pp. London and New York.

The food and parasites of fish in some deep basins of northern L. Päijänne

P. Bagge & L. Hakkari

Hydrobiology Research Center, University of Jyväskylä, SF-40100 Jyväskylä 10, Finland

Keywords: deep basins, fish food, metazoa parasites, pollution

Abstract

The composition of the fish stock, food and 'macroparasites' were studied in eleven basins (22–100 m) of Lake Jyväsjärvi and North Päijänne in August–September 1976. The fishing was done by means of a series of nets (meshes 15, 21 and 35 mm) laid on the bottom overnight. No fish were found in the two northernmost basins owing to bad oxygen conditions caused by waste waters. Smelt and burbot were the most abundant fish in the catches in other basins but the vendace was rare. Relict crustaceans and some 'deep water' copepods (e.g. *Heterocope borealis*) played an important role in the food of fish in all basins.

Parasites spreading via relict crustaceans were abundant in smelt and burbot, but absent in vendace and ruffe. The most common parasite in smelt was *Cystidicola farionis*, in burbot *Echinorhynchus borealis*, in ruffe *Triaenophorus nodulosus* and in vendace the gill parasite *Ergasilus sieboldi*. Females of a relatively rare copepod *Salmincola lotae* were found in the oral cavity of five burbot (at depths of 50 and 100 m).

Introduction

Relict amphipods (especially *Pallasea quadrispinosa*) and semipelagically living *Mysis relicta* and *Chaoborus flavicans* are known to occur relatively abundantly in the zoobenthos of many deep basins of L. Päijänne (Särkkä, 1979; Bagge, unpubl.). In many large lakes as in L. Ladoga, studied a.o. by Jääskeläinen (1917) glacial relicts are shown to be very important food items for deep living fish species and the observations of Savolainen (1975) confirm that this is also the case with the relict fourhorn sculpin *(Myoxocephalus quadricornis)* in many lakes of the Finnish lake district.

In order to evaluate the role of zoobenthic animals in the food of deep living fish, we have fished using a series of three gill nets in eleven deep basins (22–100 m) in Northern L. Päijänne and L. Jyväsjärvi in August–September 1976. Since the fish were analysed immediately after fishing, rough observations (by eye) could also be made on the occurrence and site of macroparasites living in different fish species. The concentration of oxygen of the water was measured in different fishing sites; thus it is possible to discuss the effects of pollution on the composition of fish in the basins. Moreover, the present material could be compared with the results of test fishings carried out by Tuunainen in shallow parts of the study area in 1969–70.

The study area and methods

The Northern part of Lake Päijänne (Fig. 1) consists of a chain of basins including the Ristiselkä, which is one of the deepest places (104 m) in the Finnish lake district. During the last few decades, the area has become badly polluted by sewage from the town Jyväskylä and neighbouring settlements and by waste waters from pulp mills situated some 30–40 km upstream of the lake (cf. a.o. Tuunainen 1971; Granberg 1973; Eloranta

Hydrobiologia 86, 61–65 (1982). 0018-8158/82/0861–0061/$01.00.

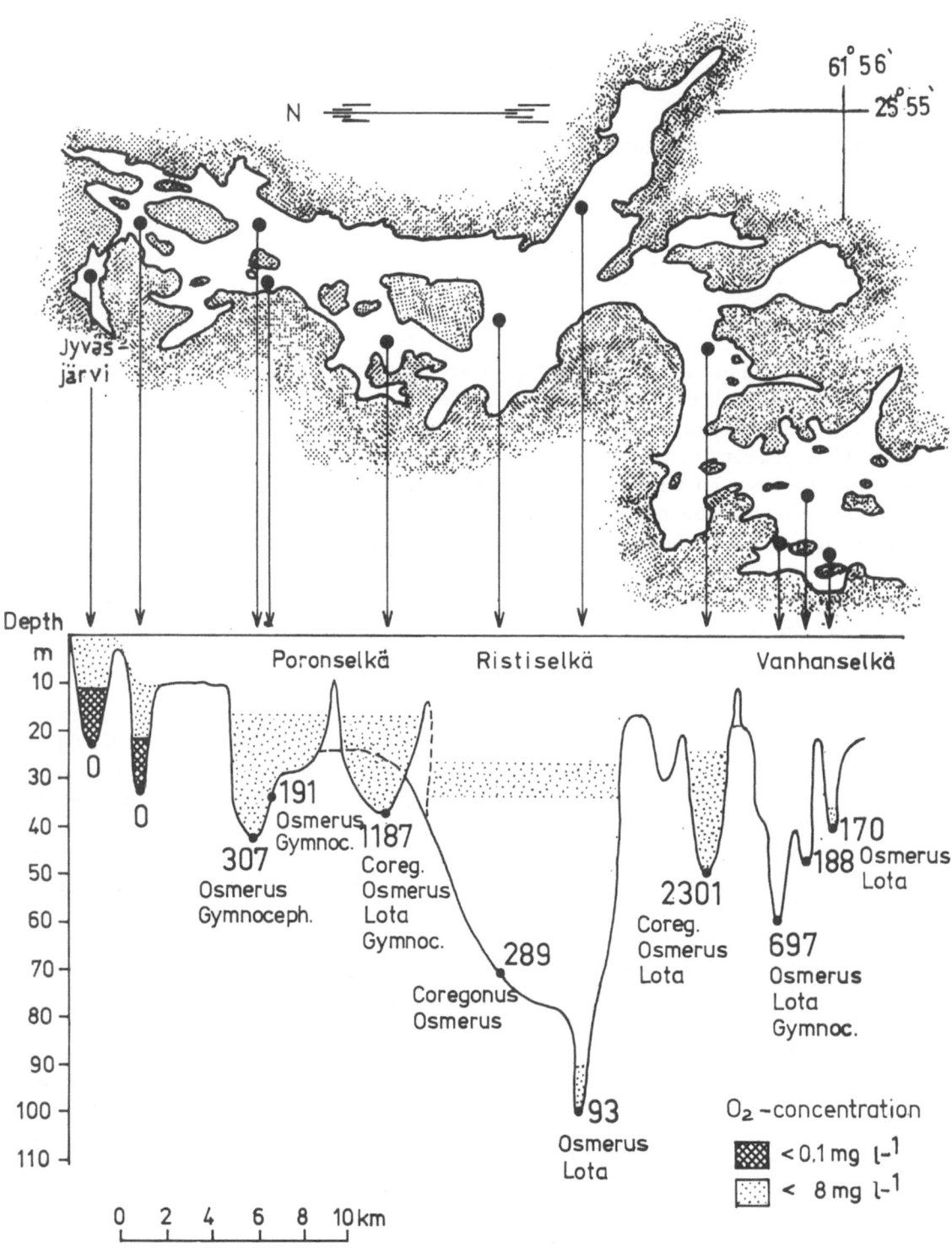

Fig. 1. The concentration of oxygen and the fish species and their fw. biomass (g catch^{-1}) in the deep basins of North Päijänne.

1976; Hakkari 1978 and Särkkä 1979). Although the quality of the water has been somewhat improved since a new purifying plant was built in 1976, the effects of waste waters are still seen on all trophic levels, especially in the northernmost parts of the area. Thus the primary production is still relatively high and the oxygen concentrations in the hypolimnion very low, especially during the late summer and late winter.

Fishing was done by means of a series of gill nets (15, 21 and 35 mm) which were laid on the bottom overnight. The fishing sites are presented in Fig. 1. The catches were measured immediately after fishing and a rough preliminary analysis (by eye) of the largest food organisms and parasites as well as their site in the fish was made before preserving the stomach contents in 70% ethanol. In the food analyses, both the frequency of occurrence and the cases when a certain food item was volumetrically most abundant were recorded. The Spearman rank correlation test was used in order to calculate the similarities in the composition of food of different species. The oxygen content of the water in each fishing site was determined by means of the Winkler method.

Results

Oxygen conditions and fish fauna of the basins

The flow of waste waters in the area is indicated by very low O_2-concentrations of the water in the hypolimnion of the northernmost basins (Fig. 1). More southwards this waste water with clearly improved O_2-conditions but with still lower O_2-concentration than that observed in the other water masses is seen to flow in intermediate depths. The numbers and generic names of fish (presented below the basins) show the biomass (f.w.) and composition of species in single catches. As was expected, the fish fauna of the basins was relatively poor and only four fish species could be observed. The most abundant species were the burbot *(Lota lota)* and the smelt *(Osmerus eperlanus)*, which attains a relatively large size in L. Päijänne and thus has been considered to be at least partly of marine origin (cf. Koli 1969). These two species were found to a depth of 98 m.

Also, the ruffe *(Gymnocephalus cernua)* occurred

rather evenly at depths of 33–60 m, while the vendace *(Coregonus albula)* was found only sporadically. The results of the distribution of these fish species accords well with those observed by Tuunainen (1971) in the area in 1969–71. When considering the whole fish material caught in 1969–70 with that caught by us, great differences are seen, especially in the northernmost parts of the study area. Thus, according to Tuunainen (1971), the biomass of the fish was almost threefold in L. Jyväsjärvi and in the northernmost areas of L. Päijänne compared with more southern areas and the bulk of the biomass was comprised of cyprinids. In our material, which originates from greater depths, no fish was found in these northernmost basins because of the lack of oxygen. Also, in these areas the quality of water and sediment in shallower water is not good enough for the normal reproduction of the vendace (cf. Nyrönen & Hakkari 1976; Bagge & Hakkari, unpubl.).

Food

The occurrence of different food organisms and the cases when a certain food item was volumetrically most abundant in the stomachs of fish, are presented in Table 1. Here, as well as in Table 3, which presents the occurrence of parasites, the whole fish material has been considered, since the material was too scanty for spatial comparisons. In general, the number of food items used by fish in deep water was rather low, consisting mainly of copepods, relict amphipods and *Mysis relicta* and some dipteran larvae and pupae. Cladocerans, which according to Hakkari (1978) are common food items of the smelt and the vendace in the pelagial zone of L. Päijänne, were rare in the material from the deep basins.

Relict crustaceans and deep water copepods, especially *Heterocope borealis,* seem to play an important role in the food of most fish caught in the basins; only the ruffe seemed to favour dipteran larvae. Though the diets of the fish species compared were relatively different, similarities in the diets could be observed when the frequency values were tested by means of the Spearman rank correlation test (Table 2). Significant similarity could be seen between the diets of vendace and small smelt, and almost significant coefficients between large smelt and other fish species. In these cases the high

Table 1. Frequency of occurrence (F %) of the food organisms in the fish studied and cases (D %) when the food was a principal item in the stomach contents.

	Vendace		Smelt		Smelt		Burbot		Ruffe	
Length (cm)	16.8–25.7		<15		>15		15.5–43		10.2–13.8	
	F	D	F	D	F	D	F	D	F	D
Pisidiae	0	0	0	0	0	0	0	0	6	0
Ostracoda	0	0	0	0	4	4	0	0	0	0
Daphnia sp.	0	0	0	0	0	0	7	0	0	0
Limnocalanus macrurus	44	11	11	0	0	0	0	0	0	0
Eurytemora sp.	11	0	0	0	0	0	0	0	0	0
Heterocope borealis	67	0	89	56	81	7	50	3	24	0
H. appendiculata	11	0	0	0	0	0	0	0	0	0
Cyclops spp.	100	89	50	0	48	0	7	0	76	0
Mysis relicta	22	0	39	28	48	7	53	17	35	18
Pallasea quadrispinosa	33	0	11	11	74	63	70	53	29	29
Gammaracanthus lacustris	0	0	0	0	22	15	20	10	0	0
Chaoborus flavicans	11	0	6	0	0	0	20	0	12	6
Chironomidae	0	0	0	0	0	0	0	0	100	47
Pisces	0	0	0	0	0	0	13	7	6	0
No. of stomachs studied	9		18		27		30		18	

Table 2. The Spearman rank correlation coefficients between the frequencies of food items of different fish species.

	Smelt <15 cm	Smelt >15 cm	Burbot	Ruffe
Vendace	0.893**	0.618*	0.367	0.460
Smelt <15 cm	–	0.742*	0.406	0.433
Smelt >15 cm		–	0.624*	0.606*
Burbot			–	0.497

similarity values depend especially on the rich occurrence of copepods in the stomachs of fish.

Parasites

The occurrence of metazoa parasites in the fish studied is presented in Table 3. The parasites observed belong to four different taxonomical groups, consisting of five species of cestoids, two acanthocephalan species, four nematodes and two copepods. Although the parasites were picked up

Table 3. The frequency of occurrence (%) of parasites in the fish studied.

	Vendace	Smelt	Burbot	Ruffe
No. of fish studied	9	45	30	18
Length of fish studied (cm)	16.6–25.7	11.2–22.2	15.5–43	10.2–13.8
Triaenophorus nodulosus (Pall.)	0	20	37	56
Eubothrium rugosum (Batsch.)	0	0	3	0
Diphyllobothrium ditremum (Creplin)	0	9	0	0
Proteocephalus cernuae (Gmelin)	0	0	0	11
P. longicollis (Zeder)	0	2	0	0
Acanthocephalus lucii (Müll.)	0	0	40	0
Echinorhynchus borealis (Linst.)	0	4	67	0
Nematoda sp.	0	0	7	6
Raphidascaris acus (Bloch)	0	0	27	6
Cystidicola farionis Fisch.	0	84	0	0
Camallanus lacustris (Zoega)	0	0	0	6
Ergasilus sieboldi Nordm.	33	40	0	6
Salmincola lotae (Olsson)	0	0	17	0

when alive, the list is very incomplete owing to the rough study method. Common parasites found in most fish species studied were the cestoidean *Triaenophorus nodulosus,* whose encysted plerocercoids were especially abundant in the livers of ruffe, and the gill parasite *(Ergasilus sieboldi).* On the other hand, the parasite fauna of different fish species were relatively specific. The most common parasite in the smelt was *Cystidicola farionis,* which often occupied the swim bladders in great numbers (40–50 specimens/fish). Similarly, the acanthocephalan worm *Echinorhynchus borealis* was very abundant in the intestine and blind sacks of the burbot. Since both species use amphipods as their intermediate hosts, their great abundance in deep water fish could be expected. *Salmincola lotae,* which occupied the oral cavity of five burbots (at depths of 50–100 m), has hitherto only been known to occur in eastern Finland (in Rantasalmi: Gadd 1904 and in Juva Luonterinselkä: Bagge, unpubl.).

Conclusions

Though the fish material obtained from the deep basins of North Päijänne is rather small and the fishing method was selective, the following conclusions could be made:

1. The fish stock of the deepest parts of the basins was relatively poor, consisting mainly of smelt, burbot, ruffe and a few vendace.

2. No fish were caught in the two northernmost basins owing to pollution and very bad oxygen conditions in the hypolimnion during the late summer stagnation. This is the opposite of results obtained by Tuunainen (1971) in the shallower parts of the area.

3. The diet of the fish from the deep areas consisted of only a few food items.

4. Relict crustaceans and deep water copepods, especially *Heterocope borealis,* seemed to play an important role in the food of fish caught in the basins; only the ruffe seemed to favour dipteran larvae.

5. Though the diets of the fish species were in general relatively different, significant or almost significant similarities in the diet could be observed, especially between vendace and small smelt, and between larger smelt and other fish species (Table 3).

6. The parasite fauna of fish caught in the deep water consisted of several cestoideans, acanthocephalans, nematodes and copepods, which except for *Salmincola lotae* are known to be relatively common fish parasites in Finland.

7. Fish lice *(Argulus* spp.) were absent in the material though both *A. foliaceus* and *A. coregoni* are known to be relatively abundant fish parasites at least in shallow water in the large lakes of Central Finland (Bagge, unpubl.).

8. Parasites spreading via relict crustaceans were abundant in smelt and burbot but absent in vendace and ruffe.

References

Eloranta, P., 1976. Phytoplankton and primary production in situ in the lakes Jyväsjärvi and North Päijänne in summer 1974. Biol. Res. Rep. Univ. Jyväskylä 2: 51–66.

Gadd, P., 1904. Parasit-Copepoder i Finland. Acta Soc. Fauna Flora fenn. 26 (8): 1–58.

Granberg, K., 1973. The eutrophication and pollution of Lake Päijänne, Central Finland. Ann. bot. fenn. 10: 267–308.

Hakkari, L., 1978. On the productivity and ecology of zooplankton and its role as food for fish in some lakes of Central Finland. Biol. Res. Rep. Univ. Jyväskylä 4: 1–87.

Jääskeläinen, V., 1917. Pohjois-Laatokan kaloista ja kalastuksista. Suomen Kalatalous 4: 217–302.

Koli, L., 1969. Eräistä kalastomme taksonomisista kysymyksistä. Luonnon Tutkija 3: 93–105.

Nyrönen, J. & Hakkari, L., 1976. Pohjois-Päijänteen kalastossa tapahtuneista muutoksista ja niihin johtaneista tekijöistä. (Summary: On the factors affecting the species composition of fish in Northern Päijänne). Hydrobiol. Res. Centre, Univ. Jyväskylä, Rep. 72: 87–133.

Savolainen, E., 1975. Distribution and food of Myoxocephalus quadricornis (L.) (Teleostei, Cottidae) in fresh waters of eastern Finland. Ann. zool. fenn. 12: 271–274.

Särkkä, J., 1979. The zoobenthos of Lake Päijänne and its relation to some environmental factors. Ann. zool. fenn. 160: 1–46.

Tuunainen, P., 1971. Observations on the composition and abundance of fish fauna in Lake Päijänne, Central Finland. (Preliminary report). Jyväskylän hydrobiol. tutkimuslaitos. Tiedonantoja 15: 1–19.

Size and structure of crayfish (Astacus astacus) populations on different habitats in Finland

K. Westman & M. Pursiainen

Finnish Game and Fisheries Research Institute, Fisheries Division, P.O. Box 193, SF-00131 Helsinki 13, and Evo Inland Fisheries and Aquaculture Research Station, 16970 Evo, Finland

Keywords: freshwater crayfish, *Astacus astacus*, population size, population structure, catching methods

Abstract

The many inland waters in Finland make crayfish production an important potential resource. The rational utilization and management of this resource requires knowledge of the size and structure of the crayfish populations. The difficulties often encountered in catching crayfish complicate population studies. Mark–recapture and electric fishing have been used in the studies. The number of adult crayfish measuring more than 70 mm in a 4-ha lake was estimated at 620, and the number in a 13-ha lake at 3 480. In the lakes, the density of adult crayfish was around $0.6–1.4$ m^{-2} and in one stream studied about 2.5 m^{-2} rising to several individuals per m^2 in the best biotopes.

Introduction

Astacus astacus, the only endemic crayfish species in Finland, occurs up to about 65° N in eastern and about 67° N in western Finland. Some isolated, self-perpetuating crayfish populations also exist further north (Westman 1973).

There are some 60 000 lakes in Finland, with a total area of 31 613 km^2, and owing to their irregular shores, their total shoreline is very long, measuring about 130 000 km. The total length of the numerous Finnish rivers exceeds 20 000 km. Consequently, the extent of the littoral zone, which is the habitat of the crayfish, is remarkably great, and Finland is particularly well-suited by nature for large-scale crayfish production.

Since 1967, the Finnish Game and Fisheries Research Institute has conducted studies on the bionomics, life history and population dynamics of crayfish in different habitats. The aim of this paper is to review some of the results obtained concerning the size and structure of crayfish populations in lake and river habitats. This kind of information is needed for evaluation of the potential for crayfish production in Finnish waters and planning the rational utilization and management of this resource.

Study areas

The studies were conducted in 1979 in two lakes with isolated crayfish populations and in 1970 in one river. Lake Slickolampi (Pohja 60° 01′N, 23° 34′E) lacks inlets and outlets and has an area of 4.2 ha. The length of the shoreline is 1 000 m, and the greatest depth 5 m. In the littoral zone, 64% of the bottom consists of mud, 24% of sunken trees, twigs, vegetation and litter, and 12% of rock, stones and gravel. The *Astacus* population disappeared from the lake in the 1960s, probably due to the crayfish plague, but since the end of that decade a new population has been developing. The plague-resistant American crayfish, *Pacifastacus leniusculus*, introduced in Lake Slickolampi in 1971, has been reproducing since the mid 1970s, and the population is slowly increasing despite the much stronger *Astacus*

Hydrobiologia 86, 67–72 (1982). 0018-8158/82/0861-0067/$01.20.

68

population (Westman & Pursiainen 1979). The crayfish population was unexploited until 1979.

Lake Vuorijärvi (Kuhmoinen, 61°39′N, 24°49′E) is also without inlets and outlets. Its area is 13 ha, and the length of the shoreline is 2 000 m. The greatest depth is 25 m. Of the bottom in the littoral zone, 50% consists of sunken trees, twigs, litter and vegetation, 27% of mud and 23% of rock, stones and gravel. The native *Astacus* population has been exploited yearly by catching crayfish that have reached the minimum body length of 10 cm stipulated in Finnish regulations.

The River Raudanjoki (Loppi, 60°40′N, 24°10′E) is a small river between Lake Keritty and Lake Punelia. The study area comprised a rapid stretch, about 80 m long and 2–4 m wide, and 15 m of the littoral of the pool lying above the rapids. The depth of the rapid area did not exceed 1 m. The bottom was stony, providing abundant hiding places for crayfish. The vegetation was abundant. The native crayfish population had been heavily exploited.

Methods

The crayfish populations in the two lakes were investigated by means of trapping, electric fishing and the mark–recapture method. In the river only electric fishing was used to sample crayfish. A new cylindrical trap model (Evo-trap), which prevents crayfish from escaping, was used (Westman *et al.* 1979a). The mesh size was 7 mm, which catches crayfish over the size of about 70 mm. The traps, baited with fresh fish (roach), were fastened at 5 m intervals to a floating nylon line located about 2.5 m from the shoreline.

The density of the catchable crayfish population was calculated by means of the capture–recapture method. As the behavioural dominance of the males and different moulting cycles for male and female crayfish influence the yield in the traps, each sex was treated separately (cf. Abrahamsson 1966). The crayfish that were caught were marked with waterproof pencil or by electric cauterization (Abrahamsson 1965) and returned to the same place from which they had been taken. Recapture was made 2 weeks later (for further details see Westman & Pursiainen 1979).

To obtain an idea of the structure, sex ratio and

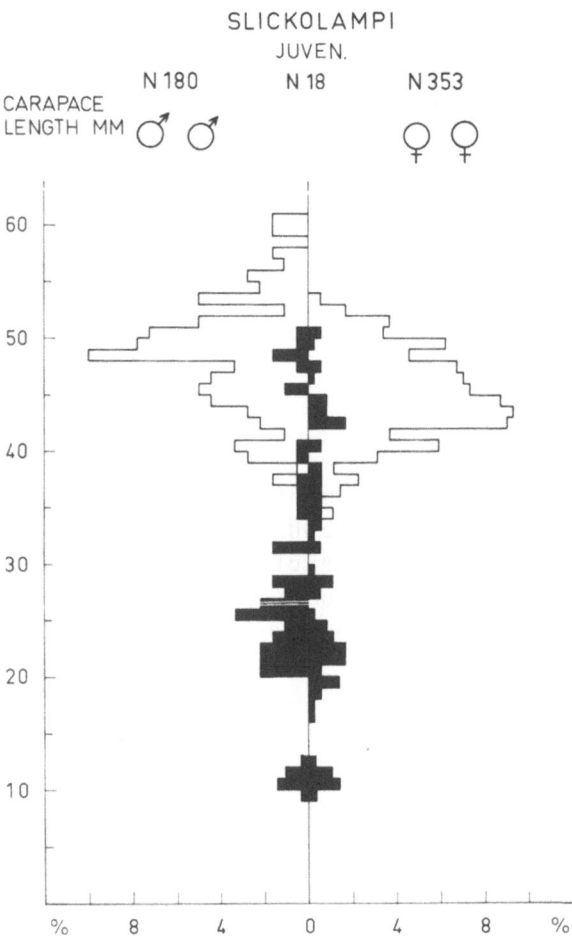

Fig. 1. Size-frequency distribution of crayfish (*Astacus astacus*) collected by trapping (white columns) and electric fishing (black columns) from Lake Slickolampi in August 1979.

density of the whole crayfish population, including small juveniles and individuals that were inactive for some reason and did not enter the traps, electric fishing was conducted in the study areas using portable equipment that produced a balanced current. The technique and equipment has been described elsewhere (Westman *et al.* 1979b).

The length index used in defining the size classes was total length in the River Raudanjoki, but cephalothorax length elsewhere. The latter, which is roughly half of the total length, is a more accurate index of size as the cephalothorax is fixed and rigid, whereas the abdominal joints are flexible. Maturity was determined by inspection of the cement glands located under the abdomen (Abrahamsson 1971).

Results

Lake Slickolampi

The density of the catchable adult crayfish population (cephalothorax length ≥ 35 mm), was estimated during a period of high activity for both males and females (7–22 August, 800 trap-nights). The number of trappable males was estimated at 186 ± 44 (S.E.) and that of the females at 433 ± 67 (S.E.). Thus the adult *Astacus* population at Slickolampi consisted of about 620 individuals, i.e. about 0.6 crayfish per metre of shoreline.

Electric fishing performed in three littoral areas with bottom of different types but suitable for crayfish (total area 80 m²) yielded 122 crayfish that were at least 2 summers old and 18 first summer juveniles. The length distribution of the crayfish is presented in Fig. 1.

The density of crayfish aged 2 summers or more in the three different electric fishing areas was 1.2, 1.3 and 2.2 m⁻². The mean density of all the crayfish was 1.5 m⁻² and that of the adult crayfish was 0.6 m⁻².

The sex ratio of the trapped crayfish was 32% males and 68% females. The uneven sex distribution may to some extent result from the increased activity of the females that had moulted recently after the hatching of the juveniles. Of the catchable females, about 69% were mature, i.e. nearly 1/3 of the mature females were not spawning in autumn 1979.

Lake Vuorijärvi

Marking and recapture of the catchable crayfish population was carried out on 13–29 August (600 trap-nights). The size of the catchable adult population of *Astacus* was estimated at about 1 433 ± 415 (S.E.) males and 1 352 ± 296 (S.E.) females, giving a total of 2 785 individuals. In 1979, about 1 500 m of the total shoreline of 2 000 m was fished. As the unfished shoreline had about the same type of bottom as the fished, about 25% should be added to the population size obtained, i.e. the catchable crayfish population in Vuorijärvi in August 1979 comprised about 3 480 individuals. This means that there were about 1.7 crayfish per metre of shoreline, which is about three times the value obtained in Lake Slickolampi. The estimate made in Vuorijärvi

Table 1. The size of the catchable crayfish (*Astacus astacus*) population in Lake Vuorijärvi estimated by means of the mark-recapture method, and the legal-size (≥ 10 cm) crayfish removed from the lake yearly.

Year	Catchable population ≥ 70 mm	Crayfish ≥ 10 cm removed	
		N	% of the population
1975	3 600	399	11.1
1976	4 100	271	6.6
1977	3 000	305	10.2
1978	2 700	570	21.1
1979	3 500	293	8.4

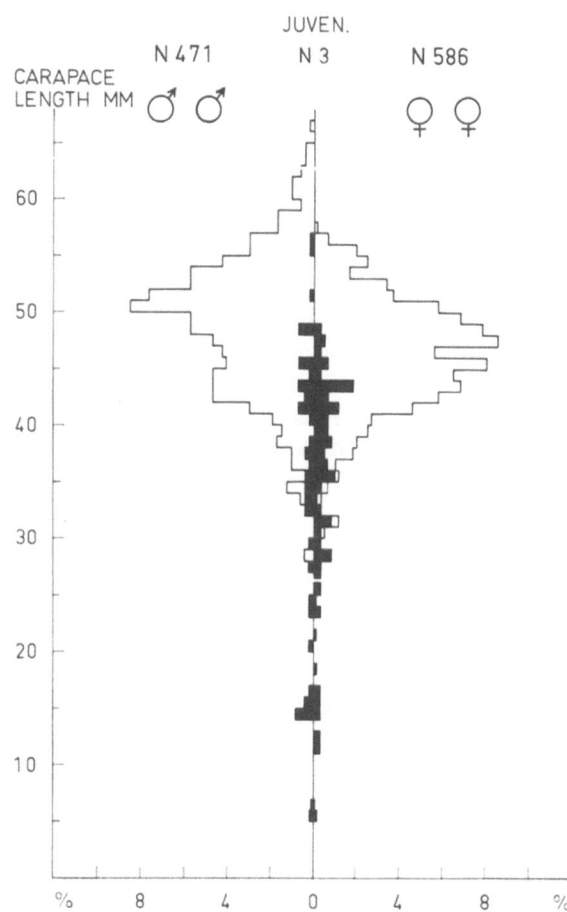

Fig. 2. Size-frequency distribution of crayfish (*Astacus astacus*) collected by trapping (white columns) and electric fishing (black columns) from Lake Vuorijärvi in July–August 1979.

in 1979 is in good agreement with the calculations made by the capture-recapture method in earlier years, when the whole shoreline was fished with

70

traps at 5 m intervals (Table 1). These figures seem to indicate that the population of catchable crayfish in Lake Vuorijärvi has settled at a fairly constant level.

Electric fishing was performed on 23–26 July in five littoral areas (total area 64 m²). The catch was 142 crayfish 2 or more summers old, and 3 juveniles. The length distribution of the crayfish collected has been presented in Fig. 2. The small number of juveniles may be the result of the early sampling time.

The density of crayfish that were at least 2 summers old in the different areas varied from 0.3 to 7.7 m⁻². The mean density of all the crayfish was 2.3 m⁻² and that of adult crayfish was 1.4 m⁻².

The sex ratio in the crayfish caught with traps was 47% males and 53% females. Of the trapped females the proportion that was mature was as high as 95% in 1979. In 1975–1978 the proportion of mature females varied from 76 to 86%.

The crayfish population in Vuorijärvi has been exploited intensively, and in 1975–1979 a total of 1 838 crayfish was removed from the lake. An average 11% of the adult, catchable population has been taken from the lake each year (Table 1). This does not seem to have caused any harm to the population or its renewal. At the present price level, the wholesale value of the yearly catches has varied from about 1 100 to 2 300 Finnish marks (FMK).

The River Raudanjoki

The total number of crayfish collected with electric fishing was 1 123 specimens in, at least, their 2nd summer with a total length range of 25–105 mm, and 190 1st summer juveniles measuring 14–17 mm (Fig. 3). About 65% of the crayfish caught were juveniles (⩽70 mm in total length). Due to the heavy exploitation, the proportion of crayfish that had reached the minimum fishing size was small, only 0.2%. The mean density of crayfish aged 2 summers or more was approximately, 4.1 m⁻². The density of adult crayfish was approximately 2.5 m⁻². The sex ratio was 1:1. Of the sexually mature females, only 39% were mature.

There is no reliable method for determining age directly from a crayfish. Consequently, the year classes have to be estimated from the length frequency distribution. To obtain reliable information by this method, a large sample is needed in which all

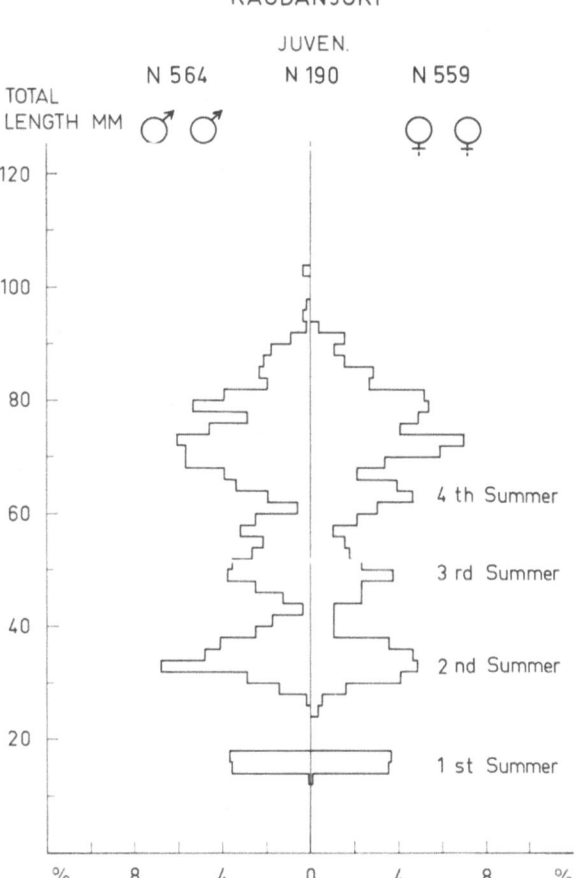

Fig. 3. Size-frequency distribution and visually estimated age groups of crayfish (*Astacus astacus*) collected by electric fishing in August 1970 in the River Raudanjoki.

the length- and age-groups are well represented. The length distribution of the crayfish collected at River Raudanjoki (Fig. 3) may be used as a rough basis for separating the age groups 0+ years (1st summer), 1+ years (2nd summer), 2+ years (3rd summer) and perhaps also 3+ years (4th summer).

According to Abrahamsson (1971), the males are sexually mature when 60 mm in total length and the females when 80 mm. Consequently, in the River Raudanjoki the males seem to be 4 summers old when they reach sexual maturity, and the females 5–6 summers old. According to a very rough estimate, the males reach the minimum size limit of 10 cm in body length when 8–9 summers old and the females when 9–10 summers old. From the 2nd year onwards the males seem to grow faster than the females.

Discussion

A special problem in sampling crayfish, as opposed to fish, is the great variation in activity and catchability caused by moulting. Moreover, the crayfish is a bottom-dwelling creature, and requires special sampling techniques. Sampling with baited traps is the method most commonly used in research on crayfish. However, Brown & Brewis (1979) found that the use of trapping for marking and recapture resulted in the underestimation of the population size of *Astacus* (*Austropotamobius*) *pallipes*, and they strongly recommended that trapping should be used only as an auxiliary sampling method in population studies. The results obtained in the present studies seem, however, to indicate that relatively reliable population estimates can be obtained from trapping results by using a trap model from which the crayfish are not able to escape, by treating each sex and preferably also the different size groups separately, by standardizing trapping techniques and by using sufficient catching effort during a period of high activity for both males and females.

The main limitation of sampling with traps is that only adult crayfish, over 60–70 mm, actively entering the traps can be caught, even with fine-meshed traps. To examine the whole population, trapping must be supplemented with a method suitable for sampling small crayfish as well. Where conditions were suitable for it, electric fishing proved to be a very practical and useful method for obtaining representative samples of the whole crayfish population (cf. Westman *et al.* 1979b).

The numbers of trappable *Astacus astacus* in Lake Slickolampi and Lake Vuorijärvi (about 0.6 and 1.7 m^{-1} of shoreline) were much smaller than those reported by Niemi (1977) from the River Pyhäjoki (64°30′N). Also using the mark–recapture method, Niemi obtained values ranging from 0.8 to 11.5 m^{-1} of shoreline, or 0.1 to 1.6 m^{-2} in different sampling areas. In the River Varisjoki (64°24′N) the numbers were 8.6 m^{-1} of shoreline, or 0.6 m^{-2} (Jäppinen 1976).

The crayfish densities obtained with electric fishing differed considerably not only between different water bodies but also between different areas within the same water. It was observed that the numbers of *Astacus* can be remarkably high in littoral areas and on river bottoms suitable for crayfish. In Lake Vuorijärvi the highest density recorded for crayfish aged 2 summers or more was about 8 m^{-2}, and the sexually mature animals numbered about 3.8 m^{-2}. These values are smaller than the true densities, as some of the crayfish observed in the study areas could not be caught.

The greatest numbers of crayfish in the two lakes were found in biotopes with hard bottoms of rock, stones, or gravel, and on bottoms suitable for burrowing, covered with sunken tree trunks, twigs, litter and vegetation, which provide shelter for the animals.

According to electric fishing conducted in Lake Slickolampi and the River Raudanjoki, about 69% and 65% of the crayfish sampled were ≤70 mm in total length. Due to difficulties in catching the 1st summer juveniles, however, their proportion in the population is underestimated. According to Abrahamsson (1966), individuals less than 3 years old (≤75 mm) represented about 75% of the *Astacus* population examined with electric fishing in the Rögle ponds, South Sweden. In a small lake in the Lithuanian SSR, Cukerzis (1975) observed that the individuals which were sexually immature constituted 90% of the population.

In the material sampled from the River Raudanjoki, the relatively small differences in the strength of the age groups older than 1 summer seem to indicate that after the first year the natural mortality of *Astacus* is quite low. Accordingly, it would perhaps be advisable to use older juveniles for stocking, rather than the newly hatched crayfish at present stocked in great numbers in some European countries.

According to the material collected in the River Raudanjoki, the growth of *Astacus* seems to be rather slow compared with that reported from, for example, Eesti SNT (Estonia) (Järvekülg 1958) and South Sweden (Abrahamsson 1966, 1971). For example in South Sweden the males were sexually mature in the 3rd summer and the females in the 4th summer, one year earlier than in the River Raudanjoki. The slower growth in the present study may chiefly be ascribed to differences in the length of the growing season, but perhaps also to differences in the food and water quality.

The females in the River Raudanjoki seem to reach the minimum size limit of 10 cm when they are 9–10 summers old at the earliest. It is thus possible for females to reproduce 4–5 times before reaching the legal catching size. However, not all of the sexually mature females in the study localities spawned every year. Similar observations have been made in Sweden by Abrahamsson (1972), who found that the proportion of sexually mature females spawning varied from 53% to 97% in different localities. Water temperature and the food supply seemed to influence reproductivity.

These observations seem to indicate that the present minimum size limit stipulated in the Finnish regulations is justified. According to the observations made in Lake Vuorijärvi (Table 1), due to the size limit even intensive exploitation of crayfish seems not to be harmful to the crayfish population.

Acknowledgements

The authors wish to express their gratitude to Mr. J. Louhimo for valuable assistance, to Mrs. A. Damström for checking the English manuscript and to Mr. O. Ranta-aho for drawing the illustrations.

References

Abrahamsson, S., 1965. A method of marking crayfish Astacus astacus Linné in population studies. Oikos 16: 228–231.

Abrahamsson, S., 1966. Dynamics of an isolated population of the crayfish Astacus astacus Linné. Oikos 17: 96–107.

Abrahamsson, S., 1971. Density, growth, and reproduction in populations of Astacus astacus and Pacifastacus leniusculus in an isolated pond. Oikos 22: 373–380.

Abrahamsson, S., 1972. Fecundity and growth of some populations of Astacus astacus Linné in Sweden. Rep. Inst. freshw. Res. Drottningholm 52: 23–37.

Brown, D. J. & Brewis, J. M., 1979. A critical look at trapping as a method of sampling a population of Austropotamobius pallipes (Lereboullet) in a mark and recapture study. Freshwater Crayfish 4: 159–164. Institut National de la Recherche: Agronomique, Thonon-les-Bains, France.

Cukerzis, J., 1975. Die Zahl, Struktur und Produktivität der Isolierten Population von Astacus astacus L. Freshwater Crayfish 2: 513–527. Lousiana State University, USA.

Järvekülg, A., 1958. Joevähk Eestis. Eesti NSV Teaduste Akadeemia, Tartu.

Jäppinen, R., 1976. Varisjoen ravusta ja ravustuksesta. Kalamies (6): 1–4.

Niemi, A., 1977. Population studies on the crayfish Astacus astacus L. in the River Pyhäjoki, Finland. Freshwater Crayfish 3: 81–94. University of Kuopio, Finland.

Westman, K., 1973. The population of the crayfish, Astacus astacus L. in Finland and the introduction of the American crayfish Pacifastacus leniusculus Dana. Freshwater Crayfish 1: 41–55. Studentlitteratur, Lund, Sweden.

Westman, K. & Pursiainen, M., 1979. Development of the European crayfish Astacus astacus L. and the American crayfish Pacifastacus leniusculus (Dana) populations in a small Finnish lake. Freshwater Crayfish 4: 243–250. Institut National de la Recherche Agronomique, Thonon-les-Bains, France.

Westman, K., Pursiainen, M. & Vilkman, R., 1979a. A new folding trap model which prevents crayfish from escaping. Freshwater Crayfish 4: 235–241. Institut National de la Recherche Agronomique, Thonon-les-Bains, France.

Westman, K., Sumari, O. & Pursiainen, M., 1979b. Electric fishing in sampling crayfish. Freshwater Crayfish 4: 251–256. Institut National de la Recherche Agronomique, Thonon-les-Bains, France.

Seasonal and spatial distribution of humus fractions in a chain of polyhumic lakes in southern Finland

V. Pennanen
Department of Limnology, University of Helsinki

Keywords: humus, gel filtration, particulate, colloidal, dissolved fractions, stratification

Abstract

Moderately concentrated, highly coloured natural water was fractionated on Sephadex G-100 columns with distilled water. The applied fractionation procedure produces elution profiles where colloidal and truly dissolved materials are clearly distinguished. The particulate fraction which cannot penetrate the dextran gel bed was calculated from the difference between the predicted and observed yield of gel filtration. The coloured material of polyhumic waters investigated was thus separated into particulate and colloidal (light scattering, iron containing), and truly dissolved (no light scattering, fluorescent) fractions with characteristic distribution patterns in creeks and lakes. The vertical distribution of particles, colloids, and dissolved material in the main lake basin indicated a heterogenous stratification of the three fractions.

Introduction

Humic substances are probably the most widely distributed natural products on the earth's surface, occurring in soils, lakes, and the sea (Schnitzer & Khan 1972). Visibly high concentrations of dissolved or dispersed humic matter are found in streams and small lakes in regions of podzolic soil (Jackson 1975). Few observations have been presented about temporal and spatial distribution of aquatic humus, and there is no general hypothesis concerning the behaviour of humic substances in natural waters. Sharp (1973) has analysed the vertical distribution of several organic fractions, including colloidal, in sea water using ultrafiltration techniques. Wetzel & Otsuki (1974) reported annual variation of particulate and dissolved organic carbon in a marl lake. Differences in the vertical distribution of two gel-fractionated coloured fractions obtained from lake water of varying humus concentration were observed in some lakes (Pennanen 1975). Lock *et al.* (1977) separated a colloidal organic fraction from running waters by centrifugation. Seasonal variation in the concentration and in the composition of fulvic acids in a eutrophic humus-rich polder lake has been reported (de Haan & de Boer 1979; de Haan *et al.* 1981). Recently Stewart & Wetzel (1980) proposed that fluorescence/absorbance ratios might be useful in delineating seasonal and depth distribution patterns of dissolved humic material of low and high molecular weight.

The methodological difficulties in fractionating aquatic organic matter including humus are well known. At present no general method has been adopted for monitoring quantitative and qualitative fluctuations of aquatic humus (see Schnitzer & Khan 1972; Gjessing 1976; Wetzel & Likens 1979). Because many organic substances, e.g. from pulp mills effluents and domestic waste waters, disturb the identification of humic material, a study area without interference from related organic substances was chosen. In Finland there are thousands of polyhumic lakes, ponds, rivers, and creeks without industrial or domestic waste water inputs. To understand the effects of expanding activities in

Hydrobiologia 86, 73–80 (1982). 0018–8158/82/0861–0073/$01.60.
© Dr W. Junk Publishers, The Hague.

74

forest and peat land draining and recently even peat mining requires knowledge of the normal distribution patterns of humus in lakes and rivers.

Material and methods

Improved techniques of gel fractionation (cf. Pennanen 1975) were used in the present study to observe qualitative and quantitative distribution of coloured humic material in a chain of polyhumic lakes. Relative quantifications based on optical monitoring are presented including particulate, colloidal, and dissolved fractions of coloured matter.

The gel fractionation method has been used in several humus studies since the early sixties (Povoledo 1964; Gjessing 1965; Söchtig 1966; Shapiro 1967; Ghassemi & Christman 1968; Cameron *et al.* 1972; de Haan 1972; Pennanen & Sederholm 1974; Salo & Saxen 1974; Mantoura & Riley 1975; Means *et al.* 1977; Stabel 1978; Tambo & Kamei 1978; Stewart & Wetzel 1980). The results of gel fractionation are difficult to compare because of differences in the procedures. Treatments before and after concentrating the humus, the gel type chosen, elution system used, etc. all influence the results of elution when such a multimolecular and diffuse material as natural humus is fractionated.

The humus concentrates to be gel-fractionated are in most cases filtered or centrifuged in order to obtain a stable filtrate and high recovery in fractionation yield. There are some disadvantages in such procedures. It is probable that the sample to be fractionated contains a decreased portion of colloids after filtration or centrifugation because some colloids are retained on the filter or sedimented together with particles and precipitates. In addition, if the quantification of removed fractions is inadequate or arbitrary, the final results concerning the distribution of several humus fractions becomes disproportionate. When the distribution of humus in natural waters is being investigated, the total amount of humic material must be taken into consideration.

Study area and sampling

The area studied is in the catchment of Kokemäenjoki in Southern Finland. The main lake

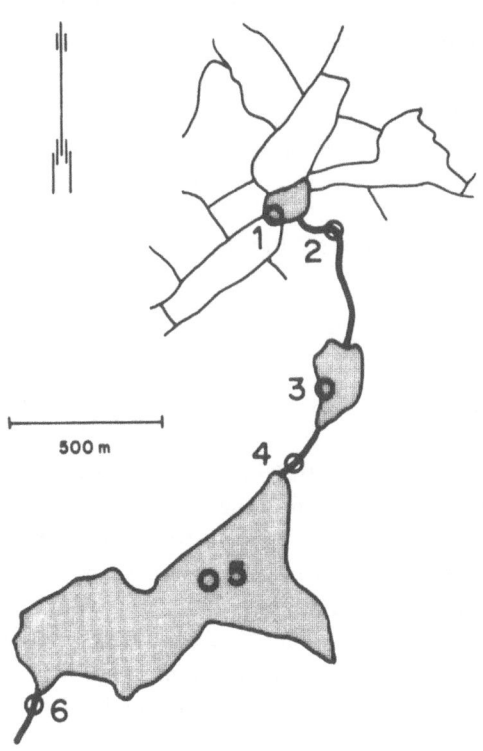

Figure 1. Sampling stations. Lake Pikku Hakojärvi (1), creek from L. Pikku Hakojärvi (2), Lake Keskinen Hakojärvi (3), incoming water to L. Hakojärvi (4), deepest point of Lake Hakojärvi (5), exit from L. Hakojärvi (6).

studied, Lake Hakojärvi (location 61°15′N, 25°12′E) is a well documented dimictic lake with high humus concentration (about 20 to 35 mg l^{-1}) and low salinity (spesific conductivity 3.4 to 4.0 mS m^{-1}) (Lehmusluoto & Ryhänen 1972). The drainage area of the Lake Hakojärvi (area 0.17 km², volume 840 000 m³, mean depth 5 m) is 1.8 km². The surrounding marshland of the smallest lake (Lake Pikku Hakojärvi) was drained in 1965–1967. The lakes are surrounded by coniferous forests.

Samples were collected monthly at six sampling stations (Fig. 1). At stations 1–4 and 6, samples of epilimnion (about 30 cm from the surface) were taken. A vertical profile was collected in Lake Hakojärvi (station 5).

Gel-fractionation procedure

Water samples from lakes and creeks were collected in polyethylene bottles and stored 1–5 days at +4 °C protected from light. Zooplankton and large fibres were removed by prefiltration through a nylon net (60 μm) before concentration. A 300 ml portion of water was evaporated to 30 ml in a rotatory evaporator (Rotavapor, Büchi, Switzerland) under reduced pressure at +40 °C. 10 ml portions of the concentrate were chromatographed immediately after concentration through a dextran gel bed column of width 2.5 cm and length 40 ± 1 cm (Sephadex G-100, Pharmacia, Sweden). Distilled water was used as eluant. Two portions of each sample were chromatographed simultaneously using identical columns. The fresh humus water concentrate was pipetted on top of the column through an open application cylinder (cuvette) with a nylon net (60 μm) bottom. When the concentrate was adsorbed into the gel bed the cuvette was rinsed thoroughly with distilled water until the sample had moved about 2 cm inside the gel bed. The adaptor cuvette was removed and the water flow adjusted. The retention of particles on the adaptor cuvette and on the top of the column was controlled. When necessary, brown particles on the top of the column were removed after chromatographing. The microbial condition of the gel bed was controlled by microscopal observation. The gel bed was autoclaved after eluting about ten samples.

The effluent was automatically collected with a siphon collector (Hans Hösli V 150, Switzerland). Chemical and physical measurements were performed on each subfraction of elution (about 16 tubes of 15 ± 1 ml) as well as on the initial unfractionated water sample.

Absorbance (at 420 nm) measurements were performed with a spectrophotometer (Spectronic 200 UV, Shimadzu/Bausch & Lomb, Japan) against distilled water in 1 cm quartz cuvettes. Fluorescence emission (excitation at 350 nm, emission at 455 nm) and scattered light (350/350 nm) readings were scanned with an Aminco Bowman (USA) spectrofluorometer. Quinine sulphate in 0.1 N H_2SO_4 was used for calibrating the apparatus. Total iron was analyzed with a Perkin Elmer (USA) atomic absorption spectrophotometer (290 B). Total organic carbon analyses were performed on samples of the date 1975-09-08 using the apparatus developed by Salonen (1979).

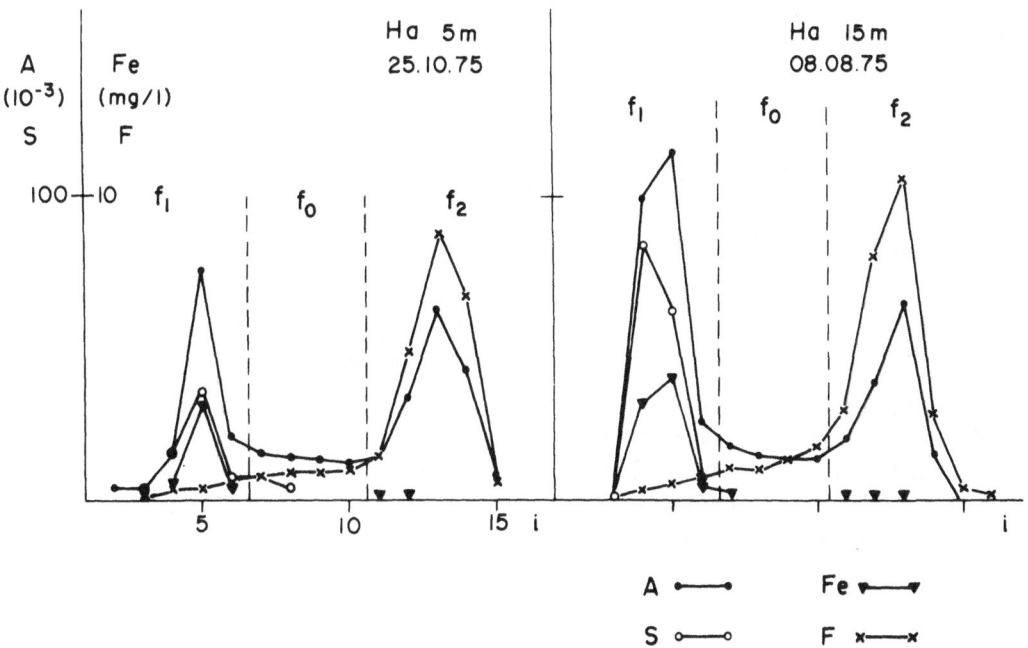

Figure 2. Elution profiles from Lake Hakojärvi epilimnion (Ha 5 m) and hypolimnion (Ha 15 m). Sephadex G-100, distilled water as eluant. Absorbance (A 420 nm), Scattered light (S 350/350 nm), Fluorescence (F 350/455 nm) total iron (Fe) were measured on each test tube (i). Dotted lines indicate how the fractions (f_1, f_0, f_2) were made up.

Results

The elution profile and nomenclature of the fractions

The elution profile of humus water on Sephadex G-100 consists of two peaks measureable at 420 nm (Fig. 2). The fast eluting (large) molecules are called the f_1-fraction, slowly eluting (small) molecules the f_2-fraction. The elution profile between the main fractions f_1 and f_2 is summarized separately as the fraction f_0. Each three fractions correspond the total absorption caused by the absorbing substances within the dotted lines in Fig. 2. The contours of the fractions varied to some extent with the elution profile in each experiment, but the results of the elutions can be summarized as follows:

$$f_1 = \sum_{i=4}^{6} A_i V_{ei} \qquad (1a)$$

$$f_0 = \sum_{i=7}^{10} A_i V_{ei} \qquad (1b)$$

$$f_2 = \sum_{i=11}^{15} A_i V_{ei} \qquad (1c)$$

where A_i = the absorbance of a subfraction
V_{ei} = the volume (15 ml) of a subfraction.

The observed recovery (R_o) of an elution is the sum of the fractions:

$$R_o = f_1 + f_0 + f_2. \qquad (2)$$

As one knows the initial absorbance of the unfractionated sample, the concentration coefficient used (10 ×), and the portion volume (10 ml) applied on the top of the column, the theoretical (100%) fractionation yield (R_t) of a portion can be calculated:

$$R_t = V_I A_I \qquad (3)$$

where V_I = calculated volume (100 ml) of the fractionated portion

A_I = measured absorbance of the unfractionated portion.

As a result of comparison between R_t and R_o it was noticed that for absorbance (and total iron) $R_t > R_o$. In other words some coloured (iron containing) material was lost during elution. Visual observations indicated that the main part of this loss is retained on the application cuvette and sometimes even on the top of the gel bed.

By calculating the loss of material one gets the fraction f_x:

$$f_x = R_t - R_o. \qquad (4)$$

There were two consistent replicates of each sample available, and the results in this study were calculated using the mean values of two elutions. In this paper the quantifications of the fractions are presented based on absorbance measurements only.

General properties of the fractions

Optical measurements (absorbance, fluorescence, scattered light) and total iron analyses were used for identifying the fractions in elution profile (Fig. 2). The general properties of the coloured

Table 1. General properties of the fractions.

	f_x	f_1	f_0	f_2
Fluorescence	Very weak	Very weak	Moderate	Very high
Scattered light	High	High	Very low	Very low
Total iron	High	High	Occasional	Occasional
Organic carbon	Uncertain	Noticeable	Noticeable	High

Table 2. Total organic carbon (TOC) content (C/mg 1^{-1}) of the unfractionated sample (R_t) and of the fractions. The observed recovery (R_o) is the sum of the elution profile $(f_1 + f_0 + f_2)$. See equations 1–4.

Station	Depth/m	R_t	R_o	f_x	f_1	f_0	f_2
3	0	20.0	20.7	–	3.6	4.4	12.7
4	0	18.7	19.6	–	3.6	4.3	11.6
6	0	11.8	12.4	–	2.0	2.2	8.1
5	1	11.7	12.8	–	2.4	2.2	8.3
	5	12.7	13.1	–	2.2	2.4	8.4
	9	13.4	13.2	(0.2)	2.3	2.9	8.0
	15	15.0	13.6	(1.4)	3.5	2.9	7.2
	16	15.4	15.6	–	4.5	3.1	8.0

Table. 3. Annual ranges and mean values (A 10^{-2}) of the fractions at stations 1-6.

Station	Depth/m	f_x:range	mean	f_1:range	mean	f_0:range	mean	f_2:range	mean
1	0	0.1-1.8	1.0	1.6-3.3	2.3	1.2-2.0	1.6	2.8-4.2	3.7
2	0	0.1-1.7	0.9	0.9-3.1	2.1	1.0-2.0	1.6	3.0-6.5	4.1
3	0	0.2-2.1	1.2	1.3-3.0	2.1	1.3-3.1	1.8	1.2-4.2	3.8
4	0	0.1-1.7	0.9	1.5-3.0	2.1	1.1-2.0	1.6	1.7-4.2	3.5
6	0	--1.6	0.7	1.1-1.9	1.5	0.6-1.6	0.9	2.0-3.0	2.4
5	1	0.2-1.4	0.9	1.0-1.7	1.4	0.5-1.5	0.8	2.1-3.6	2.5
	5	0.0-1.1	0.6	1.4-1.8	1.7	0.6-1.3	0.9	2.0-2.8	2.3
	9	0.0-2.0	0.8	1.1-2.2	1.7	0.6-1.7	1.0	1.8-2.9	2.3
	15	0.1-4.3	1.8	2.1-4.4	3.1	0.6-1.1	0.9	2.1-2.5	2.3
	16	0.1-9.9	4.2	2.3-7.5	4.9	0.6-1.2	1.0	2.0-2.4	2.2

fractions are summarized in Table 1. Details in optical properties (absorption coefficients, absorbance/fluorescence and absorbance/iron relations of the fractions) will be demonstrated and discussed elsewhere.

Total organic carbon (TOC) was analysed from the samples of 1975-09-08. The results are summarized in Table 2. The carbon content of the fraction f_x remained uncertain in this experiment while the other fractions had remarkable concentrations of organic carbon.

Horizontal and vertical distribution of the fractions

The horizontal distribution of the fractions can

be divided into two groups (epilimnion sampling at stations 1-4, and 5-6) as demonstrated in Table 3. Vertical observations were made at the deepest point of Lake Hakojärvi. The annual mean values of the fractions indicated that fractions f_0 and f_2 were almost uniformly distributed in the water column and fractions f_x and f_1 were regularly stratified (Fig. 3). The stratification profiles of f_x and f_1 differed from each other to some extent (Table 3).

Seasonal variation of the fractions

On the basis of monthly observations during the year 1975, the seasonal distribution of the fractions (Fig. 4) can be examined in the following groups:
i)   stations 1-4 (strong terrestrial influence),
ii) station 5: 1-9 m (free water column in Lake Hakojärvi),
iii) station 5: 15-16 m (near-bottom water in Lake Hakojärvi).

Fractions f_x and f_1 varied strongly both temporally and spatially in all groups. Seasonal variation of the fractions f_0 and f_2 was noticeable at stations 1-4. In Lake Hakojärvi the seasonal variation of the fractions indicated a different response to stratification: f_x seemed to stratify more efficiently than f_1, especially under ice, while f_0 and f_2 were almost uniformly distributed in the water column all the year. The winter and summer periods are discernible at stations 1-4 while the dimictic nature of Lake Hakojärvi is clearly demonstrated at station 5.

There are interesting similarities and exceptions in the seasonal variation patterns at 'paired sampling' stations 1 and 2, 3 and 4, 5 and 6 respectively (see Fig. 1). The annual mean values of

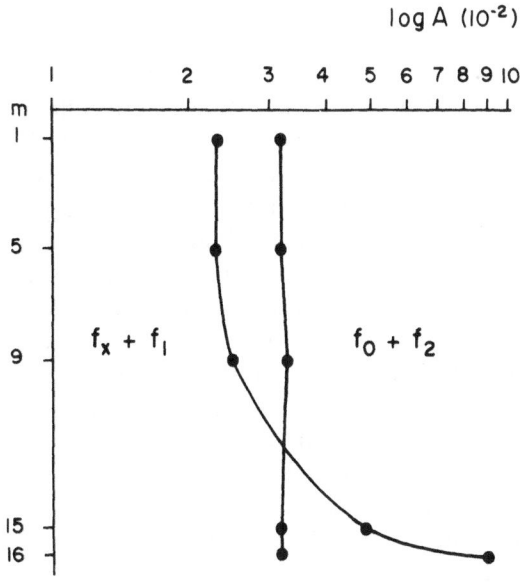

Figure 3. Annual mean values of stratifying fractions ($f_x + f_1$) and uniformly distributed fractions ($f_0 + f_2$) at the deepest point of Lake Hakojärvi.

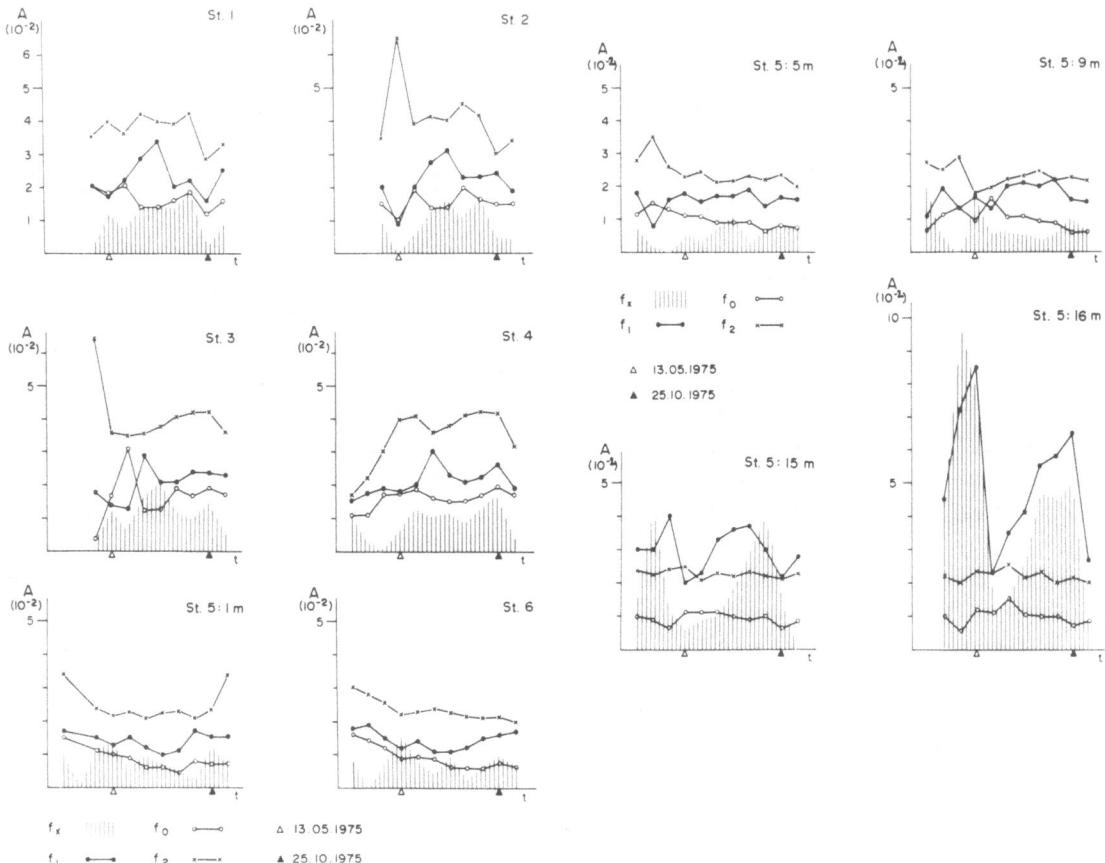

Figure 4. Seasonal variation of the fractions (f_x, f_1, f_0, f_2) at sampling stations 1–6. Vernal (△) and autumnal (▲) overturns are marked on the time scale.

the fractions are very similar (Table 3) while the distribution patterns of the fraction f_1 characterize these stations:

i)   stations 1 and 2: summer maximum,
ii)   stations 3 and 4: shorter summer maximum,
iii) stations 5 and 6: summer depression.

Discussion

After summarizing the properties available the following hypothesis is set: fraction f_x is comparable to a particulate fraction, fraction f_1 to a colloid fraction, and fraction $f_0 + f_2$ to dissolved fractions of polyhumic water.

Particles (f_x) and colloids (f_1) were quite similar on the basis of chemical and physical analyses. Their observed quantities correlate strongly although there are exceptions to this correlation. In

the hypolimnion of Lake Hakojärvi particles reach their maximum value one month before colloids under ice but one month later than colloids in summertime. The vertical distribution of particles indicated that this fraction might include components from plankton in the epilimnion.

The absorptivity of iron-containing fractions (f_x and f_1) is high compared to dissolved fractions (f_0 and f_2). These highly coloured complexes contain iron which is tightly bound with an organic ligand while the rest of the iron is inorganic (cf. Ghassemi & Christman 1968). Both iron (Shapiro 1967) and organic macromolecules (Koenings & Hooper 1976) are essential in these colloids.

The carbon content of the colloid fraction was about 20% of total organic carbon. Lock *et al.* (1977) reported that up to 50% of filterable carbon was colloidal in Canadian rivers when examined by centrifugation. In the present study it was possible to separate a fairly homogenous macromolecular fraction (f_1) by choosing a gel type with a higher exclusion limit.

The main part (about 80%) of the total organic carbon was in the truly dissolved state together with dissolved salts (cf. Pennanen 1975). The gel-filtration behaviour (e.g. Urano *et al.* 1980) and the fluorescence of this fraction indicate that it includes aromatic units. This fraction might be comparable to the fulvic acid fraction which is the most water soluble part of terrestrial humus (cf. Schnitzer & Khan 1972).

There is some relation between colloids and dissolved components. At station 3 one can note a possible transformation of f_0 to f_1. Furthermore, at the main lake basin the amount of the fraction f_0 is decreased while the concentration of iron complexes (f_1) has increased.

Conclusion

In order to trace actual distribution of humic substances in aquatic systems a relevant fractionation method is needed. After adjusting estimates for absorption coefficients for each fraction, the interactions between particulate, colloidal, and dissolved fractions can be examined.

Acknowledgements

This study was supported by the Academy of Finland. I am grateful to Professor R. Ryhänen for his interest and support during the work. Thanks are due to Mrs. Mirja Heikkinen and Mrs. Tuula Lepojärvi for their careful assistance.

References

Cameron, R. S., Swift, R. S., Thornton, B. K. & Posner, A. M., 1972. Calibration of gel permeation chromatography material for use with humic acid. J. Soil Sci. 23: 342–349.

Ghassemi, M. & Christman, R. F., 1968. Properties of the yellow acids of natural waters. Limnol. Oceanogr. 13: 583–597.

Gjessing, E. T., 1965. Use of 'Sephadex' gel for the estimation of molecular weight of humic substances in natural water. Nature 208: 1091–1092.

Gjessing, E. T., 1976. Physical and Chemical Characteristics of Aquatic Humus. Ann Arbor Science, Michigan.

de Haan, H., 1972. Some structural and geological studies on soluble humic compounds from Tjeukemeer. Verh. int. Ver. Limnol. 18: 685–695.

de Haan & de Boer, T., 1979. Seasonal variation of fulvic acids, amino acids, and sugars in Tjeukemeer, The Netherlands. Arch. Hydrobiol. 85: 30–40.

de Haan, H., de Boer, T. & Halma, G., 1981. Curie point pyrolysis mass-spectrometry of fulvic acids from Tjeukemeer, The Netherlands. Freshwat. Biol. 9: 315–317.

Jackson, T. A., 1975. Humic matter in natural waters and sediments. Soil Sci. 19: 56–69.

Koenings, J. P. & Hooper, F. F., 1976. The influence of colloidal organic matter on iron and iron-phosphorus cycling in an acid bog lake. Limnol. Oceanogr. 21: 684–696.

Lehmusluoto, P. O. & Ryhänen, R., 1972. Lake Hakojärvi, a polyhumic lake in Southern Finland. Verh. int. Ver. Limnol. 18: 403–408.

Lock, M. A., Wallis, P. M. & Hynes, H. B. N., 1977. Colloidal organic carbon in running waters. Oikos 29: 1–4.

Mantoura, R. F. C. & Riley, J. P., 1975. The use of gel filtration in the study of metal binding of humic acids and related compounds. Analyt Chem. Acta 78: 193–200.

Means, J. L., Crerar, D. A. & Amster, J. L., 1977. Application of gel filtration chromatography to evaluation of organometallic interactions in natural waters. Limnol. Oceanogr. 22: 957–965.

Pennanen, V., 1975. Humus fractions and their distribution in some lakes in Finland. In: Povoldeo, D. & Golterman, H. L. (eds.) Proc. Int. Meet. Humic Substances, Nieuwersluis 1972, Pudoc, Wageningen. pp. 207–215.

Pennanen, V. & Sederholm, H., 1974. The fluorescence activity of gelfractionated lake waters of different colour values. Aqua fenn.: 3–7.

Povoledo, D., 1964. Some comparative physical and chemical studies on soil and lacustrine organic matter. Mem. Ist. Ital. Idrobiol. 17: 21–32.

Salo, A. & Saxen, R., 1974. On the role of humic substances in the transport of radionuclides. Report SFL – A 20. Institute of Radiation Physics, Helsinki.

Salonen, K., 1979. A versatile method for the rapid and accurate determination of carbon by high temperature combustion. Limnol. Oceanogr. 24: 177–183.

Schnitzer, M. & Khan, S. U., 1972. Humic Substances in the Environment. Marcel Dekker, New York.

Shapiro, J., 1967. Yellow organic acids of lake water: differences in their composition and behaviour. In: Golterman, H. L. & Clymo, R. S. (eds.), Chemical Environment in the Aquatic Habitat, pp. 202–216. Proc. of an I.B.P. Symp. Amsterdam and Nieuwersluis 1966, Amsterdam.

Sharp, J. H., 1973. Size classes of organic carbon in seawater. Limnol. Oceanogr. 18: 441–447.

Stabel, H-H., 1978. Zur Molekulargewichtverteilung gelöster organischer Moleküle in verschiedenen Oberflächengewässern. Arch. Hydrobiol. 82: 88–97.

Stewart, A. J. & Wetzel, R. G., 1980. Fluorescence: absorbance ratios – a molecular weight tracer of dissolved organic matter. Limnol. Oceanogr. 25: 559–564.

Söchtig, H., 1966. Zur Fraktionierung von Humusstoffen durch Gelfiltration. Landbauforsch. Völkenrode 16: 25–30.

Tambo, N. & Kamei, T., 1978. Treatability evaluation of general organic matter. Matrix conception and its application for a regional water and waste water system. Water Res. 12: 931–950.

Urano, K., Katagiri, K. & Kawamoto, K., 1980. Characteristics of gel chromatography using Sephadex gel for the fractiona-tion of soluble organic pollutants. Water Res. 14: 741–745.

Wetzel, R. G. & Likens, G. E., 1979. Limnological analyses. W. B. Saunders, Philadelphia.

Wetzel, R. G. & Otsuki, A., 1974. Allochtonous organic carbon of a marl lake. Arch. Hydrobiol. 73: 31–56.

The role of nitrogen as a growth limiting factor in the eutrophic Lake Vesijärvi, southern Finland

J. Kanninen, Lea Kauppi & E.-R. Yrjänä
National Board of Water, Water Research Institute, P.O. Box 730, 00101 Helsinki 10, Finland

Keywords: bioassay, nitrogen, nitrogen fixation

Abstract

The significance of nitrogen for algal growth was studied in Lake Vesijärvi in 1979 and 1980 by algal bioassay, using *Selenastrum capricornutum* and *Anabaena cylindrica* as test organisms. Nitrogen limited the growth of *Selenastrum* for the major part of the investigation period, while phosphorus seemed to be the most limiting factor for *Anabaena*. This difference was reflected in the *in situ* succession of phytoplankton. As the ratio of inorganic nitrogen to phosphate phosphorus became smaller, nitrogen-fixing blue-green algae became dominant. Nitrogen fixation was greatest at the beginning of July, coinciding with maximum heterocyst numbers.

Introduction

In many eutrophic or hypereutrophic lakes nitrogen may limit algal growth (Gerloff & Skoog 1957; Goldman & Wetzel 1963; Goldman & Armstrong 1969; Maloney *et al.* 1972; Forsberg & Claesson 1974; Goldman 1976; Cleasson & Ryding 1977). This often leads to an increase in the biomass of nitrogen-fixing blue-green algae, which are independent of mineral nitrogen in the water.

The aim of this study was to evaluate the significance of nitrogen for algal growth and composition and to determine the amount of nitrogen fixed by blue-green algae in the eutrophic Lake Vesijärvi.

Study area

Lake Vesijärvi is situated in southern Finland, north of the city of Lahti. The lake is shallow and has three basins, of which Enonselkä is the most eutrophic. The most important hydrological and morphological characteristics are:

	Whole lake	Enonselkä
Drainage area	515 km^2	84 km^2
Surface area	110 km^2	26 km^2
Volume	663 × 10^6 m^3	176 × 10^6 m^3
Maximum depth	40 m	33 m
Mean depth	6.0 m	6.8 m
Theoretical retention time	5.4 yr	5.6 yr

The city of Lahti discharged its sewage into the Enonselkä basin until 1975, after which it was led to the River Porvoonjoki. The sewage was treated in a biological sewage works before discharge to the lake. The population of the city had increased from less then 3000 in 1905 to 95 000 in 1978. During the final years when sewage was discharged to the lake, the nutrient load was about 150 kg day^{-1} phosphorus and 800 kg day^{-1} nitrogen (Keto 1976).

The sampling station was E4 in the Enonselkä basin (Fig. 1).

Hydrobiologia 86, 81–85 (1982). 0018-8158/82/0861-0081/$01.00.
© Dr W. Junk Publishers, The Hague.

82

Fig. 1. Map of Lake Vesijärvi.

Materials and methods

Water samples were taken every two weeks as integrated samples from the euphotic zone in Enonselkä basin. The analyses of total N, NH_4-N, NO_3-N, NO_2-N, total P and PO_4-P were carried out using methods presented by Erkomaa *et al.* (1977). The composition of phytoplankton was determined by the Utermöhl technique. The number of heterocysts was also recorded.

The samples for algal tests were deep-frozen. Before the test they were rapidly thawed under hot tap water (cf. Forsberg *et al.* 1975) and filtered through a Gelman membrane filter (pore size 0.2 μm). After this the above mentioned chemical analyses were performed again. The nutrient addi-

tions used in the tests were 0, 140 and 700 μg l^{-1} nitrogen and 0, 32 and 160 μg l^{-1} phosphorus, in all combinations. The sample volume was 30 ml. In 1979 only one test alga, *Selenastrum capricornutum* Printz., was used. This alga is of the same origin as the one used by Skulberg (1968) and Forsberg (1972). In 1980 a heterocystous, nitrogen-fixing blue-green alga, *Anabaena cylindrica* Lemm. (ATCC Nr 27899), isolated by S. B. Chu from pond-water in Cambridge in 1939 (Fogg 1942), was also used. The test conditions were as follows:

	Temp-erature	Illum-ination	Gas exchange	Incubation time
S. capricornutum	20°C	5000 lx	Shaken	14 days
A. cylindrica	20°C	2000 lx	daily	21 days

Table 1. Nitrogen and phosphorus concentrations (μg l^{-1}) as well as the limiting nutrient on the basis of bioassay and nutrient concentrations in Lake Vesijärvi in 1979 and 1980. – = Not done.

Date	Depth (m)	Tot.N	NO$_3$-N+ NO$_2$-N	NH$_4$-N	Tot.P	PO$_4$-P	N$_{tot}$:P$_{tot}$	N$_m$:PO$_4$-P	Limiting nutrient Concentrations	Bioassay
22.5.1979	0–5	590	–	–	42	–	14		N,P	N
19.6.	0–4	1000	7	13	55	11	18	1.8	N	N
26.6.	0–4	800	2	27	38	3	21	9.7	N	–
10.7.	0–4	1300	1	20	56	7	23	3.0	N	N
26.7.	0–3	920	6	23	51	11	18	2.6	N	–
1.8.	0–4	820	–	–	43	–	19		N,P	–
14.8.	0–3	1000	4	47	56	4	18	13	N	N,P
21.8.	0–4	1400	9	140	310	–	4.5		N	–
12.9.	0–4	930	23	110	110	16	8.4	8.3	N,P	N
4.6.1980	0–6	540	1	8	34	6	16	1.5	N	N
18.6.	0–3	580	0	5	42	5	14	1.0	N	N
2.7.	0–3	740	3	58	62	9	12	6.8	N	N
18.7.	0–3	–	1	6	50	9		0.8	N	N
30.7.	0–3	780	1	33	53	6	15	5.7	N	N
13.8.	0–4	840	1	6	42	8	20	0.9	N	n
27.8.	0–4	700	1	42	50	17	14	2.5	N	N
8.9.	0–4	570	6	57	42	14	14	4.5	N	N

The biomass of *S. capricornutum* was measured with an electronic particle counter (Coulter Counter model Z$_B$) and as turbidity (Hach). The biomass of *A. cylindrica* was examined microscopically in a haemocytometer, and turbidity was also measured.

In 1980 the rate of nitrogen fixation was measured by the acetylene reduction method (Burris 1972; Vuorio 1977). The samples were taken from six depths (0.2, 0.5, 1, 2, 4, 6 m) and concentrated 100 times. The diurnal variations were studied once (30–31 July) by taking samples every 3 h. Sub-samples of 5 ml were injected into the serum bottles, and three duplicate sub-samples and one control sample were prepared. The acetylene concentration was 20%. The samples were incubated for 2 h at the depth from which they had been taken. Metabolic activity was stopped by adding Lugol's solution. The ethylene concentration was measured using a Perkin Elmer gas chromatograph.

The primary production of phytoplankton was measured from the same depths as nitrogen fixation using the ^{14}C method described by Vollenweider (1974). The water temperature and illumination were measured during sampling at each depth.

Results

Mineral nitrogen, especially nitrate, was almost

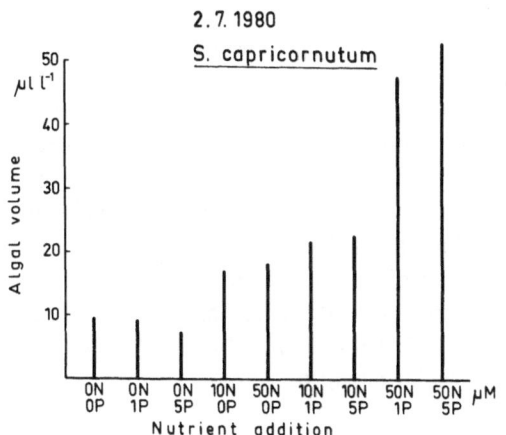

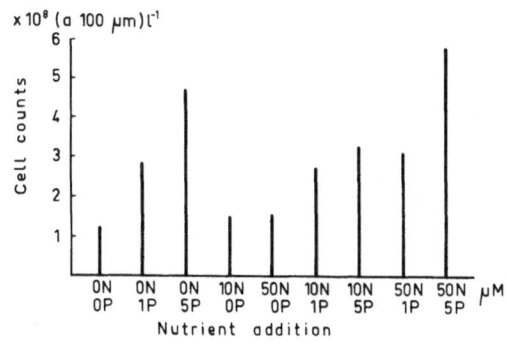

Fig. 2. Bioassay results from 2 July 1980 during the period of maximum nitrogen fixation.

84

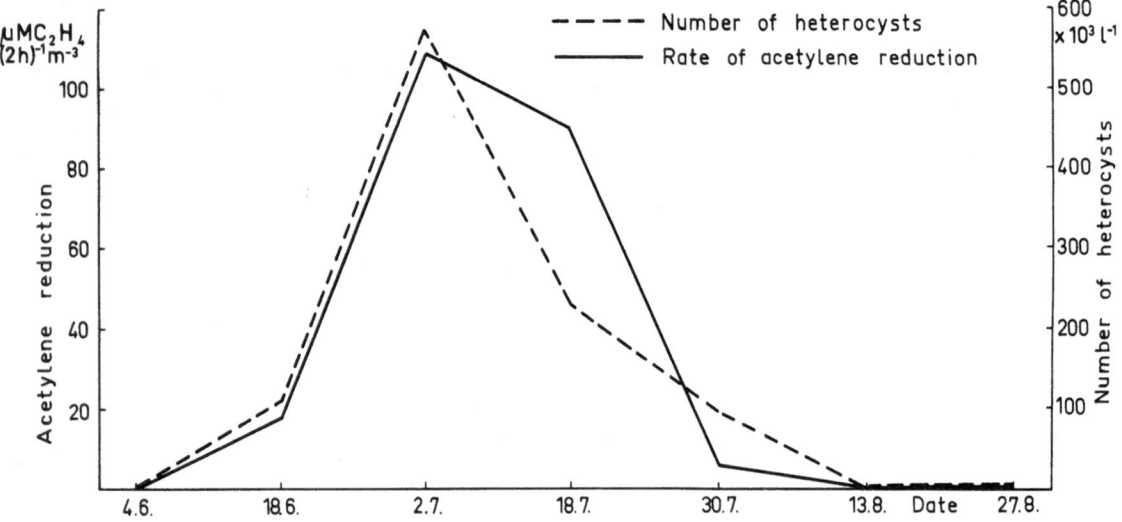

Fig. 3. The rate of acetylene reduction (μM C$_2$H$_4$ (2h)$^{-1}$m^{-3}) and the number of heterocysts at a depth of 1 m in Lake Vesijärvi in 1980.

exhausted in June and July, whereas there was always enough phosphate to allow the growth of algae (Table 1).

In both years nitrogen limited the growth of *Selenastrum capricornutum* from May to the middle of August, according to the bioassay. For *Anabaena cylindrica* the situation was the reverse: nitrogen increase had almost no influence on growth, while addition of phosphorus resulted in increased biomasses. This can clearly be seen in Fig. 2, which shows the situation on 2 July, when the greatest nitrogen fixation was measured. This is accounted for by the ability of *Anabaena cylindrica* to fix molecular nitrogen.

In 1979 the phytoplankton biomass varied from 2 to 22 mg l^{-1}. The biomass was highest in July and August, when blue-green algae were dominant. 23–34% of the total biomass consisted of heterocystous blue-green algae *(Anabaena circinalis* Rbh., *A. planctonica* Brunnth. and *Aphanizomenon flosaquae* (L.) Ralfs.). The number of heterocysts reached 220–650 × 10^3 l^{-1} but decreased at the end of August, although blue-green algae were still dominant.

The phytoplankton counts for 1980 were not complete at the time of writing. A bloom of heterocystous blue-green algae occured at the beginning of July. The maximum number of heterocysts, 572 × 10^3 l^{-1}, was observed on 2 July. The dominant heterocystous blue-green algae were *Anabaena flos-aquae* (Lyngb.) Breb., *A. planctonica* and *Aphanizomenon flos-aquae*.

Greatest nitrogen fixation, about 120 μM C$_2$H$_4$(2h)$^{-1}$m^{-3} in the three uppermost layers, was observed at the same time as the heterocyst maximum (Fig. 3). Most nitrogen fixation occurred in the uppermost 3 m (Fig. 4). In August the nitrogen fixation was almost negligible, although the maximum chlorophyll-a concentration was observed on August 7. The dominant species was *Microcystis flos-aquae* (Wittr.) Kirchn.

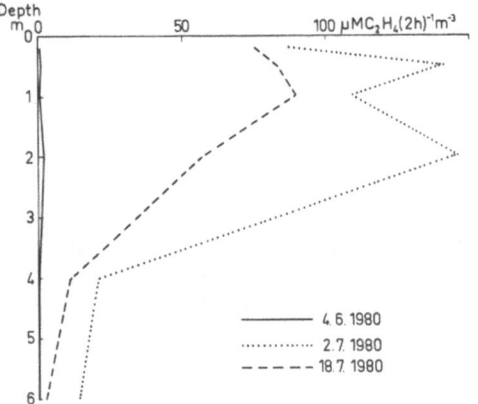

Fig. 4. Vertical variation in the rate of acetylene reduction in Lake Vesijärvi in June and July 1980.

Discussion

The results from bioassay and chemical analyses were in good agreement. The low ratios of $N_{tot}: P_{tot}$ and $N_m: PO_4\text{-}P$ suggest nitrogen limitation (Table 1). According to Tarkiainen et al. (1974) nitrogen limits algal growth in the Helsinki sea area, if $N_{tot}: P_{tot}$ is greater than $N_m: PO_4\text{-}P$.

The deficiency of nitrogen was clearly reflected in the succession of phytoplankton. In 1979, when nitrogen apparently limited algal growth throughout the summer, heterocystous blue-green algae were abundant from the end of June to the middle of August. In 1980 the succession of phytoplankton was very rapid. The bloom of nitrogen-fixing algae lasted for about 2–3 weeks in July. The chlorophyll-a maximum was, however, observed in August and was due to a bloom of Microcystis flos-aquae.

The rate of acetylene reduction measured in Vesijärvi in July was greater than those measured by Vuorio et al. (1978) in the Helsinki sea area. Rinne et al. (1979) observed acetylene reduction rates of the same order of magnitude as those in Lake Vesijärvi at one station in the Gulf of Finland in August 1977. Rusness & Burris (1970) measured an acetylene reduction rate of 108 $\mu M(2h)^{-1}$ m^{-3} in Lake Mendota in 1968 in the surface water.

Acknowledgements

We should like to thank Mr. Juha Keto and all the other persons in the Municipal Laboratory of the City Lahti and in the Helsinki Water District Laboratory who have contributed to this study. Mrs. Eeva Tarkiainen-Rinne (Helsinki City Water Laboratory) kindly provided the culture of Selenastrum capricornutum and Mr. Timo Vaara (University of Helsinki, Institute of Microbiology) the culture of Anabaena cylindrica.

References

Burris, R. M., 1972. Measurements of biological N_2 fixation with 15N_2 and acetylene. In: Sorokin, J. & Kadota, M. (Eds.) Techniques for the Assessment of Microbial Production and decomposition in fresh waters, pp. 3–14. IBP Handbook No. 23, Blackwell Scientific Oxford.

Claesson, A. & Ryding, S.-O., 1977. Nitrogen – a growth limiting nutrient in eutrophic lakes. Prog. Wat. Tech. 8, 4/5: 291–299.

Erkomaa, K., Mäkinen, I. & Sandman, O., 1977. Methods of water analyses used by authorized and Water Authority laboratories. National Board of Waters, Helsinki, Finland, Report No. 121. (In Finnish).

Fogg, G. E., 1942. Studies on nitrogen fixation by blue-green algae – I. Nitrogen fixation by Anabaena cylindrica Lemm. J. exp. Biol. 19: 78–87.

Forsberg, C., 1972. Algal assay procedure. J. Wat. Pollut. Control Fed. 44: 1623–1628.

Forsberg, C. & Claesson, A., 1974. Naturvårdsverkets rr-undersökning – III. Algtest med vatten från rr-undersökningens sjöar, Augusti–Oktober 1972. Vatten 30: 84–95. (In Swedish).

Forsberg, C., Ryding, S.-O. & Claesson, A., 1975. Recovery of polluted lakes. A Swedish research program on the effects of advanced waste water treatment and sewage diversion. Wat. Res. 9: 51–59.

Gerloff, G. C. & Skoog, F., 1957. Nitrogen as a limiting factor for the growth of Microcystis aeruginosa in southern Wisconsin lakes. Ecology 38: 556–561.

Goldman, C. R. & Armstrong, L., 1969. Primary productivity studies in lake Tahoe, California. Verh. int. Ver. Limnol. 17: 49–71.

Goldman, C. R. & Wetzel, R. G., 1963. A study of the primary productivity of Clear Lake, California. Ecology 44: 283–294.

Goldman, J. C., 1976. Identification of nitrogen as a growth limiting nutrient in wastewaters and coastal marine waters through continuous culture algal assays. Wat. Res. 10: 97–104.

Keto, J., 1976. Eutrophication process of Lake Vesijärvi. Ympäristö ja Terveys 7: 299–308. (In Finnish).

Maloney, T. E., Miller, W. E. & Shiroyana, T., 1972. Algal responses to nutrient additions in natural waters – 1. Laboratory assays. In: Likens, G. E. (Ed.) Nutrients and Eutrophication: The Limiting-Nutrient Controversy, pp. 134–140. American Society of Limnology and Oceanography, Special Symposia Vol. 1.

Rinne, I., Melvasalo, T., Niemi, Å. & Niemistö, L., 1979. Nitrogen fixation (acetylene reduction method) by blue-green algae in the Baltic Sea in 1975 and 1977. Publ. Wat. Res. Inst., Fin 1. 34: 88–107.

Rusness, C. & Burris, R. H., 1970. Acetylene reduction (nitrogen fixation) in Wisconsin lakes. Limnol. Oceanogr. 15: 808–813.

Skulberg, O. M., 1968. Studies on eutrophication of some Norwegian inland waters. Mitt. int. Ver. Limnol. 14: 187–200.

Tarkiainen, E., Rinne, I. & Niemistö, L., 1974. On the chemical factors regulating the primary production of phytoplankton in the Baltic Proper. Merentutkimuslait. Julk./Havforskningsinst. Skr. 238: 39–52.

Vollenweider, R. A., 1974 (ed.). A manual on methods for measuring primary production in aquatic environments. IBP Handbook No. 12, 2nd edn. Blackwell Scientific, Oxford.

Vuorio, H., 1977. Fixation of molecular nitrogen by planktonic cyanobacteria in the Helsinki sea area in 1974 (English summary). Rep. Water Conversation Lab., City of Helsinki 9(1): 1–58.

Vuorio, H., Rinne, I. & Sundman, V., 1978. Nitrogen fixation of planktonic blue-green algae in the Helsinki sea area determined as acetylene reduction. Aqua fenn. 8:47–57.

Seasonal succession of phytoplankton in an ice-free pond warmed by a thermal power plant

Pertti Eloranta

Department of Biology, University of Jyväskylä, SF-40100 Jyväskylä 10, Finland

Keywords: phytoplankton, thermal pollution, eutrophy

Abstract

In a pond receiving warmed cooling waters from a thermal power plant, the physical and chemical properties of the water, phytoplankton, periphyton and zooplankton were monitored on a weekly sampling schedule. In winter the phytoplankton growth was limited by poor light conditions. In mid-February a rapid phytoplankton growth started, simultaneously with increasing light energy, high nutrient concentrations and small herbivorous zooplankton populations. The increase of phytoplankton biomass was stopped by lack of free nutrients and silica at the end of March. From May until August the phytoplankton standing crop was mainly regulated by herbivorous zooplankton. The autumnal maximum of phytoplankton occurred with decreasing zooplankton populations, increasing nutrient concentrations, a turbulence favourable for diatoms and high water temperature.

Introduction

The monitoring of annual fluctuations of plankton may yield results of low utility because of infrequent sampling, poor counting methods or failure to monitor important environmental factors, such as light conditions, water temperature, nutrient concentrations, water currents and zooplankton grazing. The effects of these factors on the phytoplankton succession varies widely between seasons and therefore the correlations between phytoplankton standing crop and environmental factors for longer periods are usually not significant. Short interval monitoring is necessary to understand and explain the dynamics of phytoplankton.

Changes in phytoplankton standing crop depend on the ratio between production and loss of cells. The primary factors for production are light energy, carbon, other nutrients and water temperature, while sedimentation, grazing, wash-out and parasites are the main factors causing loss of cells. The result of this 'competition' is seen as fluctuating standing crop.

The studied pond Vasikkalampi is, due to cooling waters of a thermal power plant, almost ice-free throughout the year (Eloranta 1980a, b, 1981). The spring overturn occurs in March and fall overturn in September. In summer the pond is well stratified. In winter the water temperature is rather even from the surface to bottom, but because the warmed waters are discharging to the water surface and taking from the depth of 3–4 m, the circulation occurs only in the surface layers (0–4 m). The pond Vasikkalampi is a good natural laboratory for studies concerning plankton dynamics and the relations between plankton and environmental factors because it is a natural water body, moderately eutrophic without ice during the critical spring months. Further, it has no larger discharges and only a small outlet and it does not receive any effluents or sewage.

Hydrobiologia 86, 87–91 (1982). 0018-8158/82/0861–0087/$01.00.

Methods

The samples for chemical and plankton analyses were taken bimonthly in winter 1978–1979 but since spring 1979 almost weekly. The weekly radiation energy was measured with a LI-COR pyranometer at the pond. The water samples were transported immediately after sampling to a nearby laboratory for chemical analyses. Colour, turbidity, pH, alkality, conductivity, dissolved oxygen, reactive silica, total phosphorus, phosphate phosphorus and C.O.D. were measured from the samples. Colour and turbidity were measured spectrophotometrically at 420 and 520 nm respectively. Total phosphorus was determined after digestion in an autoclave with $K_2S_2O_8$. The liberated orthophosphate was transformed to the molybdenum blue complex with ascorbic acid. The blue colour was measured at 880 nm. Reactive silica (SiO_2) was determined by reducing a silico-polymolybdate complex to the molybdenum blue complex by reducing the complex with ascorbic acid. The blue colour was measured at 820 nm.

The samples for quantitative phytoplankton analyses (microscopical counting and chlorophyll analyses) were taken with a 2 m tube-sampler from the water layers of 0–2, 2–4 and 4–6 m. The samples for microscopical analyses were preserved in field with IKI solution with sodium acetate; later 1 ml of conc. formalin was added.

The phytoplankton counting was done with an inverted microscope and phase contrast optics. The small species were counted on one transverse strip at a magnification of 600× and the large and scarce species were counted on half the chamber floor at a magnification of 150×. For the biomass calculations 1 500–10 000 algal units were counted per sample. The size of settled samples was 50 ml in midwinter and 10 ml during other seasons. The average cell volumes of dominant species were calculated after cell measurements and the volumes were checked in different seasons.

The primary production *in situ* was measured with the ^{14}C-method 25 times in 1979. The borosilicate glass bottles were exposed from morning to morning horizontally at depths of 0.2, 0.5, 1.0, 2.0, 3.0 and 5.0 m; 1 ml of $NaH^{14}CO_3$ solution with an activity of ca. 1 μCi was added to each sample. The samples were transported after incubation to the laboratory in a dark box without the previous addition of formaldehyde and were filtrated immediately with 0.45 μm membrane filters. The activity of filters was measured with a Frieseke Höpfner thin-window beta counter.

The zooplankton samples were taken concurrent with the phytoplankton and water samples with a 1 m tube-sampler with a volume of 6.36 l. The sample water was filtrated with a 50 μm mesh nylon net and the samples were fixed with conc. formalin (5 ml/100 ml). The zooplankton counting was made with an inverted microscope. The volumes of different rotifer species were calculated after size measurements according to formulae of Ruttner-Kolisko (Bottrell *et al.* 1976) and the volumes of crustacean species were calculated according to regressions given in Bottrell *et al.* (1976). The dry weight/wet weight ratios used were 10% for copepods and rotifers (except *Asplanchna,* 3.9%) and 15% for *Cladocera* (Bottrell *et al.* 1976).

The periphyton growth was measured from artificial substrata (150 × 100 × 1 mm celluloid plates, exposed vertically with upper limit 20 cm below the water surface). The chlorophyll content of the periphyton material after 2 weeks was determined with hot 90% methanol (Iwamura *et al.* 1970).

Results and discussion

The low incident radiation energy was the primary limiting factor for primary production in January and February when the pond was covered by thin ice and snow. At this time the primary production was less than 2 mg $C_{ass}m^{-3}day^{-1}$. The optimum radiation for primary production in cold water (+4–5 °C) (see Talling 1957; Hitchcock & Smayda 1977) was reached by the beginning or mid-February depending on the weather and ice conditions. After this critical point the phytoplankton biomass started to increase, slowly at first but later in March very fast to the first spring maximum (Figs. 1 and 2; see also Eloranta 1980a, b, 1981). This maximum arose at the time of low pressure by herbivorous zooplankton and when the concentrations of free nutrients were high (Fig. 2). The ratio between the production and loss of algal cells was high even though the production was still limited by low water temperature (Fig. 1).

The increase in phytoplankton biomass was stopped by the lack of free reactive silica and

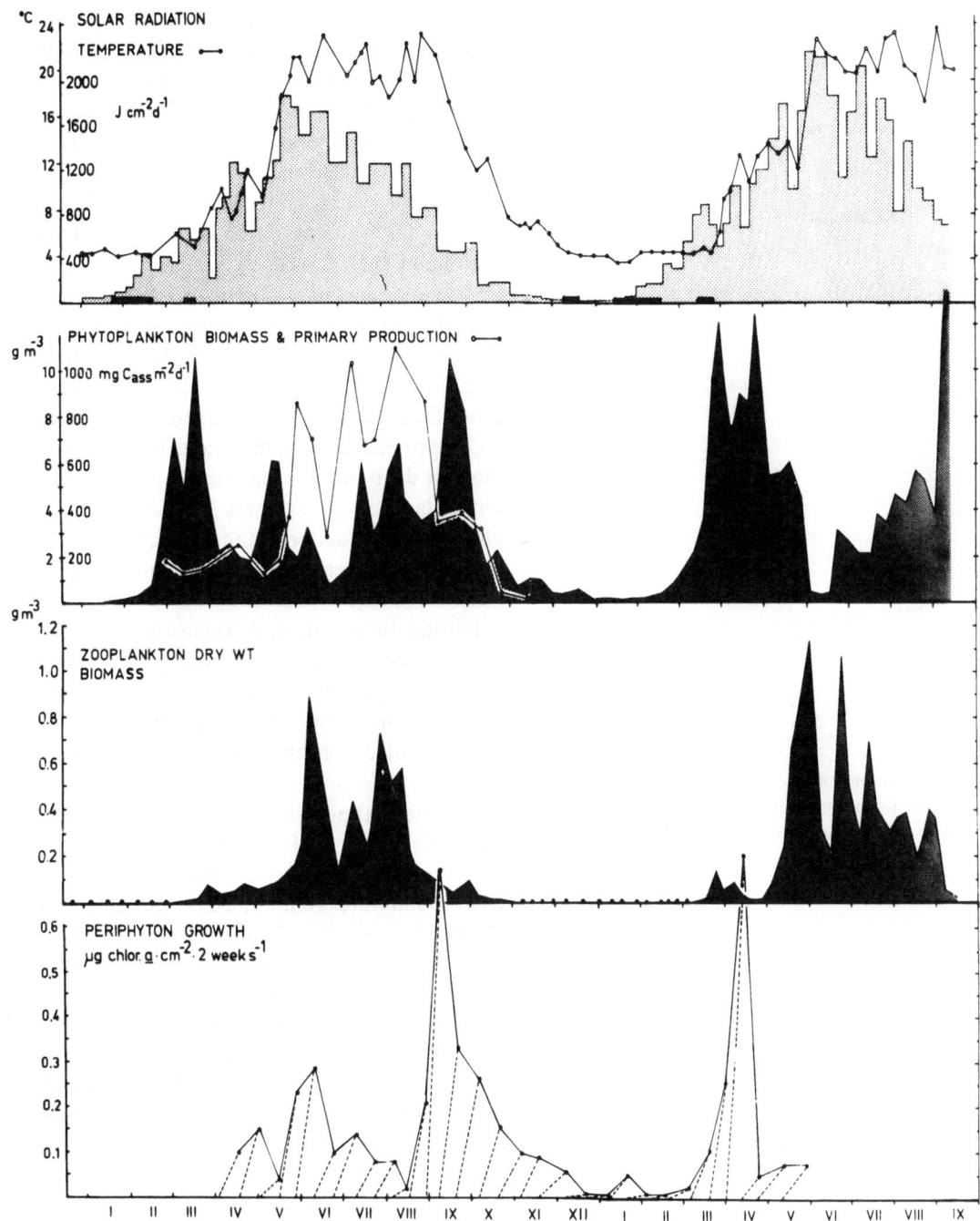

Fig. 1. The yearly successions of water temperature (1 m), total incident radiation, phytoplankton fresh weight biomass (0–2 m, shading), primary production *in situ*, zooplankton biomass (0–2 m) and periphyton growth on artificial substrata in 1979 and 1980 (the black vertical lines indicate the period of ice cover).

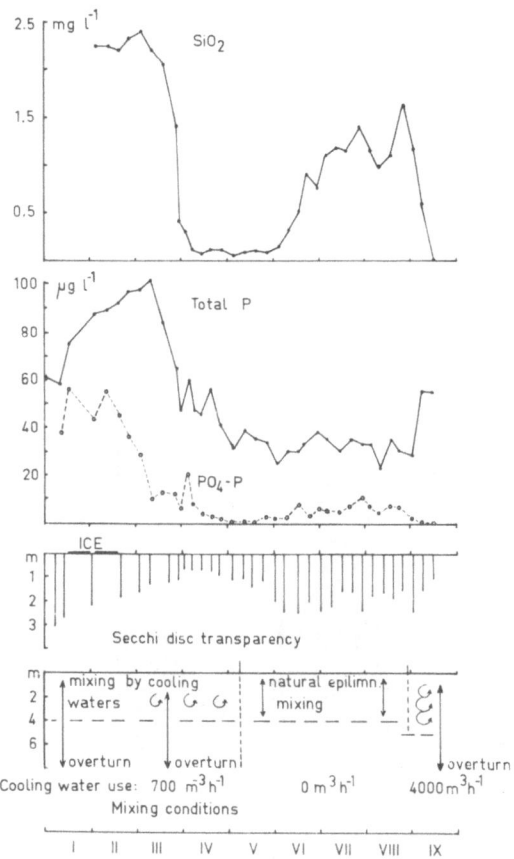

Fig. 2. The concentrations of free reactive silica, total and phosphate phosphorus in the epilimnetic water, Secchi disc transparency, and mixing conditions in the studied pond in 1980.

In spring 1980 the radiation conditions were more uniform than in 1979 and the high diatom maximum lasted until May. The maximum diatom growth was found when total incident radiation was between 630 and 840 J cm^{-2}day^{-1} (photosynthetically available radiation below the water surface ca. 290–335 J cm^{-2}day^{-1}) (Fig. 2).

The water temperature increased to +20 °C by the end of May in both 1979 and 1980. This rapid increase of water temperature induced a rapid increase in zooplankton production and biomass (dominant species were in June *Asplanchna priodonta* and *Bosmina longirostris*; Fig. 1). The phytoplankton biomass decreased rapidly with increasing grazing pressure from herbivorous zooplankton. Thus the loss of cells exceeded the production of new cells which resulted in a decreasing phytoplankton biomass and production until the first drop in the zooplankton population. The phytoplankton growth was not limited by phosphate phosphorus during the summer months which was seen as increased phosphate concentrations at this time (Fig. 2).

During June, July and August the phytoplankton biomass fluctuated according to the fluctuations of zooplankton biomass and the phytoplankton standing crop increased stepwise simultaneously with decreasing zooplankton biomass (Fig. 1).

The thermal power plant was started in the last week of August and used at this time ca. 4 000 m^{3}h^{-1} of cooling water. Thus theoretically the whole pond water volume passes through the power plant once per day. This circulation moved the metalimnion from 4 m to 5 m, increasing the nutrient concentrations in the epilimnion (Fig. 2). The water temperature rose to 20–23 °C in September, when the solar radiation decreased to 840 J cm^{-2}day^{-1} (Fig. 2). At this time the phytoplankton autumn maximum was started by rapid growth of diatoms, especially of *Cyclotella meneghiniana* (vol. 5 300–8 500 μm^3; see also Eloranta 1980b). The diatom growth was fast in spite of the low primary production (Fig. 1) because the loss of cells was low. The favourable turbulence of the water decreased the loss of cells by sedimentation and the grazing pressure by zooplankton was weak because of the large size of the diatom cells and the reduced zooplankton population. At the beginning of the diatom maximum the radiation had decreased to

decreased phosphorus concentration (Fig. 2) simultaneously adverse light conditions caused by the selfshading of dense phytoplankton and zooplankton populations (Fig. 2). Subsequently the phytoplankton biomass fluctuated according to the weather conditions. Periods of high radiation increased biomass but during periods with low radiation energy, the biomass decreased due to the lower ratio between production and loss of cells.

The concentration of total phosphorus in the water decreased during the spring months due to the sedimentation of planktonic diatoms but also due to nutrient fixation by periphytic algae and littoral macrophytes. The maximum of the periphyton growth was in March, when macrophyte growth also started (Eloranta 1980a).

the optimum level and the silica and phosphorus concentrations were increased. The silica concentration decreased during the diatom growth from 1.62 to 0.01 mg SiO_2 l^{-1} in 3 weeks, while the phosphate phosphorus concentration dropped to zero (Fig. 2).

The autumn phytoplankton maximum decreased with decreasing water turbulence, decreasing temperature, radiation energy and lack of free nutrients. After October the phytoplankton biomass was mainly limited by low radiation energy (total incident radiation less than 210 J $cm^{-2}day^{-1}$) together with low water temperature.

Summary

The phytoplankton growth is primarily limited by light in midwinter, also in open waterbodies in Finnish latitudes (60–70 °N). The phytoplankton growth started in the studied pond at incident subsurface radiation of 200–300 J $cm^{-2}day^{-1}$ by cryptomonads and diatoms but later the primary production was restricted, in spite of increasing radiation energy, by low water temperature (4–6 °C) during the first spring phytoplankton maximum. The lack of ice cover extends 3 months the period of phytoplankton spring succession, when normally the length of the period after ice melting to summer conditions is only one months. The summer minimum of phytoplankton is a result of decreased supply of free nutrients, the heavy pressure of zooplankton production simultaneously with the period of rapid changes in environmental conditions as in water currents and temperature. This period is also a boundary in seasonal succession of different phytoplankton groups. The phytoplankton biomass was during the summer strictly regulated by herbivorous zooplankton. The strong current of cooling water in August–September made possible the high maximum of big diatom *Cyclotella meneghiniana*. The mixing minimized the loss of cells by sinking and increased the supply of silica and other nutrients from hypolimnetic layers.

References

Bottrell, H. H., Duncan, A., Gliwicz, Z. M., Grygierek, E., Herzig, A., Hillbricht-Ilkowska, A., Kurasawa, H., Larsson, P. & Weglenska, T., 1976. A review of some problems in zooplankton production studied. Norw. J. Zool. 24: 419–456.

Eloranta, P., 1980a. Winter phytoplankton in a pond warmed by a thermal power station. Ann. bot. fenn. 17: 264–275.

Eloranta, P., 1980b. Annual succession of phytoplankton in one heated pond in central Finland. Acta hydrobiol. 22: 421–438.

Eloranta, P., 1981. Yearly succession of the phytoplankton in an ice-free pond in central Finland. Schweiz. Z. Hydrol. 42 (in press).

Hitchcock, G. L. & Smayda, T. J., 1977. The importance of light in the initiation of the 1972–1973 winter-spring diatom bloom in Narragansett Bay. Limnol. Oceanogr. 22: 126–131.

Iwamura, T., Nagai, H., Ichimura, S., 1970. Improved methods for determining contents of chlorophyll, protein, ribonucleic acid, and deoxyribonucleic acid in planktonic populations. Int. Rev. ges. Hydrobiol. 55: 131–147.

Talling, J. F., 1957. Photosynthetic characteristics of some freshwater diatoms in relation to underwater radiation. New Phytol. 56: 29–50.

The annual flood regime as a regulatory mechanism for phytoplankton production in Kainji lake, Nigeria

Stig-Göran Karlman
Outokumpu Oy, Kokkola Works, P.O. Box 26, 67101 Kokkola 10, Finland

Keywords: phytoplankton, primary production, Africa, man-made lakes, hydrology, transparency, subsurface light, solar radiation

Abstract

Phytoplankton production was measured *'in situ'* in Kainji lake from December 1970 to September 1972 using the oxygen light and dark bottle technique. Seasonal variations in solar radiation, transparency, temperature, and composition of subsurface light were also measured. Oxygen production per unit area varied from 220 to 4500 mg O_2 m^{-2} day^{-1}, the maximum production rate from 95 to 400 mg O_2 m^{-3} h^{-1}. Seasonal mixing of lake water and river water of varying turbidity changed the optical properties of the lake water and consequently affected phytoplankton production. The annual flood pattern was found to be an important factor regulating phytoplankton production in the lake.

Introduction

Estimates of phytoplankton production in African waters (e.g. Talling 1957, 1965a, b; Prowse & Talling 1958; Lemoalle 1969, 1973; Viner 1970; Aleem & Samaan 1969, 1971; Ganf 1972, 1974; Karlman 1973) cover a wide range of both tropical and subtropical lakes. The earlier works were geographically concentrated on East and North-East Africa while the West African lakes have not drawn attention in this respect until the last decade. The present interest in the African lakes and reservoirs can probably be attributed to the development of the large African man-made lakes (Kariba in 1958, Volta in 1964, Nasser in 1964, and Kainji in 1968). The multipurpose development of these new resources put an emphasis on research on a wide spectrum encompassing studies on sociology, public health and fisheries, aimed at socio-economic development of the regions concerned.

For the study of fisheries, the lack of data concerning production at all trophic levels led to a demand for specific studies of the production problems in tropical waters. This paper is based on a two year study of phytoplankton primary production in Kainji lake and discusses the mechanisms regulating primary production in Kainji lake.

Study area

The geographical position of the Niger river gives it a rather peculiar flood pattern. The intertropical front with the rain-bearing monsoon covers most of the drainage area at the same time of the year. The run-off at the sources in the Fouta Djallon Highlands travels 2700 km before reaching Nigeria six months later. The water from the upper drainage area is comparatively clear when it reaches Nigeria, having deposited its silt in the swampy areas around Timbuctu. Because of the clear water this flood is locally called the *black flood*. The second drainage area starts downstream of Niamey. The local flood carries more silt than the black flood and the water has a milky appearance, and for this reason it is called the *white flood*. The two different floods are hereafter referred to by these names. The annual inflow of the Niger river with corresponding lake

Hydrobiologia 86, 93–97 (1982).   0018–8158/82/0861–0093/$01.00.
© Dr W. Junk Publishers, The Hague.

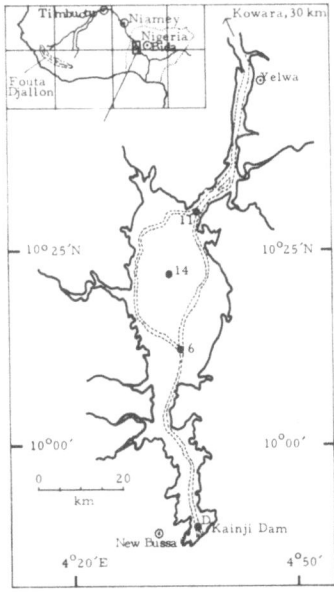

Fig. 1. Map of West-Africa and Kainji lake with the sampling stations (●). The former river channel is indicated with dotted lines.

level is shown in Fig. 2.

Kainji lake (Fig. 1) was created in 1968 when the dam at Kainji Island was completed across the Niger river. At the highest level, the lake has a maximum length of 136 km, and width of 24 km. The maximum depth, close to the dam site, is about 60 m, the mean depths are 11.9 and 7.5 m at the highest and lowest levels, respectively. The total volume is 15 km³, area 1260 km², and the length of the shoreline is 716 km. The annual drawdown is about 10 m. Kainji lake has an exceptionally high through-flow ratio of 4:1. This gives the lake the character of a wide, slow-flowing river rather than a lake.

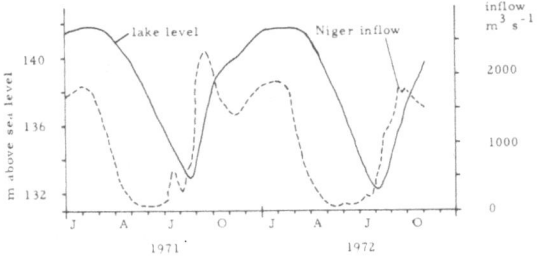

Fig. 2. Niger inflow measured at Kowara, 9 km north of Kainji lake. The lake level has been observed every third day at the dam site.

The lake can easily be divided into three parts: the northern part, representing riverine conditions; the middle part consisting of the two river paths and a shallow part in between, which partially emerges during low water; and the southern part which is narrow and deep. The four sampling stations indicated on the lake map were selected to represent the different conditions in the various parts of the lake.

The rainy season lasts from May to August in the Kainji area, the heaviest rainfall occuring in June – August. The local rains drain through some small tributaries into the lake. The major local flood, though, is caused by the rivers Sokoto and Malendo north of Kainji lake. The clear water black flood rises slowly from August to September, reaching peak inflow in January–February.

Material and methods

Temperature profiles were obtained using a bathythermograph with a depth range down to 60 m. Transparency was measured with a 25 cm diameter white-painted Secchi disc. Solar radiation was recorded weekly with a bimetallic actinograph. The readings were obtained in cal cm^{-2} week^{-1}. The actinograph was available from March 1972. Earlier values for solar radiation were approximated from Gunn-Bellani recordings in the nearby towns of Yelwa and Bida. The underwater illumination and vertical light transmission were measured with a submarine photometer. The unit had one submersible and one deck photocell. The filters used had peak transmission of 440 nm (blue), 530 nm (green), and 700 nm (red).

Data on the regulation of the lake were kindly supplied by the National Electricity and Power Authority, the agent responsible for operating the power station and lake regulation.

The phytoplankton primary production was measured 'in situ', using the oxygen light and dark bottle technique. Water samples were obtained with a non-metallic van Dorn water sampler equipped with a mercury thermometer. The whole procedure was always carried out in the shade to avoid light shocks. The incubation bottles, 250 ml clear Kimax bottles, were suspended horizontally with one light and one dark bottle for each depth. Duplicate bottles were thought unnecessary because of the close vertical spacing. Five to seven pairs of bottles were normally used. The exposure time was 1000–1400 h. The photosynthetic activity was stopped immediately after exposure by adding Winkler's reagents to the bottles, and the subsequent titrations were carried out in the field with a precision of ± 0.04 mg of O_2 l^{-1}. The oxygen production was measured as the difference between the light and the dark bottles (gross production). The measured production was converted to daily production estimates using a factor of 1.7. This factor was obtained empirically by determing the daily production with 1 h intervals.

Results

The surface water temperatures varied from 22 to 32 °C. The lake is subject to total water mixing in December–January due to nocturnal cooling and is homothermic at about 27 °C. Thermal stratification develops in February and lasts until December in the deepest part. During stratification, oxygen is depleted in the hypolimnion, and hydrogen sulphide is developed.

The annual variation of transparency for all stations is shown in Fig. 3. The lowest annual transparency at station 11 in August–September marks the peak of the white flood. The movement of the turbid water mass can be traced through the lake with simple measurements of transparency. The peak of the white flood gradually reaches the dam site about six weeks later.

The annual solar radiation in Kainji lake was 678 kJ cm^{-2}. This value is in good agreement with Talling's (1965b) value of 645 kJ cm^{-2} for lake Victoria. The mean weekly value for 1972 was 13 kJ cm^{-2}, the annual variation ± 25%. In contrast to Lake George (Ganf 1974), Kainji shows a distinct seasonal variation in insolation with peak radiation during the dry months (February, March, April) and a marked decrease during the rainy season (July–August). The available radiant energy during the rainy season (mean value 11 kJ cm^{-2} week^{-1}) ought to be high enough for maintaining a high photosynthetic activity.

The vertical light transmission was found to be affected in two ways by different flood conditions. The light transmission was directly correlated to transparency as measured with a Secchi disc. The

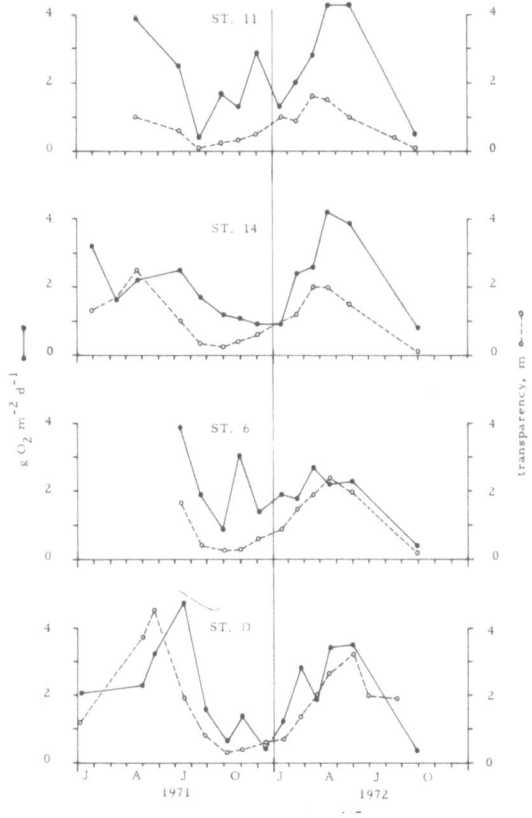

Fig. 3. Seasonal variation of daily phytoplanktonic production, ΣP, and transparency for all stations.

spectral composition also changed. During the white flood, the red part of the spectrum (see Fig. 4) penetrated furthest, while the green part somewhat dominated when transparency was over 2 m. The situation with red light predominant is, according to Ganf (1974), normal in lakes with either a dense crop of phytoplankton or, as in Kainji lake, turbid water.

The seasonal variations of both phytoplankton production and transparency are shown in Fig. 3. For tropical conditions the range of ΣP is wide, with a tenfold variation for stations 11, 6 and D, and a twentyfold difference between the highest and lowest recorded values. The rather constant temperature and illumination suggest that the subsurface light conditions strongly affect production. The phytoplankton abundance has been shown by Adeniji (1975) to vary with flood conditions; hence both production and abundance are influenced in

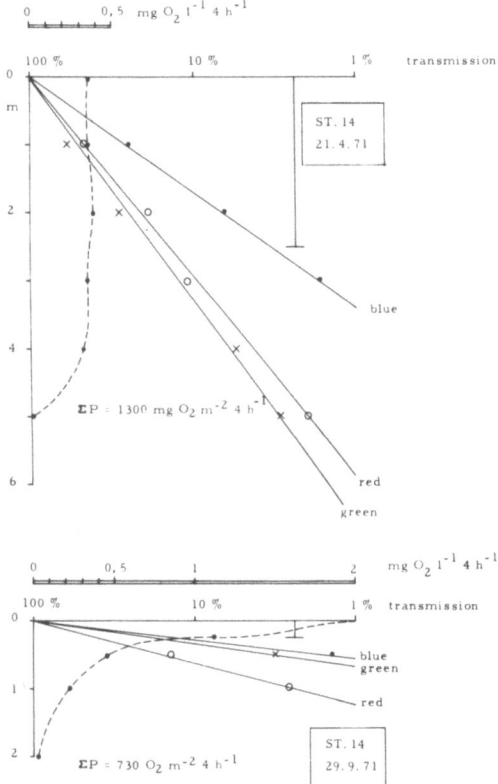

Fig. 4. Typical responses of the production integral (broken line), vertical light transmission, and transparency to different flood conditions. Black flood conditions with high transparency are pictured above, and white flood with turbid water below.

the same way. In Fig. 4 the effect of the two flood conditions is demonstrated. The transparent water of the black flood has the 1% transmission level at about 7 m. The resulting production integral is deep and narrow. During the white flood, with highly turbid water, the light attenuates to 1% within 1 m. The resulting integral is shallow and wide at the surface. The loss of production at depth is thus partially compensated by an increased production rate at the surface.

Discussion

The pre-impoundment studies of the Kainji lake area (White 1965; Imevbore & Visser 1969; Imevbore 1970) contain valuable background data for a better understanding of present lake conditions. Imevbore (1970) noted that the Niger river had a

very low mineral content compared with other major African rivers. The river water was especially poor in nutrients, such as phosphate and nitrate. Imevbore also found that the phytoplankton biomass was considerably lower during the white flood than during the black flood.

After the creation of Kainji lake, this observation was confirmed by the early studies of phytoplankton abundance by Adeniji (1975). The transition from riverine to lacustrine conditions hence does not seem to have changed the phytoplankton abundance pattern. This is probably due to the high throughflow ratio. From the preimpoundment studies it was possible to predict a rather low production level for the lake because of the low mineral content of the river water. On the ot'er hand, it could be expected that a potentially hi￼h turnover rate and an uninterrupted production period could offset the impact of limited nutrient supply. This proved not to be the case. Karlman (1973) compared the phytoplankton production in Kainji lake with 24 other tropical and subtropical lakes both in Africa and India. He concluded that both the range of daily production and the annual mean value for Kainji were among the lowest recorded. The surprisingly low production for a lowland tropical lake was attributed to a low nutrient content in the water and low transparency limited by fine inorganic silt.

Melack (1976) discussed primary production comparing data on mostly the same lakes as Karlman, quoting a value of 6.5 mg of O_2 m^{-2} day^{-1} for Kainji as obtained from Henderson (1973). This early estimate, based on a single measurement, was found to be too high because of a calculation error and when corrected fell within the present range of production.

The effect of turbidity has also been observed in Lake Chad by Lemoalle (1973). As in Kainji lake the turbidity was mainly due to inorganic silt. The transparency was even more restricted than in Kainji lake with a range of 0.1–0.8 m compared to 0.1–4.5 m for Kainji. Although Lake Chad had a transparency comparable to the northern part of Kainji lake, the daily production could be twice that in Kainji. The comparison with lake Chad indicates that although transparency regulates primary production in accordance with flood conditions, it is the nutrient supply that sets the limits for production in Kainji lake.

References

Adeniji, H. A., 1975. Some aspects of the limnology and the fishery development of Kainji lake, Nigeria. Arch. Hydrobiol. 75: 253–262.

Aleem, A. A. & Samaan, A. A., 1969a. Productivity of Lake Mariut, Egypt. part 1. Physical and chemical aspects. Int. Rev. ges. Hydrobiol. 54: 313–355.

Aleem, A. A. & Samaan, A. A., 1969b. Productivity of Lake Mariut, Egypt. Part II. Primary production. Int. Rev. ges. Hydrobiol. 54: 491–527.

Ganf, G. G., 1972. The regulation of net primary production in Lake George, Uganda, East-Africa. In: Kajak, Z. & Hillbricht-Ilkowska, A. (eds.) Productivity Problems of Fresh Waters, pp. 693–708. PWN Warsaw-Krakow.

Ganf, G. G., 1974. Incident solar irradiance and underwater light penetration as factors controlling the chlorophyll *a* content of a shallow equatorial lake (Lake George, Uganda). J. Ecol. 62: 593–609.

Henderson, F., 1973. A limnological description of Kainji lake, 1969–1971. FAO Technical Report Fi: DP/NR/66/524/10, 47 pp.

Imevbore, A. M. A., 1970. The chemistry of the River Niger in the Kainji reservoir area. Arch. Hydrobiol. 67: 412–431.

Imevbore, A. M. A. & Visser, S. A., 1969. A study of microbiological and chemical stratification of the Niger river within the future Kainji lake area. In: Obeng L. E. (ed.) Man-made Lakes. The Accra Symposium, Accra, Ghana Univ. Press.

Karlman, S-G., 1973. Pelagic primary production in Kainji lake, Nigeria. FAO Technical Report No. 3, Fi: SF/NIR 24, 59 pp.

Lemoalle, J., 1969. Premières données sur la production primaire dans la région de Bol (avril-octobre 1968) (Lac Tchad). Cah. ORSTOM, sér. Hydrobiol. 3: 107–119.

Lemoalle, J., 1973. L'énergie lumineuse et l'activité photosynthétique du phytoplankton dans le Lac Tchad. Cah. ORSTOM, sér. Hydrobiol. 7: 95–116.

Melack, J. M., 1976. Primary productivity and fish yields in tropical lakes. Trans. Am. Fish. Soc. 105: 575–580.

Prowse, G. A. & Talling, J. F., 1958. The seasonal growth and succession of plankton algae in the White Nile. Limnol. Oceanogr. 3: 223–238.

Talling, J. F., 1957. Diurnal changes of stratification and photosynthesis in some tropical African Waters. Proc. Roy. Soc. B 147: 57–83.

Talling, J. F., 1965a. The photosynthetic activity of phytoplankton in East African lakes. Int. Revue ges. Hydrobiol. 50: 1–32.

Talling, J. F., 1965b. Comparative problems of phytoplankton production and photosynthetic productivity in a tropical and a temperate lake. Mem. Ist. Ital. Idrobiol. 18 Suppl.: 399–424.

White, E., 1965 (ed.). The first scientific report of the Kainji Biological Research Team. Liverpool. Mimeographed.

Viner, A. B., 1970. Hydrobiology of Lake Volta, Ghana. II. Some observations on biological features associated with the morphology and water stratification. Hydrobiologia 35: 230–248.

The effect of land use on the diatom communities in lakes

J. Meriläinen[1], P. Huttunen[2] & K. Pirttiala[1]
[1] Karelian Institute, Section of Ecology, University of Joensuu, P.O. Box 111, SF-80101 Joensuu 10, Finland

[2] Department of Biology, University of Joensuu, P.O. Box 111, SF-80101 Joensuu 10, Finland

Keywords: diatoms, lakes, land use, sediment

Abstract

Data on recent diatom community structure and relevant environmental characteristics from the lakes and their catchments have been collected from 151 oligotrophic lakes in eastern Finland. The pattern of frequency distribution of diatoms as a function of environmental variables, including land use in the catchment, differs between diatom taxa and indicates the optimum conditions and amplitude of occurrence for particular species. This kind of study should lead to increased understanding of the environmental requirements of diatom species and will be useful in the interpretation of historical changes in lakes as well as in forecasting possible future changes.

Introduction

Land use affects lakes in different ways. After forest clearance, small lakes in particular become more exposed to the wind so that depth of the epilimnion increases, and conditions for phytoplankton improve during summer stratification. The erosion connected with tilling of soil increases the inflow of nutrients to lakes. The progression from slash-and-burn culture to field cultivation, to more frequent tilling and more intensive fertilization generally led to a more eutrophic lake phase (e.g. Huttunen & Tolonen 1977; Vuorinen 1978).

Lakes themselves could also be used by man. Severe changes were caused by retting of fibre plants in many small lakes in Finland (e.g. Tolonen 1978; Huttunen 1980). Dissolved oxygen was depleted by decomposing organic matter from the bottom-most water layer during the winter and summer stratifications. This was succeeded by leakage of nutrients from the bottom sediments and eutrophication of the lakes. Study of the effects of land use on a number of lakes within one area may be complicated. Essen-

tially similar land use may produce different phenomena in different lakes, depending on the morphometry of the lake, and topography, bedrock, soil, and vegetation of the watershed. If organic soil (peatland) is drained and cultivated, the increased leaching of allochthonous humic matter to the lake may lead to decreased species diversity, similar to when nutrient loading increases (e.g. Patrick 1961; Meriläinen 1967). However, advantageous addition of humic matter in combination with the nutrients available in the lake may cause an increase in species diversity. This group of problems thus makes it difficult to interpret the structure of communities from species diversity indices alone.

This paper presents first results from a study aimed at analysing the dependence of diatom community structure in Finnish lakes on land use in their catchments.

Study area and methods

The present paper is based on data from 151 lakes in eastern Finland located partly on granite–gneiss

Hydrobiologia 86, 99–103 (1982). 0018-8158/82/0861-0099/$01.00.

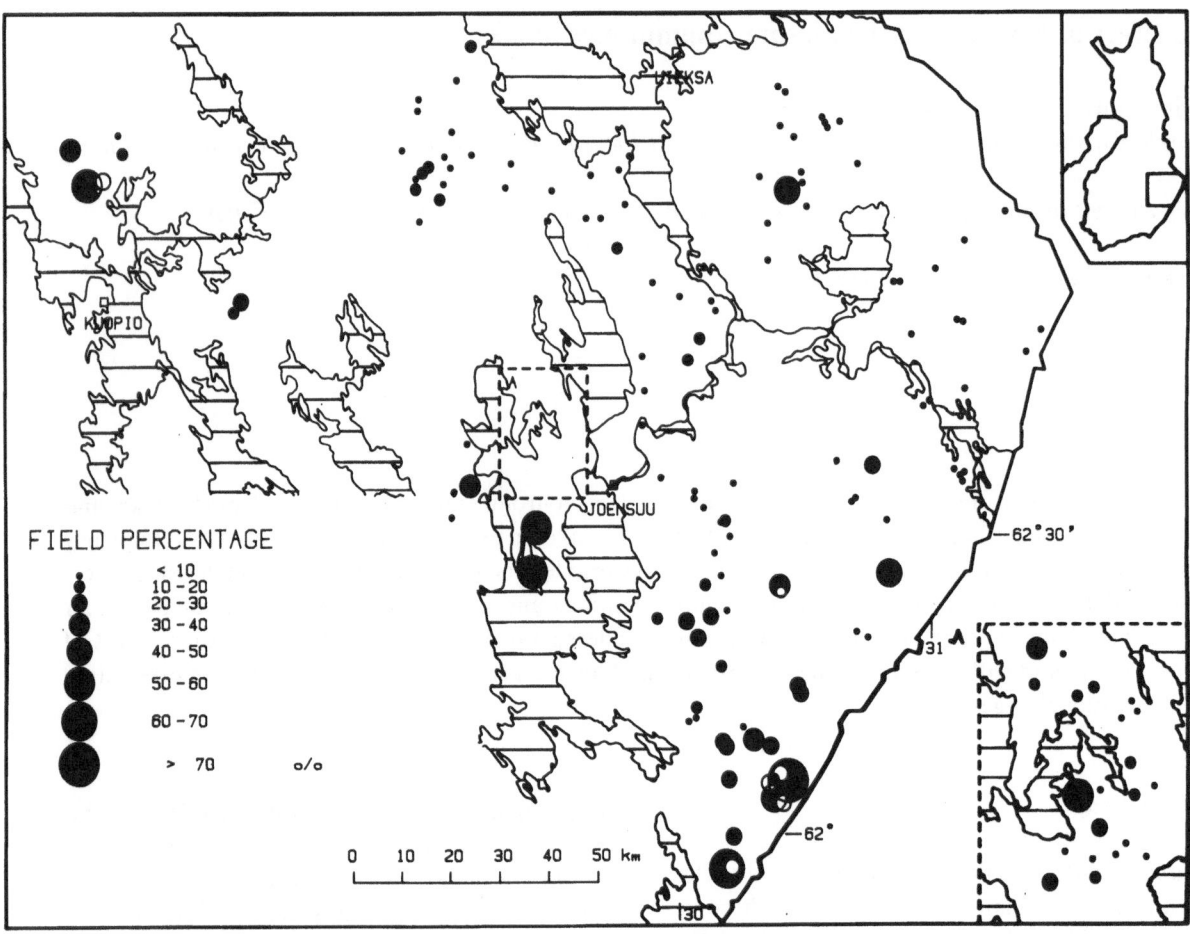

Fig. 1. Location of experimental lakes. The area of circles indicates the proportion of arable land of the drainage area.

complex, partly on Karelidic schists with some on apatite bedrock (Fig. 1). In planning the project (established 1978) attention was paid to other studies with related research strategies (Meriläinen 1967; Brugam 1980; Crisman 1978; Ilmavirta 1980). More than 60 physico-chemical parameters were measured and analysed from the catchments, lake waters and bottom sediments. The size of the lakes ranged from 5 to 1500 ha. In addition to size and productivity of the lakes, such characteristics as land use and percentage of peatland in the catchment were used as criteria to select the study sites. River-like lakes with short residence time as well as shallow unstratified lakes were avoided. The research strategy of the investigation has been presented earlier (Huttunen *et al.* 1980). Most of the sites were oligotrophic lakes, but covering a range of abiotic and biotic properties: transparency 1–11 m, pH 4.5–7.5, conductivity 1.1–17.2 mS m^{-1},

alkalinity 0.01–0.6 mval l^{-1}, total P 2–61 μg l^{-1}, chlorophyll *a* 0.5–24.5 μg l^{-1}. The diatom analyses were based on sediment samples collected in late winter through the ice using a Kajak sediment sampler. Five subsamples were obtained from the deepest part of each lake and the upper 3 cm of their sediments were homogenized. Water samples were taken during the autumnal circulation and during the sharp thermal stratification in summer. The vernal circulation period was avoided because it is short and often incomplete in these lakes.

Results and discussion

For further characterization of the study basins some lake parameters are described as a function of field percentage and peatland percentage in the catchment (Fig. 2). In lakes without or with only

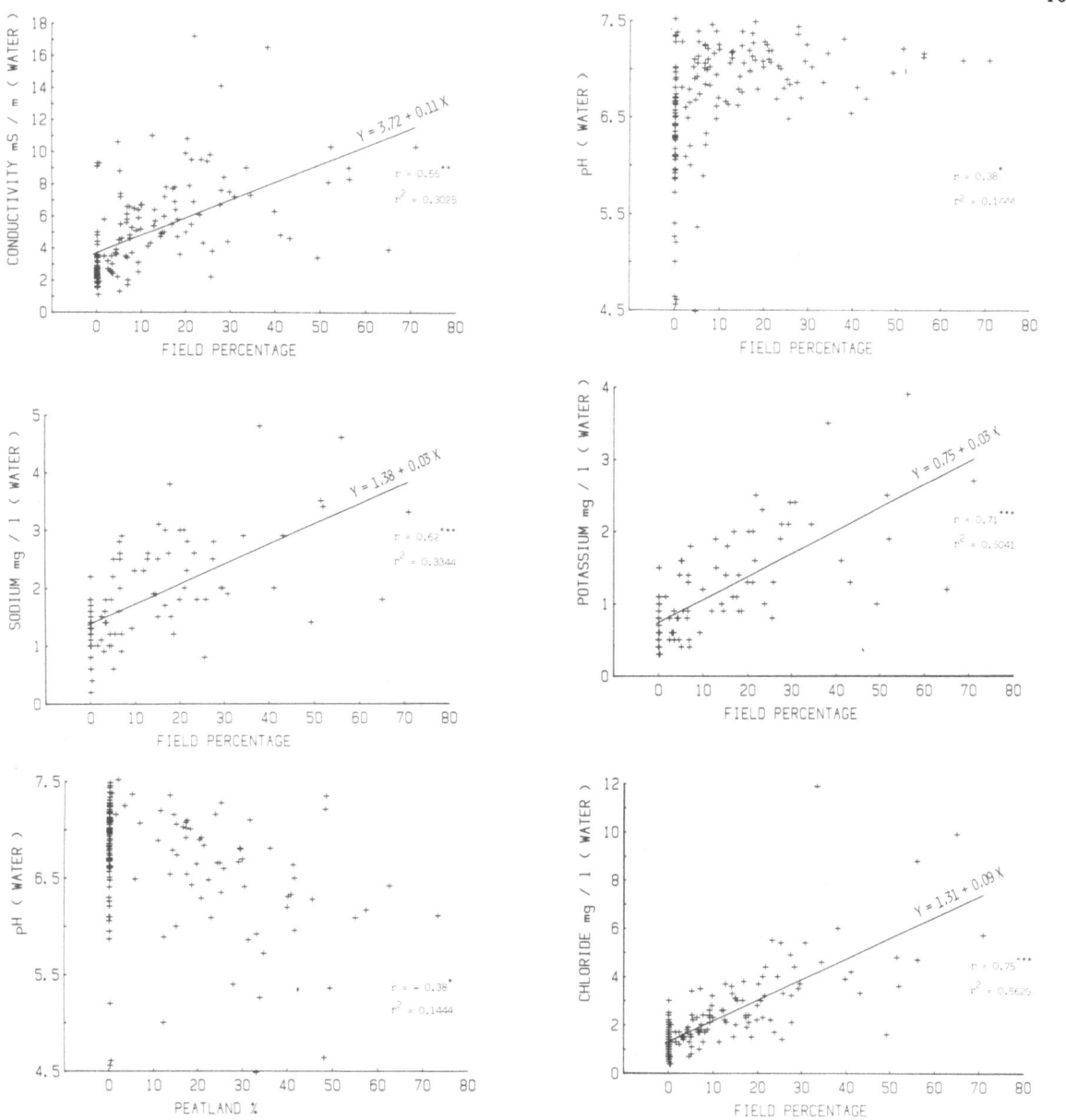

Fig. 2. Water characteristics plotted against the proportion of peatland or arable land on the drainage area.

small fields in the surroundings (< 10% of the catchment) the pH covers the whole pH range (4.5–7.5) during the autumnal full circulation. But in the lakes which are surrounded by larger cultivated areas (> 10% of the catchment) pH is at least 6.5. Positive correlations exist between field percentage and conductivity, sodium, potassium, calcium, and magnesium concentrations of the water. A highly significant positive correlation was found between field percentage and chloride of the water, probably an effect of artificial fertilizers.

Kerekes (1974) reported very significant positive correlation between the chloride content and conductivity of water in oligotrophic lakes. This is confirmed also by our results. As expected, the pH of the water decreases with increase of peatland in lake surroundings. This negative correlation has, however, only a significance of 95%.

At present the diatom communities of only 21 lakes have been investigated, and such preliminary results allow only limited generalization. However, the first step towards community interpretation can

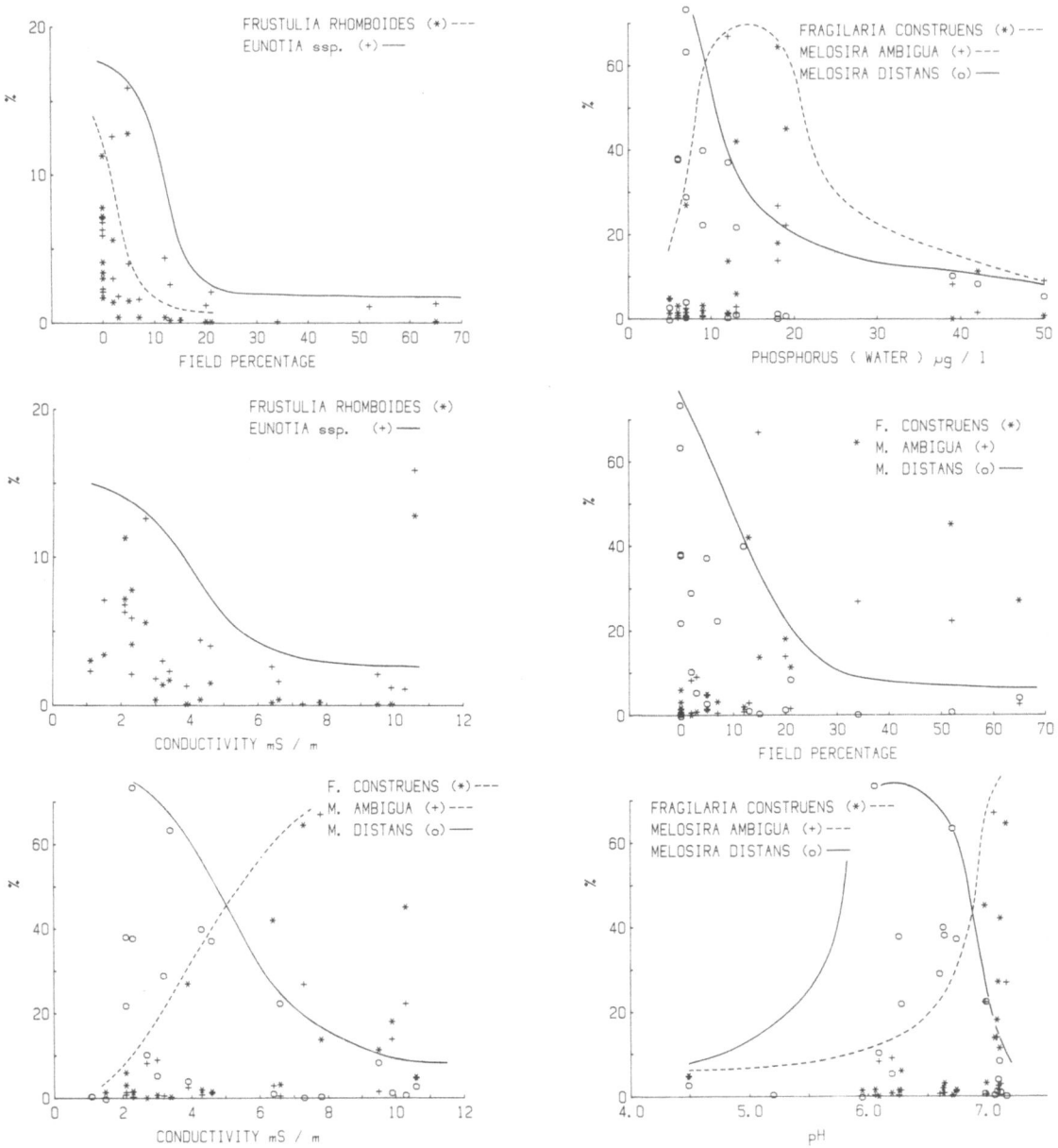

Fig. 3. The relative distribution of some groups or taxa of diatoms plotted against water and drainage area parameters in 21 lakes. The tentative range of the taxa is shown as an area between axes and the curve.

be taken by plotting the distribution patterns against particular environmental parameters. Initially we have calculated results on a percentage basis, but the data would also allow the use of absolute values.

For this paper some common diatom taxa and groups of taxa were chosen to discuss their relative frequencies as a function of pH, conductivity, and total phosphorus of the water (measured during the autumnal full circulation) and the field percentage in the catchment.

Diatom distributions as a function of environment

Proportion of field in drainage area

Melosira distans (s.s.), *Frustulia rhomboides* (incl. *F. r.* var. *saxonica*), and *Eunotia* spp. (excl. *E. arcus),* all common diatoms in oligotrophic lakes, rarely occur in lakes surrounded by fields. The highest representation of *Fragilaria construens* (s.l.) is from lakes with more field in the surroundings. The distribution pattern of *Melosira ambigua* is more 'indifferent' (Fig. 3).

pH

The proportions of *Frustulia rhomboides* and *Eunotia* spp. decrease with increasing pH. *Melosira distans* concentrates in the lakes with the pH between 6.0 and 6.7, whereas *Fragilaria construens* and *Melosira ambigua* are more frequent in the waters with higher pH (Fig. 3).

Conductivity

The proportions of *Frustulia rhomboides* and *Eunotia* spp. as well as *Melosira distans* decrease with increasing conductivity of the water, whereas the distribution patterns of *Fragilaria construens* and *Melosira ambigua* are the reverse (Fig. 3).

Total phosphorus

Frustulia rhomboides, Eunotia spp., and *Melosira distans* show the same trend with the increasing phosphorus as with the conductivity, but the peaks of *Fragilaria construens* and *Melosira ambigua* are situated at higher concentrations (Fig. 3).

Conclusions

The pattern of frequency distribution of diatoms as a function of environmental variables, including the changes of lake surroundings caused by land use, differs from one taxon to another. It also seems likely that the interpretation of the effects of land use on the lakes can be better based on indicator communities than on indicator species. First results suggest that this research strategy can produce ecological information from diatom communities which is useful in the interpretation of both prior lake phases and present conditions, as well as in forecasting changes in the future. Later, when a greater number of distribution patterns of diatoms have been analysed, their differences can be used for more sophisticated community analyses. Clustering of diatom assemblages, calculation of indices, and stepwise multiple regression analysis between background characteristics and the diatom species of communities will also be carried out.

References

Brugam, R. B., 1980. Postglacial diatom stratigraphy of Kirchner Marsh, Minnesota. Quat. Res. 13: 133–146.

Crisman, T. L., 1978. Algal remains in Minnesota lake types: a comparison of modern and late-glacial distributions. Verh. Int. Ver. Limnol. 20: 445–459.

Huttunen, P., 1980. Early land use, especially the slash-and-burn cultivation in the commune of Lammi, southern Finland, interpreted mainly using pollen and charcoal analyses. Acta bot. fenn. 113: 1–45.

Huttunen, P., Meriläinen, J. & Pirttiala, K., 1980. Research strategy for the ecology of diatoms. Environmental parameters, data handling, and preliminary diatom information. Nordic meeting of diatomologists 1980. Lamni Biological Station, Finland, May 6–7, 1980, papers 20–32.

Huttunen, P. & Tolonen, K., 1977. Human influence in the history of Lake Lovojärvi, S. Finland. Finskt Museum 1975: 68–105.

Ilmavirta, V., 1980. Phytoplankton in 35 Finnish brown-water lakes of different trophic status. Dev. Hydrobiol. 3: 121–130.

Kerekes, J., 1974. Limnological conditions in five small oligotrophic lakes in Terra Nova National Park, Newfoundland. J. Fish. Res. Bd Can. 31: 555–583.

Meriläinen, J., 1967. The diatom flora and the hydrogen-ion concentration of the water. Ann. bot. fenn. 4: 51–58.

Patrick, R., 1961. A study of the numbers and kinds of species found in rivers in eastern United States. Proc. Acad. nat. Sci. Philadelphia 113: 215–258.

Tolonen, K., 1978. Effects of prehistoric man on Finnish lakes. Pol. Arch. Hydrobiol. 25: 419–421.

Vuorinen, J., 1978. The influence of prior land use on the sediment of a small lake. Pol. Arch. Hydrobiol. 25: 443–451.

Convection in bottom sediments and its role in material exchange between water and sediments

K. M. Lappalainen
Water Protection Association of Savo, Neulamäen teollisuuskylä, SF-70150 Kuopio 15, Finland

Keywords: convection, sediment, temperature, convectional diffusion, material exchange

Abstract

In two Finnish lakes, in winter and summer, a constant temperature in the sediment was not reached until a depth of 1.5–2 m. The thermal stratification pattern in the sediment was similar to that in water. However, the lack of turbulent mixing in the sediment resulted in a thin 'episediment'. This stratification was caused by convection. Convection currents in the sediment were most prevalent during spring and autumn overturn when the density of overlying water was greater than that of the interstitial water. Convection was also possible in winter. The duration and magnitude of convection was dependent on warming and cooling rates. Such convection currents may be important in material exchange between sediment and water since material concentrations in interstitial water are 5–100 times greater than in overlying water.

Introduction

The literature contains many data on chemical, biological and hydrodynamic factors affecting transfer of materials between water and sediments (e.g. Lee 1970). However, relatively little is known about the depth to which mixing extends in sediments. It has been suggested that this depth is only in the order of millimeters and that the mechanism of material exchange is primarily diffusive (Hayes & Phillips 1958; Mortimer 1971). Other evidence indicates that mixing extends to depths of 5–15 cm (Lee 1970).

Under certain temperature conditions, convective diffusion is effective in mixing water and could also effect mixing in sediments. If so, it may promote material exchange, especially of gases. Such a hypothesis may be investigated by measurement of temperature profiles in the bottom sediments of lakes.

Materials and methods

Vertical temperature profiles in bottom sediments were obtained using an electrical thermistor attached to a rigid rod. Measurements were accurate to ±0.1 °C. Profiles were obtained from a small, eutrophic pond, Sammakkolampi (Kuopio, Central Finland), and also from a large lake, Kallavesi (Central Finland), where two sites were used, in the open water area and in Särkilahti bay.

Results and discussion

Some typical variations of winter and summer temperatures in Lake Kallavesi are presented in Fig. 1. The constant deep sediment temperature +6 °C is reached at the depth of about 2 m. Both the thermal gradient and seasonal variations of temperature are much greater in sediments in shallow littoral areas than in deeper profundal sediments.

Vertical temperature profiles in winter and

106

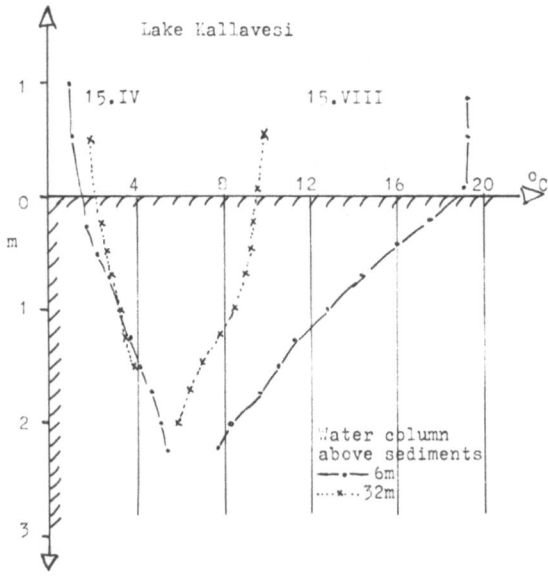

Fig. 1. Temperature profiles in lake Kallavesi in 1980.

summer for Sammakkolampi are shown in Fig. 2. Ice melted on 5 May 1980. If differences between temperatures in profiles are the result of the time interval from 5 May to 9 May 1980, about 40 cal cm^{-2} day^{-1} heat was absorbed from the water to the sediment at the water level of 2.5 m. This heat flux exceeds the normal maximal heat conduction, which is about 1–5 cal cm^{-2} day^{-1}. The solar radiation must also have a minor direct effect, through absorption in brown water. These temperature conditions are favorable for interstitial water convection. Water temperature first approaches the temperature of maximum density. When water temperature is over 4 °C but under the temperature of the deep sediment and its interstitial water, conditions for convection still exist. The small effect of salts on density is ignored. Profiles at a water depth of 4 m (Fig. 2) show less variation than at 2.5 m and conditions for convection are poor, since interstitial water temperatures are near 4 °C.

Results in Fig. 3 represent conditions suitable for autumnal sediment convection. When overlying water becomes cooler, it again becomes denser than interstitial water in the sediment. This situation continues until the water temperature is 4 °C. Estimations of heat flux from sediment to overlying water are 10–40 cal cm^{-2} day^{-1}.

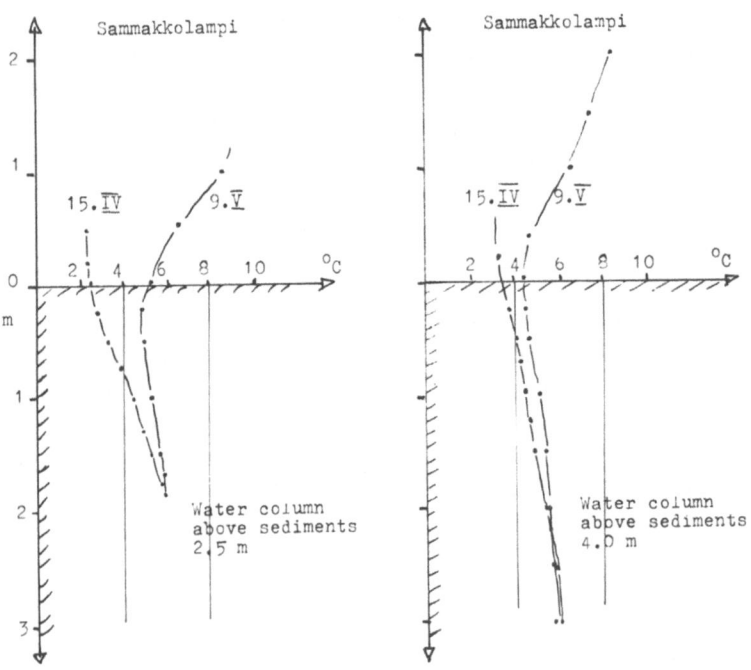

Fig. 2. Temperature profiles in the pond Sammakkolampi in the spring of 1980.

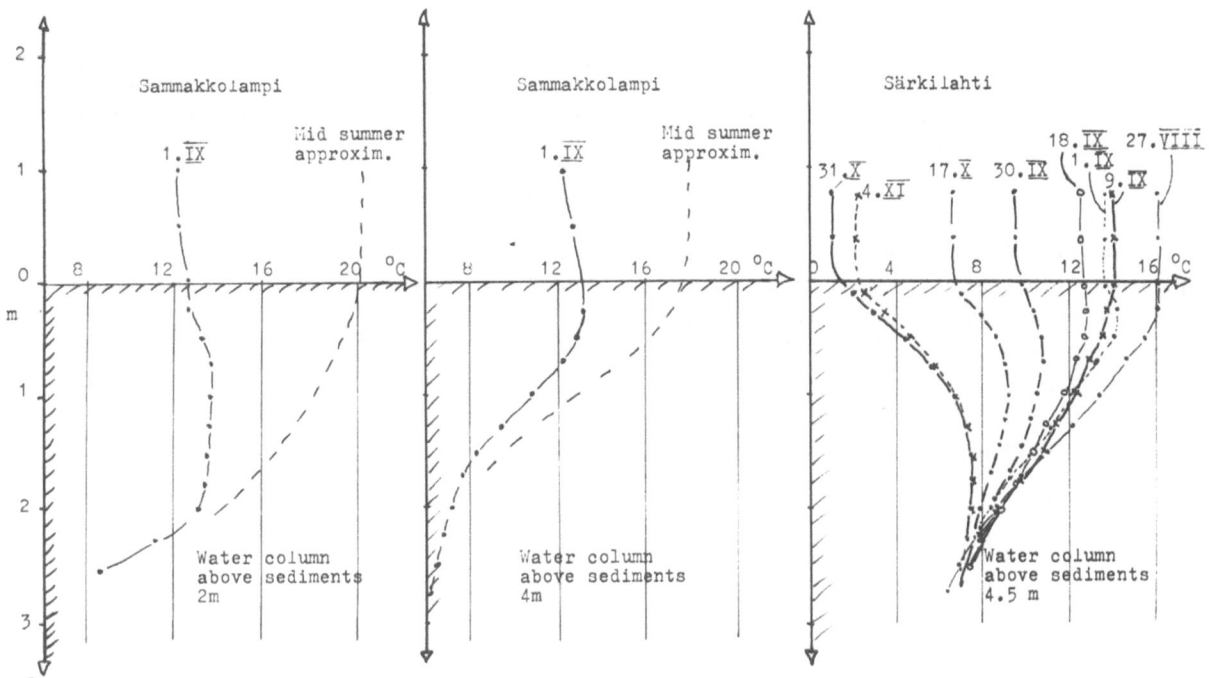

Fig. 3. Temperature profiles in the pond Sammakkolampi and in the bay Särkilahti in the autumn of 1980.

Conclusions

Figure 4 presents the principal temperature limits within which interstitial water convection is possible.

Convectional heat transfer, interstitial water currents and transfer of soluble and gaseous materials can take place in spring and autumn under the following conditions (Tw = temperature of water in winter, Ts = temperature of water in summer and Td = constant temperature of sediment and its interstitial water):

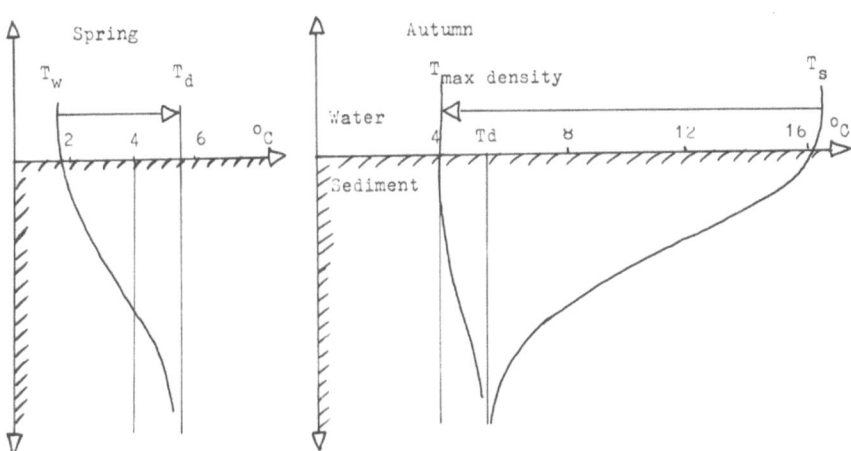

Fig. 4. The temperature variation types exposed to the convection in the interstitial water. Tw = temperature of water in winter, Ts = temperature of water in summer, Td = rather constant temperature of deep sediment.

	Spring	Autumn
$Td > 4\,^{\circ}C$	$Tw \geqslant Td$	$Ts \geqslant 4\,^{\circ}C$
$Td < 4\,^{\circ}C$	$Tw \geqslant 4\,^{\circ}C$	$Ts \geqslant Td$
$Td = 4\,^{\circ}C$	$Tw \geqslant 4\,^{\circ}C$	$Ts \geqslant 4\,^{\circ}C$

Convectional mixing has a longer and broader effect when temperature variation is large. The temperature profiles suggest that the mixing depth may be some tens of centimeters. The variations in sediment temperature are especially great in large lakes which have a thick epilimnion and which are well oxygenated. The variations in temperatures may be an important original cause of trophic state differences.

References

Hayes, F. R. and Phillips, J. E., 1958. Lake water and sediment, IV. Radiophosphorus equilibrium with mud, plants, and bacteria under oxidized and reduced conditions. Limnol. Oceanogr. 3: 459–475.

Lee, G. F., 1970. Factors affecting the transfer of materials between water and sediments. Water Chemistry, University of Wisconsin, July 1970. 50 pp.

Mortimer, C. H., 1971. Chemical exchanges between sediments and water in the Great Lakes – speculations on probable regulatory mechanisms. Limnol. Oceanogr. 16: 387–404.

Winter microbial activity in Lake Tuusulanjärvi

Timo Tamminen

Helsinki Water District, P.O.Box 278, SF-00531 Helsinki 53, Finland

Keywords: heterotrophic activity, 3-H-glucose, dark CO_2 uptake, primary productivity, ATP, winter

Abstract

Microbial heterotrophic activity, dark CO_2 assimilation, primary productivity and microbial ATP were measured monthly in the extremely eutrophic Lake Tuusulanjärvi during the winter of 1979–1980. Because of continuous water circulation caused by low temperature and artificial aeration of the lake, no winter stratification developed. Very low summertime ^{3}H-glucose turnover times of 5 h increased to a level of 10–20 h from August to January. Winter maximum of 110 h was measured in March, and turnover times returned to 10–20 h in April, before the vernal bloom of algae occured. Oxygen saturation remained over 46% during the winter.

High primary productivity was observed in November (400–500 mg C m^{-3} day^{-1}), and measurable productivity was detected under ice in January (80 mg C m^{-3} day^{-1}). Dark CO_2 assimilation increased to 14% of primary productivity in March. No correlation was found between ^{3}H-glucose turnover rate and dark CO_2 assimilation. ATP correlated slightly better with primary productivity than with turnover rate. The single concentration method proved to be sensitive for winter heterotrophic activity measurement.

Introduction

Investigation of microbial heterotrophic activity in lakes has concentrated on ice-free periods. In dimictic lakes, however, oxygen consumption by heterotrophic bacterial activity aggravates the critical effects of winter stagnation, especially in small water bodies where the ice-covered period with isolation from atmospheric oxygen is long.

Knowledge of heterotrophic processes would aid modelling of oxygen conditions under ice, but traditional methods – plate counts, microscopy – are inadequate. Some reports have suggested that the radioglucose assimilation methodology would be suitable for measuring heterotrophic bacterial activity in winter conditions (Munro *et al.* 1973; Morita *et al.* 1977).

In this study a method developed by Williams & Askew (1968) and Azam & Holm-Hansen (1973) was utilized, in which heterotrophic activity measurement is based on the assimilation of ^{3}H-labelled glucose by the natural population of heterotrophic bacteria in water. In this method, only one concentration of radioactive substrate is added into the sample in contrast to the kinetic method (Parsons & Strickland 1962), so that the technical performance is analogous with the standard primary productivity method.

Material and methods

Lake Tuusulanjärvi (surface area: 6 km²; maximum depth: 10 m; mean depth: 3.1 m) has received municipal pollution from town Järvenpää for several decades, and the non-point nutrient load is heavy (>50%). Complete oxygen exhaustion has occurred frequently, both during heavy blue-green

Hydrobiologia 86, 109–113 (1982). 0018-8158/82/0862-0109/$01.00.

algae blooms and under ice. Since 1972 the lake deep has been aerated during critical seasons (Numminen & Lemmelä 1976). In spite of aeration, oxygen concentrations in the lake deep sank below 1 mg l^{-1} during the winters of 1976–77 and 1977–78. Municipal pollution was terminated in March 1979.

Samples were taken on a monthly basis from July 1979 to May 1980. Six depths above the lake deep (0, 1, 3, 5, 7, 9 m) were sampled with a Ruttner sampler. On two dates (16.8.1979, 19.5.1980) only three depths were sampled (1, 5, 9 m). Water was transported to the laboratory in clean 1 l polyethylene bottles, and incubations normally started within 5–7 h of sampling.

Heterotrophic activity was measured with D-(6-^{3}H)glucose (specific activity 22.5 Ci mmol^{-1}, Radiochemical Centre, Amersham, England). This was added to the 0–2 m sample of 5 July 1979 in 15 concentrations (0.003–2.65 μg glucose l^{-1}). The kinetic parameters obtained are presented in Fig. 1. The concentration chosen for routine measurements with the single concentration method was 0.05 μg glucose l^{-1}. Incubations were carried out in the dark, and temperature was within $\pm 3\,°$C of the in situ temperature. Sample volume was 50 ml. Incubation time was 2 h, and incubation was terminated by adding 0.5 ml of formaldehyde (35%) into the sample. All samples were incubated in triplicate. A blank sample for each triplicate was prepared by adding the formaldehyde before the addition of radioactivity.

Samples were filtered on 0.45 μm filters (Millipore), which were oxidized with a Junitek oxidizer. Radioactivities were measured with an LKB-Wal-

lac RackBeta liquid scintillation counter. Scintillation cocktail consisted of 2.6 ml H$_2$O in 7.4 ml PCS (Amersham). Measured radioactivities were calculated into turnover times (T) and turnover rates (1/T).

Primary productivity measurements were made in vitro according to the Finnish standard SFS 3049. Incubation time was 24 h. All incubations in light were carried out in triplicate. In addition to the normal dark sample, a blank was prepared by adding the formaldehyde to the sample before the addition of radioactivity. Dark CO$_2$ assimilation was calculated by diminishing the blank value from the dark sample. Filtration, oxidation and measurement procedures were similar to heterotrophic activity measurements, and the scintillation cocktail consisted of 8 ml Carbo-Sorb and 8 ml Permafluor V (Packard).

ATP was measured with a modification of the methodology presented by Holm-Hansen & Booth (1966). Samples of 10–25 ml were filtered on 0.45 μm filters (Millipore), and filtration stopped before the filters were sucked dry. Filters were immediately transferred to boiling 0.02 M Tris-HCl buffer (0.002 M EDTA, pH 7.75 in 25 °C). Extraction time was 5 min., after which extracts were deep-frozen in glass ampoules. ATP content of extracts was measured with LKB-Wallac ATP monitoring reagents and an LKB-Wallac Luminometer 1250. Each measurement was made in triplicate and with an internal standard of ATP.

Physical and chemical analyses were made in the Helsinki Water District laboratory according to Finnish standard methods (SFS-catalogue 1055, 1981).

Results

The aeration of the lake started in early autumn, but full efficiency was not reached until 12 February 1980. The lake was frozen over at the beginning of December. Aeration broke ice over the lake deep at the end of February, and ice cover of the whole lake broke up on 29 April.

No permanent stratification appeared during the research period, and maximum thermal difference between 1 and 9 m was 3.2 °C (August 1979). From October 1979 to May 1980, the water column was thermally homogeneous. Continuous circulation

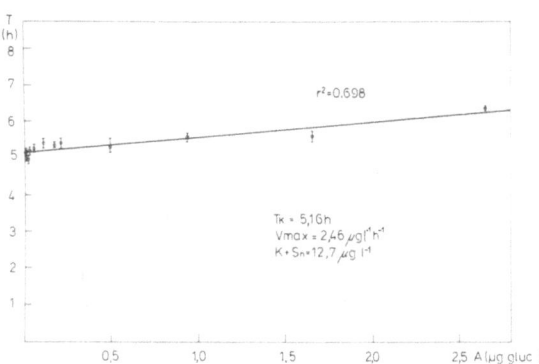

Fig. 1. Kinetic parameters in the 0–2 m sample of 5 July 1979. Vertical lines represent the standard deviation of triplicates.

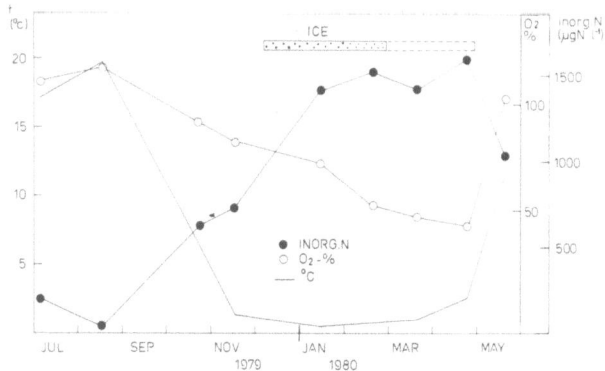

Fig. 2. The succession of temperature, oxygen saturation percentage and inorganic nitrogen ($NH_4 + NO_2 + NO_3$) in 1 m.

caused the temperature to fall to near 1 °C by November (Fig. 2).

The lack of physical stability also prevented the development of chemical or biological stratification. Only on 16 August, immediately after a heavy algal bloom, did the oxygen saturation percentage of the bottom layers sink to 5%, while at 5 m it remained at 81%. Saturation percentage decreased to a minimum of 46% compared with less than 25% during four previous winters.

Nutrient concentrations in Lake Tuusulanjärvi were high, total nitrogen ranging from 1100 μg l^{-1} to 2300 μg l^{-1}, and total phosphorus from 45 μg l^{-1} to 140 μg l^{-1}. The mineralization of organic nitrogen was a cumulative process, which proceeded until midwinter, when over 70% of total nitrogen was inorganic, and over 98% of this was nitrate-nitrogen (Fig. 3).

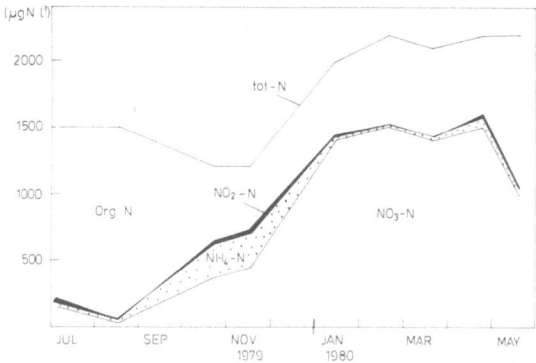

Fig. 3. The transformation of the nitrogen pool (1 m). Organic nitrogen was not measured but it is calculated as the difference between total and inorganic nitrogen.

With the exception of the summer 1979 sampling dates, biological parameters varied little with depth. Integral values representing the whole water column give therefore an adequate picture of the succession (Fig. 4). The pattern of all the measured parameters was essentially similar, showing a steady decline in biological activity during autumn and winter and rapid increase in late April and May. Correlations between biological parameters and temperature are presented in Table 1.

Discussion

The results show clearly that the sensitivity of the heterotrophic activity method is sufficient in a eutrophic lake like Lake Tuusulanjärvi. The highest measured turnover times were 300–700 h, and the method could reliably detect activities up to several thousand hours without any modifications of incubation time or substrate concentration. Unpublished data from brackish water areas suggest that wintertime turnover times of ^{3}H-glucose seldom exceed 1000 h even in the unpolluted coastal areas of the Gulf of Finland.

The summer turnover times of 5–7 h (July 1979, May 1980) are among the lowest reported (cf. Sepers 1977; Hoppe 1978). The intermediate level of 10–20 h, which prevailed from August to January and was reached again in April, also indicates remarkably high heterotrophic activity especially at temperatures below 1 °C. During this period oxygen saturation decreased steadily. The reduction of heterotrophic activity to 100–110 h (February-

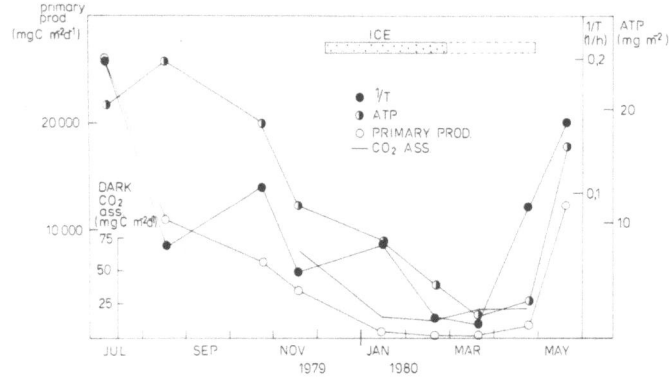

Fig. 4. The succession of biological variables of the water column. Turnover rate (1/T) is a weighted mean of six depths, and other parameters are integral values over depth (per square meter). Since the depth is 10 m, a weighted mean value of parameters (per cubic meter) is obtained by dividing results by 10.

March) undoubtly prevented the development of a serious oxygen deficit after midwinter following the termination of municipal pollution and the long aeration period during the winter.

The development of oxygen and inorganic nutrient concentrations under ice is essentially a cumulative process, whereas turnover rate, primary productivity and dark CO_2 assimilation are dynamic variables, and ATP is an indicator of bioactivity. This different nature of the parameters prevented correlation or regression analyses of the dependence of oxygen saturation on biological variables. However, the heterotrophic activity levels represented by turnover rates of glucose have a clear connection with the oxygen saturation of the water body.

The correlations between biological parameters (Table 1) are generally significantly positive, with the exception of the correlation between the two heterotrophic parameters, turnover rate and dark CO_2 assimilation. Since no fractionation proce-

dures preceded the ATP measurement, the relative portions of different groups of organisms in ATP results are not known. The ATP values of the water column (Fig. 4) resemble on several dates a relative sum value of primary productivity and turnover rate, which would lend support to the use of ATP results as an overall measure of bioactivity. ATP correlates slightly better with primary productivity than with heterotrophic parameters.

Dark CO_2 assimilation has been proposed as a measure of bacterial production (Sorokin 1965), and it has been suggested to represent 6% of total carbon assimilation of aerobic heterotrophic bacteria (Kuznetsov & Romanenko 1966). Bacterial production values calculated according to this theory are presented in Table 2. These bacterial production estimates exceed primary productivity in February and March.

The reason for the non-correlation of turnover rate of glucose and dark CO_2 assimilation is not known. Differences in incubation time and mea-

Table 1. Correlations between biological parameters and temperature. Significance symbols represent P = 0.001*** and P = 0.05*.

	n	°C	ATP	PP-net	PP-dark	CO_2-ass
ATP	45	0.730***				
Primary prod. (net)	45	0.843***	0.713***			
Primary prod. (dark)	44	0.728***	0.696***	0.740***		
Dark CO_2 ass.	28	0.162	0.454*	0.606***	0.932***	
1/T	47	0.732***	0.593***	0.803***	0.560***	−0.052

Table 2. Primary productivity and bacterial production (calculated according to Kuznetsov & Romanenko 1966) of the water column.

date	PP-net (mg C m^{-2} day^{-1})	Dark CO_2 ass. (mg C m^{-2} day^{-1})	Dark CO_2 ass. (% of PP)	Bacterial production (mg C m^{-2} day^{-1})	(% of PP)
15.11.1979	4350	68	1.6	1130	26
14. 1.1980	468	15	3.2	253	54
20. 2.1980	190	13	6.7	213	112
19. 3.1980	146	20	13.9	338	232
24. 4.1980	1040	21	2.0	348	34

surement sensitivity might produce artificial scattering, or methods might measure different, relatively independent heterotrophic processes. Since the theoretical basis of the dark CO_2 assimilation method is somewhat ambiguous (Overbeck 1979), turnover rate is considered to give more reliable information on heterotrophic processes in the lake.

Turnover rate and turnover time of a substrate describe the turnover of a certain fraction of the dissolved organic carbon pool. The results cannot be transformed into actual uptake velocity without knowing the natural substrate concentration, and even so, total heterotrophic uptake of carbon would not be known. However, because of the sensitivity and precision of the method, its theoretical clarity and simple technical performance, analogous with primary production measurement, the single concentration method is a valuable tool in ecological microbial research.

Acknowledgements

I wish to thank the field and laboratory staff and heads of the research division of Helsinki Water District, Ms. Sinikka Numminen and Ms. Riitta Niinioja; Mr. Antti Uusi-Rauva of the Isotope Laboratory in the Faculty of Agriculture and Forestry, University of Helsinki, for co-operation in radioactive and ATP measurements; and Mr. Jorma Kuparinen for valuable criticism of the manuscript.

References

Azam, F. & Holm-Hansen, O., 1973. Use of tritiated substrates in the study of heterotrophy in seawater. Mar. Biol. 23: 191–196.

Holm-Hansen, O. & Booth, C. R., 1966. The measurement of adenosine triphosphate in the ocean and its ecological significance. Limnol. Oceanogr. 11: 510–519.

Hoppe, H., 1978. Relations between active bacteria and heterotrophic potential in the sea. Netherl. J. Sea. Res. 12: 78–98.

Kuznetsov, S. I. & Romanenko, V. I., 1966. Production der Biomasse heterotropher Bakterien und die Geschwindigkeit ihrer Vermehrung im Rybinsk-Stausee. Verh. int. Verein. Limnol. 16: 1493–1500.

Morita, R. Y., Griffiths, R. P. & Hayasaka, S. S., 1977. Heterotrophic activity of microorganisms in Antarctic waters. In: Adaptations within Antarctic ecosystems. The proceedings of the 3rd SCAR Symposium on Antarctic Biology, pp. 99–113. Gulf Publishing Co., Texas.

Munro, A. L. S., Williams, G. R. & Massie, L. C., 1973. Seasonal variation in microbial activity. Bull. Ecol. Res. Comm. (Stockholm) 17: 269–270.

Numminen, S. & Lemmelä, R., 1976. Tuusulanjärven ilmastuksen tuloksista. Vesitalous 17: 18–20.

Overbeck, J., 1979. Dark CO_2 uptake - biochemical background and its relevance to *in situ* bacterial production. Arch. Hydrobiol. Beih. Ergebn. Limnol. 12: 38–47.

Parsons, T. R. & Strickland, J. D., 1962. On the production of particulate organic carbon by heterotrophic processes in sea water. Deep-Sea Res. 8: 211–222.

Sepers, A. B., 1977. The utilization of dissolved organic compounds in aquatic environments. Hydrobiologia 52: 39–54.

SFS-catalogue 1055, 1981. Finnish Standards Association, Helsinki.

Sorokin, Y. I., 1965. On the trophic role of chemosynthesis and bacterial biosynthesis in water bodies. Mem. Ist. Ital. Idrobiol. 18: 187–205.

Williams, P. J. leB. & Askew, C., 1968. A method of measuring the mineralization by micro-organisms of organic compounds in sea-water. Deep-Sea Res. 15: 365–375.

The influence of sulphite mill effluents on heterotrophic activity as measured by glucose assimilation

Jorma Kuparinen*

Tampere Water District Office, P.O. Box 297, SF-33101 Tampere 10, Finland

Present address: Tvärminne Zoological Station, SF-10850, Tvärminne, Finland

Keywords: heterotrophic activity, 3-H-glucose, natural populations, sulphite waste

Abstract

The influence of sulphite mill effluents on the heterotrophic activity of natural populations in the recipient water was studied during the summer of 1979. A single concentration method (3-H-glucose, $0.05 \ \mu g \ l^{-1}$) was used to measure the heterotrophic activity. The influence of the effluent was studied by adding it in concentrations varying from 0.001 to 10% to water samples derived from different sections of the watercourse. Adaptation to rather high concentrations of the effluent was observed near to the point of discharge. Low concentrations of the effluent had a stimulatory effect on heterotrophic activity. Short turnover times (5–20 h) of glucose and inhibition of primary productivity demonstrated the influence of effluents on planktonic populations.

The single concentration method allows rapid and simple detection of horizontal and vertical effects of effluents on natural bacterioplankton in the recipient water.

Introduction

Forest industries cause severe pollution problems in Finland, as in many other countries. The effluents from different kinds of forest industries have both inhibitory and stimulative effects on the biotope in the recipient water. Toxic effects of the effluents are being studied intensively but the extent of these effects is still poorly understood because of the large number of different toxic compounds in effluents.

A distinctive effect in the receiving waters is however caused by the load of easily degradable organic compounds and inorganic nutrients in the water. A variety of carbon sources coupled with nutrient load from the settlements and farmlands surrounding the watercourse provide a good substrate for active bacterial metabolism. As a result of vigorous consumption of oxygen during the decomposition of organic matter and oxidation of reduced compounds, oxygen concentrations dimin-

ish in water or disappear totally, causing fish deaths. Of the total BOD_7 load in Finnish waters, the share of forest industries is about 85% (Vesihallitus 1976).

The upper part of the basin of the river Kokemäenjoki (Fig. 1) was chosen for the study of the effects of effluents on heterotrophic activity and primary productivity. The paper and pulp mills (sulphite process) of G. A. Serlachius Ltd are located on the Keuruu watercourse in the town of Mänttä (M). A more detailed description of the study area, its hydrography and effluent discharge has been given by P. Eloranta (1970). Because of the heavy effluent load from the mills, the waters in the lower watercourse suffer from oxygen depletion in the deeps up to 30 km from the mills, until the point where Tarjannevesi runs into the watercourse causing dilution of effluents. In the vicinity of the mills (up to 10 km) biological processes are strongly inhibited (P. Eloranta 1970; Eloranta & Kettunen 1979).

Hydrobiologia 86, 115–119 (1982). 0018-8158/82/0862-0115/$01.00.

© Dr W. Junk Publishers, The Hague.

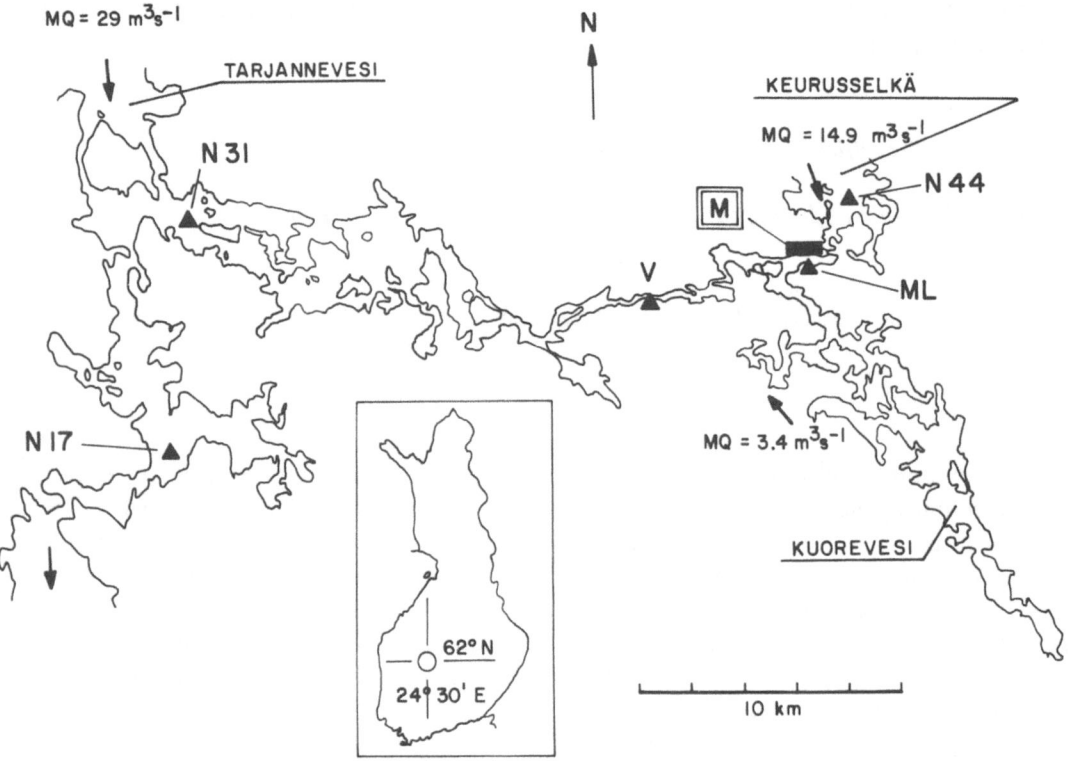

Fig. 1. Location of the study area. Sampling stations are marked as black triangles. M = location of the mills, ML = Mänttä bay. Arrows indicate the direction of flow.

Heterotrophic activity was chosen as the main parameter to be measured because heterotrophic bacteria are mainly responsible for the decomposition of organic matter and are thus the major cause of oxygen depletion in the water. Rao *et al.* (1976) and Rokosh *et al.* (1977) found aerobic heterotrophic bacteria to be the most sensitive indicators of the effects of effluents. If the numbers of aerobic heterotrophs can indicate these effects it would seem reasonable that their metabolic activity should also do so.

Material and methods

The study area and sampling stations are shown in Fig. 1. Samples were taken from depths of 0.1, 1, 3 and 5 m and also from 1 m above the sediment surface. For the effluent test a surface sample from 0 to 2 m was taken from all the stations. The effluent, which was added to the water samples from different sampling stations, was taken from a depth of 0 to 1 m in front of the mills at Mänttä bay (ML) and filtered on a 0.45 μm membrane filter (Millipore) before use. All water samples were carried to the laboratory cooled in aluminium boxes and stored overnight at $+6 \pm 2\,°C$.

Heterotrophic activity was measured with D-(6-^{3}H)glucose (Radiochemical Centre, Amersham, England) of specific activity 22.5 Ci mmol^{-1}. Labelled glucose was added to the 50 ml water samples in 100 ml serum bottles at a final concentration of 0.05 μg l^{-1}. When using such a small concentration the amount of added substrate (A) could be neglected (equation 1) and the turnover time (T) or turnover rate (T^{-1}) calculated (Williams & Askew 1968; Azam & Holm-Hansen 1973).

$$v = \frac{c}{Ct}(Sn + A), \qquad (1)$$

in which v = rate of uptake (μg substrate l^{-1} h^{-1}) at ambient substrate concentration, c = radioactivity (DPM = disintegrations per minute) measured from the sample, C = radioactivity (DPM) added to the sample, t = incubation time (h), Sn = ambient substrate concentration (μg l^{-1}), A = concentration of substrate added to the sample (μg l^{-1}). When A $\approx$ 0, then

$$v = \frac{c}{Ct} \times Sn \qquad (2)$$

$$\text{or } \frac{Sn}{v} = \frac{Ct}{c} = T \text{ (turnover time).} \qquad (3)$$

Triplicate incubations of 2–3 h were carried out for heterotrophic activity measurements at *in situ* temperatures. A blank for each sample was made by adding 0.5 ml formalin (39%) prior to adding the radioactivity. After the incubation bacteria were filtered on 0.45 μm membrane filters (Millipore) and dried at room temperature. Dry filters were oxidized in a JUNITEK oxidizer. ^{3}H-activity in a 2.6 ml water + 7.4 ml PCS (Radiochemical Centre, Amersham, England) scintillation cocktail was counted with a LKB-WALLAC RackBeta liquid scintillation counter.

Primary productivity (24 h) was measured according to the SFS-3049 standard. For counting the ^{14}C-activity, 0.45 μm membrane filters (Millipore) were wetted in the scintillation vials with 0.3 ml distilled water and soaked in 2.0 ml of dioxane (Merck p.a.) for at least 2 h. When the filters became transparent, 10.0 ml of PCS scintillation cocktail was added. The vials were kept at room temperature overnight and shaken a few times, after which the filters became totally transparent in the cocktail. Radioactivity was counted with the same instrument as above.

Chlorophyll-*a* and other parameters were measured in the Tampere Water District Office laboratory according to the Finnish standard methods (SFS standards).

Results and discussion

The extent of the influence of the paper and pulp mills can be seen in the lignin concentrations (Fig. 2). The upper watercourse had a lignin concentration of 1 mg l^{-1} from surface to bottom but 5 km from the mills at the lower watercourse (V) it was almost 20 times higher and 40 km from the mills eight times higher still. At station N31 the bottom sample had a slightly higher lignin content

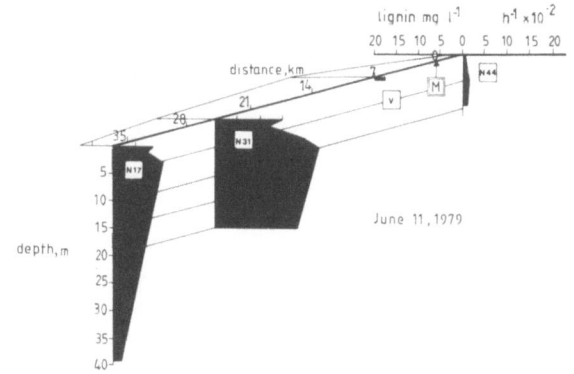

Fig. 2. Glucose turnover rate (h^{-1}) of the water column and lignin content (mg l^{-1}) at 0.5 m at stations N44, V, N31 and N17 on 11 June 1979. The location of the mills is marked M.

than the surface samples but at the other stations the concentrations were evenly distributed. Some other physical and chemical parameters also revealed the presence of effluents up to 40 km from the mills. At station N31 and N17 the pH was 4.9 and 5.6, the conductivity 8.5 and 6.4 mS m^{-1} in the surface water respectively, whereas at station N44 the pH was 6.7 and the conductivity 4.2 mS m^{-1}. The oxygen saturation in surface water from 0 to 5 m was 45% at station N31 and 79% at N17, while in the upper waterway it was 95%.

The glucose turnover rate was low, 0.009 h^{-1}, throughout the whole water column at station N44 (Fig. 2). At Vilppula rapids (V) it was 0.018 h^{-1} and at station N31 increased up to 0.2 h^{-1}. At station N17 the rate had decreased to 0.1 h$^-$ but was still considered to be high. The glucose turnover rate of 0.2 h^{-1} corresponds to a 5 h turnover time, which is among the shortest found in the literature (Sepers 1977; Gocke 1977). The effect of this increased heterotrophic activity was an oxygen depletion at a depth of 20 m at station N31.

Primary productivity had its highest values, 100 mg C m^{-3} day^{-1} (net), at station N44 and its lowest, 10 mg C m^{-3} day^{-1} (net), at station V (Fig. 3) The highest chlorophyll-*a* concentration, 14 μg l^{-1}, was detected at station N31, the other stations all having 5–6 μg l^{-1}. High chlorophyll-*a* concentration and low primary productivity showed strong inhibition at stations V and N31. These results are in good agreement with the results of V. Eloranta (1976) who found that the ^{14}C-fixation rate of *Ankistrodesmus falcatus* var. *acicularis* culture was

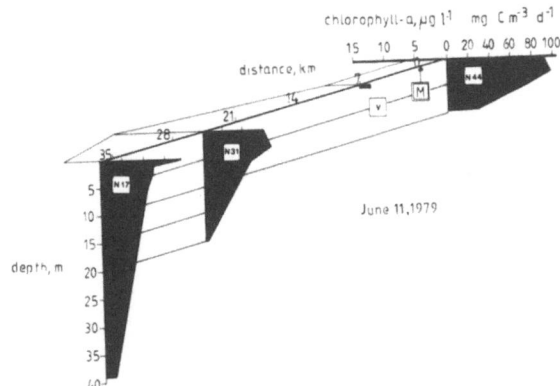

Fig. 3. Primary productivity (mg C m^{-3} day^{-1}) of the water column and chlorophyll-*a* content (μg l^{-1}) at 0.5 m at stations N44, V, N31 and N17 on 11 June 1979. The location of the mills is marked M.

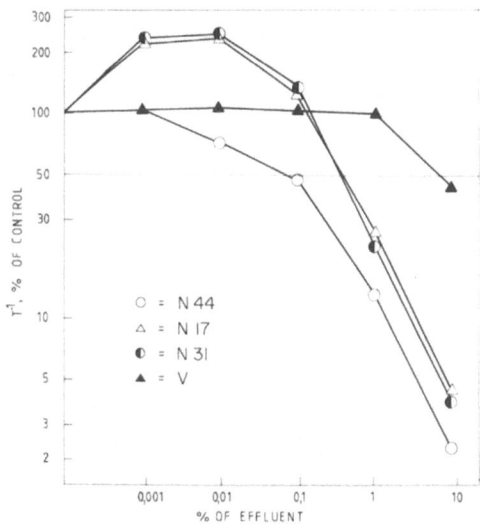

Fig. 4. The influence of effluent (a sample from 0 to 1 m from Mänttä bay) on glucose turnover rate (h^{-1}) at stations N44, V, N31 and N17.

significantly more sensitive to the black liquor and spent sulphite liquor effluents than was the chlorophyll-*a* content.

Heterotrophic activity and primary productivity measurements showed that effluents have both stimulative and inhibitory effects on the biotope. The organic nutrients, especially soluble carbohydrates such as sugars and sugar acids (P. Eloranta 1970; Poole *et al.* 1977), provide a good source of carbon and energy for micro-organisms. This was shown in glucose turnover rates at station N31 where the effluent content was still high. The inhibitory effect was shown in both glucose turnover rate and in primary productivity at the Vilppula rapids but only in productivity at station N31, suggesting that bacteria are not as sensitive to effluents as are phytoplankton. However, this may be due not to the difference in initial sensitivity but to the greater ability of bacteria to develop resistant populations.

To study the adaptation of bacterial populations a series of experiments was carried out in which effluent water was added in different concentrations to natural water samples taken from the sampling stations. The results of one of these experiments is presented in Fig. 4. When effluent was added to samples from the Vilppula rapids (V), no change in glucose turnover rate was observed until 10% concentration. In the samples from all other stations 50% inhibition was reached at between 0.01 and 1.0% concentration of effluent,

station N44 being the most sensitive. Populations at the stations N31 and N17 were able to utilize low concentrations of effluent and both reached 50% inhibition at about 0.3% concentration.

At the Vilppula rapids the bacteria are constantly exposed to high concentrations of effluents and therefore did not react to small additions of effluent in the tests. Also pH, conductivity, acidity and lignin concentration of sample water remained unaltered at up to 10% effluent addition. At the other stations chemical parameters changed at 1.0 and sometimes even at 0.1% concentration of effluent. The changes in glucose turnover rate were found to occur at one tenth or one hundredth of the concentration needed for the change in common physical and chemical parameters. Fifty per cent inhibition at 0.1% level of effluent shows that bacteria are sensitive to abrupt increases of effluent concentration in water. However, when the flow of effluent is regular, resistant populations may develop, which can maintain a considerable decomposition rate of organic matter in water.

The use of heterotrophic activity measurement has proved valuable in detecting and describing variations in the influence of effluents from paper and pulp mills. The extreme sensitivity of aerobic heterotrophic populations to organic nutrients seems to be a valuable tool along with the primary productivity and standard physical and chemical measurements in assessing water quality and trophic status.

Acknowledgements

I would like to thank Mr. Jaakko Keränen at the Tampere Water District Office for help and support during this study. I am grateful to Mr. Antti Uusi-Rauva for helping with radioactive measurements. The cooperation of the Tampere Water District Office laboratory staff is gratefully acknowledged. I also thank Mr. Michael Bailey for revising the manuscript. The study was financed by the Tampere Water District Office.

References

Azam, F. & Holm-Hansen, O., 1973. Use of tritiated substrates in the study of heterotrophy in seawater. Mar. Biol. 23: 191–196.

Eloranta, P., 1970. Pollution and aquatic flora of waters by sulphite cellulose factory at Mänttä, Finnish Lake District. Ann. bot. fenn. 7: 63–141.

Eloranta, P. & Kettunen, R., 1979. Phytoplankton in a watercourse polluted by a suphite cellulose factory. Ann. bot. fenn. 16: 338–350.

Eloranta, V., 1976. Effects of different process wastes and main sewer effluents from pulp mills on the growth and production of Ankistrodesmus falcatus var. acicularis (Chlorophyta). Biol. Res. Rep. Univ. Jyväskylä 2: 3–33.

Gocke, K., 1977. Heterotrophic activity. In: Rheinheimer, G. (ed.) Microbial Ecology of a Brackish Water Environment, pp. 198–222. Springer-Verlag, Berlin.

Poole, N. J., Parkes, R. J. & Wildish, D. J., 1977. Reaction of estuarine ecosystems to effluent from pulp and paper industry. Helgoländer wiss. Meeresunters. 30: 622–632.

Rao, S. S., Rokosh, D. A. & Jurkovic, A. A., 1976. Influence of a point source pulp mill effluent discharge on the nearshore bacterial communities in Lake Superior. In: Proceedings of the Second Federal Conf. on Great Lakes, pp. 397–408. Great Lakes Basin Commission.

Rokosh, D. A., Rao, S. S. & Jurkovic, A. A., 1977. Extent of effluent influence on lake water determined by bacterial population distributions. J. Fish. Res. Bd Can. 34: 844–849.

Sepers, A. B. J., 1977. The utilization of dissolved organic compounds in aquatic environments. Hydrobiologia 52: 39–54.

Vesihallitus, 1976. Application of water pollution control principles. Publications of the National Board of Waters, Finland 16: 1–352 (in Finnish).

Williams, P. J. Leb. & Askew, C., 1968. A method of measuring the mineralization by micro-organisms of organic compounds in sea-water. Deep-Sea Res. 15: 365–375.

Establishing the pattern of heterotrophic bacterial activity in three Central Amazonian lakes

Hakumat Rai[1] & Gary Hill[2]
[1] Max Plack Institute for Limnology, A. G. Plankton Ecology, 2320 Plön, Postfach 165, West Germany
[2] 5509 D Lennox Ave., Bakersfield, CA 93309, U.S.A.

Keywords: heterotrophic, bacteria, V_{max}, Central Amazon, primary production, multitrophic

Abstract

Three lakes of the Central Amazon were studied in depth for a year. Patterns of heterotrophic activity were seen to develop as the result of a variety of factors in the three lakes. Primary mechanisms that appear to control the heterotrophic activity were the water levels, the nutrient concentrations, including labile carbon, and the chlorophyll-a levels. Heterotrophic activity (V_{max}) in the three lakes varied from values comparable to extreme oligotrophic to eutrophic in each of the lakes studies. The large range ($0.045 - 9.644 \, \mu g \, l^{-1} \, h^{-1}$) of V_{max} is viewed as a part of the cycle of growth and renewal which is characteristic of the Central Amazonian Ria and Varzea lakes.

Introduction

There is a beautiful and natural symmetry expressed in the annual rise and fall of the Central Amazonian aquatic ecosystem. This average 10.12 m rise of the rivers floods the Ria (river) and Varzea (floodplain) lakes with energy in the form of nutrients. The myriad lakes pay for this energy with biomass in an annual cycle of growth, renewal and exchange (Hill & Rai 1980).

Patterns of heterotrophic activity (V_{max}) are therefore necessarily established during each seasonal cycle. Seasonal trends in each of the three lakes studied express patterns that could justifiably be called indicative of either an oligotrophic, mesotrophic or eutrophic system, depending entirely on what point in time they were being viewed.

The seasonal trend of V_{max} was apparently established in each of the three lakes independently and slightly differently, although in each case, primary production was of chief importance.

Patterns of V_{max} in natural aquatic systems have been studied by Wright & Hobbie (1966), Hobbie & Crawford (1969), Andrews & Williams (1971), Overbeck (1975), Gocke (1977), Rai (1978, 1979), Rai & Hill (1981b) and others. Most workers in this area of study have found rather close correlations between V_{max} and primary production and/or the densities of saprobic bacteria. Rai & Hill (1981b) found no relationship between V_{max} and saprobic bacteria in their work on Lago Tupe in the Central Amazon, but did report on an apparently controlling nutrient mechanism which was indirectly responsible for the development of the seasonal trend of V_{max}.

Location of lakes

Lago Janauari, Lago Cristalino and Lago Tupe are all located within 30 km of Manaus, Amazonas, Brazil (Fig. 1). Lago Janauari is a 79 ha mixed water Ria/Varzea lake that receives both Rio Negro and Rio Solimões waters, and is the only lake in this study which supports the growth of floating meadows. Floating meadows are composed primarily of *Paspalum repens* and *Echinochloa polystachya*. Where they exist, these plants exert

Hydrobiologia 86, 121–126 (1982). 0018-8158/82/0862-0121/$01.20.
© Dr W. Junk Publishers, The Hague.

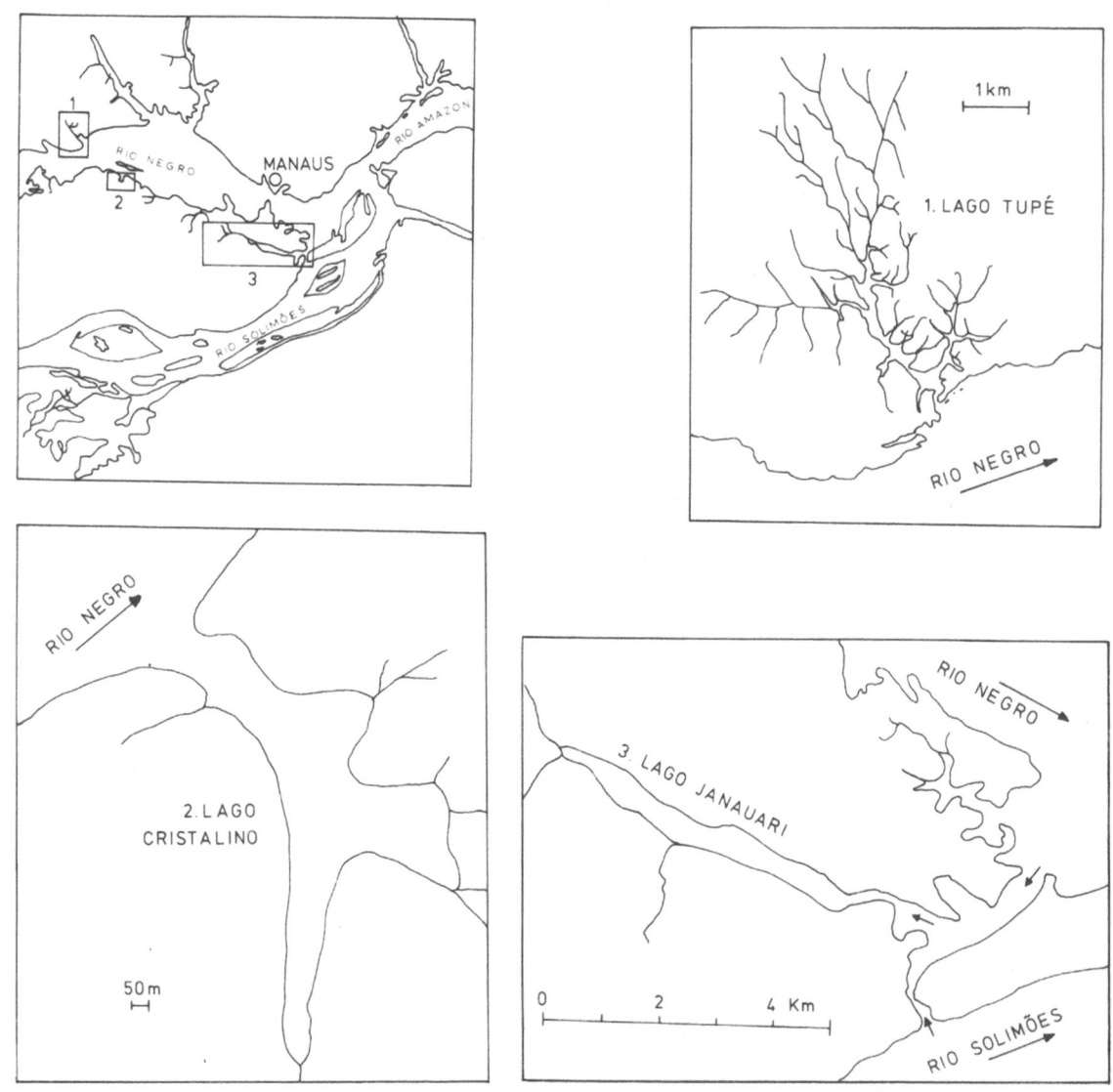

Fig. 1. A map of the study area and the three lakes.

notable effects on the chemistry and the microbiology of the lakes (Rai & Hill, unpublished data). Lago Cristalino (24 ha) and Lago Tupe (68 ha) are both black water lakes influenced only by the Rio Negro, and are situated on opposing banks of the river.

Methods

Methods and formula derivation for *in situ* heterotrophic uptake are discussed in Wright &
Hobbie (1966), Hobbie & Crawford (1969) and Rai (1978, 1979). Three parameters relating to the kinetics of the bacterial population were derived: the theoretical maximum velocity of uptake (V_{max}), the approximation of the natural substrate concentration ($Kt + Sn$), and the turnover time (Tt) for the substrate to be completely removed from solution by natural populations present in the water sample. methods for the remaining physical and chemical tests are the same as described in detail by Rai & Hill (1981a, b).

Results and discussion

Lago Janauari

Annual variations of V_{max} in Lago Janauari are shown in Fig. 2. July exhibited the lowest values throughout the water column with a standard deviation of only 0.01. This was the time of highest water levels in all three lakes. In January, when the annual rise was under way, there developed a maximum value of V_{max} at the surface and at all depths, with a standard deviation in the water column of 68.19. When the lake is oligotrophic, there is little deviation; but more eutrophic conditions result in pronounced deviations. This was also the case for the other two lakes. The trophic state seems to determine the amount of fluctuations throughout the water column.

Variations in the water levels of Lago Janauari are shown in Fig. 2 with the maximum amounts of Chl-a in the water column. These two variables are negatively reelated (r = –0.90), so that the variations in the depth of the lake can account for 81% of

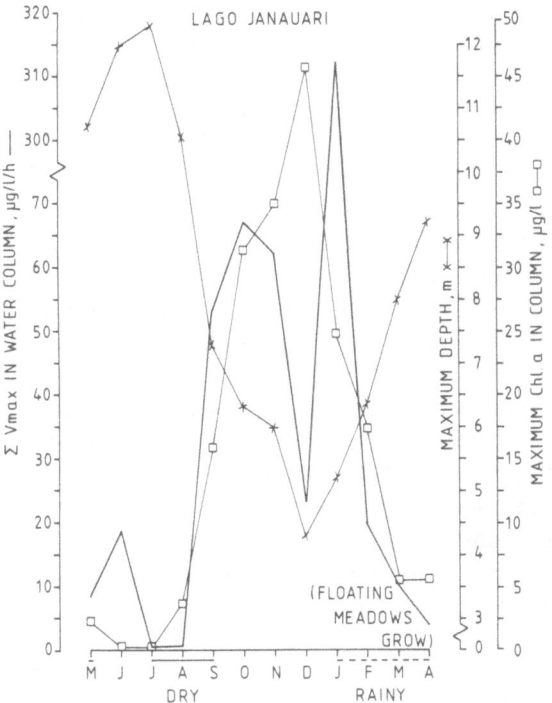

Fig. 2. Decreasing water levels in Lago Janauari permit the autochthonous growth which develops a burst of heterotrophic activcity.

the variability of Chl-a. Considering this, and the work of Sioli (1975), Schmidt (1973), Rai & Hill (1981a) and others, it can be said with some foundation that, in the lakes of the Central Amazon, autochthonous production occurs during the low water period.

Increases in Chl-a are generally (but loosely) seen in this same figure to be associated with the developing increases in V_{max}. The patterns of Chl-a and V_{max} are quite similar except for December where V_{max} decreases when Chl-a is peaking. This may be the result of the Rio Negro influence which was just beginning its rise in December. The rising Rio Negro, rich in the first washings of the caatingas, might perhaps depress the bacterial activity by bringing in new populations of bacteria which were unable to utilize glucose. Saprobic bacteria in Lago Janauari exhibit peaks in May, July, September, December, March and April (Rai & Hill, unpublished) and do not appear to be related to V_{max}.

The pattern of V_{max} in Lago Janauari appears therefore to be somewhat related to Chl-*a* bearing algae with the exception being understood in terms of shifting bacterial populations. Surface Chl-*a* and V_{max} show a correlation of +0.68. Other influences on V_{max} can be visualized in terms of the floating meadows. The rise of V_{max} throughout the water column in January to a summation of 312.03 µg glucose l^{-1} h^{-1} does not occur at high Chl-a levels and so autochthonous production does not seem to be the source of highest V_{max}.

The floating meadows are stranded on the shore under low wateer conditions where they undergo active microbial decomposition. As the waters begin to rise, the organic products of this active process are washed into the lake where the heterotrophic bacterial population reaches eutrophic levels throughout the column. Total bacteria and DOC also begin major developments at this time (Rai & Hill, unpublished), further supporting the hypothesis of a shore-lake interaction. Natural substrate concentration (Kt + Sn) is also highest in January throughout the column. It therefore appears that the seasonal pattern of V_{max} in Lago Januari is established in relation to at least two major dissolved organic sources. The first source is autochthonous algal production which results in V_{max} levels of 42.1 µg glucose l^{-1} h^{-1}. The second source is thought to be the floating meadows which results in V_{max} levels of 19.7 µg glucose l^{-1} h^{-1}.

Lago Cristalino

Lago Cristalino is a 24 ha black water Ria lake lateral to the Rio Negro. V_{max} at Lago Cristalino averaged 1.35 μg glucose l^{-1} h^{-1} at the surface with a standard deviation of 2.71 and a range that went from an oligotrophic 0.04 μg l^{-1} h^{-1} to a eutrophic 9.64 μg l^{-1} h^{-1}. The pattern at the surface was bimodal with a major and a minor peak (Fig. 3). The major peak occurred in October. The maximum in Lago Janauari occurred in January, and was attributed primarily to the floating meadows. Lago Cristalino is a black water lake and so does not support the growth of the floating meadows. A different mechanism for the development of V_{max} is therefore indicated. Different times for peak activity also suggest different origins for the heterotrophic activity.

The suggestion in Fig. 3 is that the natural substrate concentration is the leading mechanism regulating the activity of the heterotrophic bacterial population. It seems indeed to account for more than 90% of the variability of V_{max}. Andrews & Williams (1971) found a similar relationship be-

tween V_{max} and Kt + Sn in their work on the Pacific.

This same relationship holds true in vertical considerations at Lago Cristalino. Variations in V_{max} are seen to have no clear positive relation with the total bacterial fluctuations (Rai & Hill, unpublished). In October, when V_{max} peaks at the surface, total bacteria densities were 1.12 × 10^8 l^{-1} and the SPC counts were 195 × 10^3 l^{-1}; so that the saprobic bacteria represented only 0.17% of the total bacterial population. Vertically, the SPC influences V_{max} slightly in November, but not in September. The implication is that the SPC does not exert great influences and, in this lake, is sometimes benefited by the same substrates that increase V_{max} (Rai & Hill, unpublished).

What does show a consistent similarity in fluctuations to the fluctuations of V_{max} is the level of dissolved phosphates. At this low level (<1 μg l^{-1}) P can be a limiting nutrient. Accordingly, from May through October, there is a positive relationship between V_{max} and phosphorus. The relationship is however thought to be indirect since affinity for glucose must be predicated on glucose producers. Chl-a bearing algae appear, (Rai & Hill, unpublished), to be producing labile carbon which is contributing significantly to the pool of Kt + Sn. When starvation eliminated by increasing consumption of phosphates in Lago Cristalino, algae show a burst of growth which results in increased labile carbon available for the heterotrophic bacteria.

Labile carbon, estimated by the method of Zsolnay (1975), was perfectly correlated with DO <r = + 1.00). This further indicates that the source of the labile carbon was growing algae whose growth released both labile carbon and DO.

Increasing consumption of phosphates appears to develop an actively growing algal population which produces DO and labile carbon in proportional amounts. The labile carbon then adds to the natural substrate concentration which is the major controlling factor for the establishment of the pattern of V_{max} in Lago Cristalino.

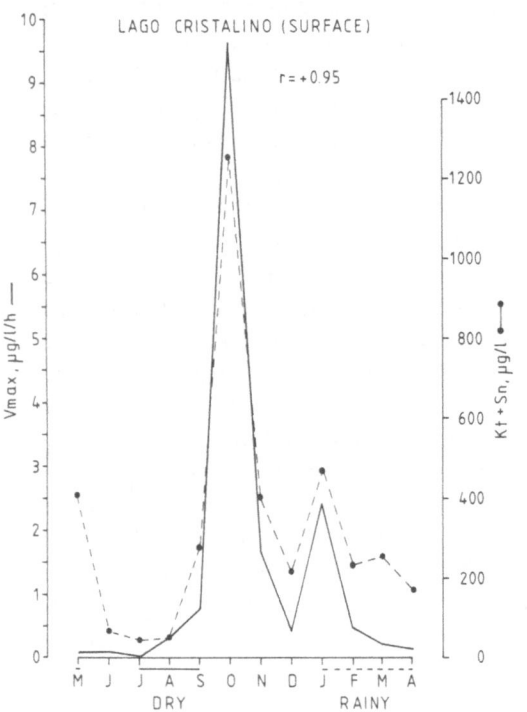

Fig. 3. The seasonal distribution of V_{max} and Kt + Sn at the surface of Lago Cristalino.

Lago Tupe

Lago Tupe is a 68 ha black water Ria lake located across from Lago Cristalino, and lateral to the Rio Negro. Comprehensive studies on Lago Tupe have been reported by Rai & Hill (1981a, b).

Heterotrophic bacterial activity in Lago Tupe at the surface ranged from 0.09 to 9.3 μg glucose l^{-1} h^{-1} with an annual average of 2.0 μg glucose $l^{-1} h^{-1}$. The seasonal pattern is shown in Fig. 4 with the nitrate nitrogen: phosphate (NO_3-N:PO_4-P) ratio. Again, the seasonal cycle develops from oligotrophy to eutrophy and back to oligotrophic conditions. Lago Tupe also exhibits maximum V_{max} during the low water productive period, just as Lago Janauari and Lago Cristalino do. However, the pattern of V_{max} seems to have been established in a manner quite different from that in Lago Janauari and along similar but slightly different paths from that in Lago Cristalino.

Highest values of V_{max} in Lago Tupe occurred concomitantly with highest Chl-a., lowest PO_4 and low water levels (Rai & Hill 1981b). Algae are most productive at low water levels when the flow is more quiescent. The Chl-a bearing primary producers are seen to be somewhat inversely related to PO_4 concentrations. It is indicated that the death of algae releases glucose. A death release is indicated in this lake, since there was a significant and inverse relationship between the labile carbon and Chl-a (r = –0.74). The release of algal glucose serves as a natural substrate for the bacteria which then increase in activity. When Chl-a is low, V_{max} is

therefore low and PO_4 high since the lack of algal growth permits PO_4 accumulation. It seems clear then that V_{max}, or heterotrophic activity, in Lago Tupe is inextricably dependent on algal production, which again depends on nutrient levels and water levels. Of nutrients, phosphorus is not the only nutrient in Lago Tupe that is of importance. Nitrates also exert great influence. This can be seen in Fig. 4 where the NO_3-N: PO_4-P ratio has a highly significant relationship with V_{max} (r = + 0.89). Nitrates and phosphates then appear to control algal growth in Lago Tupe, and the death release of dissolved organics provides the substrate for the increases in observed heterotrophic activity.

Acknowledgements

This work is the result of a cooperative effort between the Max Planck Institute for Limnology, Department of Tropical Ecology (Plön, West Germany) and the Institute Nacional de Pesquisas da Amazonia (Manaus-Amazonas, Brazil).

The authors would like to express their gratitude to Dr. H. Sioli and Dr. Kerr for providing facilities and for their interest and encouragement. Dr. Overbeck kindly provided facilities for the radioactive measurements. Miss I. Mau deserves a good measure of credit for her meticulous cpm work, and Mrs. H. Buhtz gave invaluable assistance in her excellent chemical and microbiological analyses, in addition to drawing the figures.

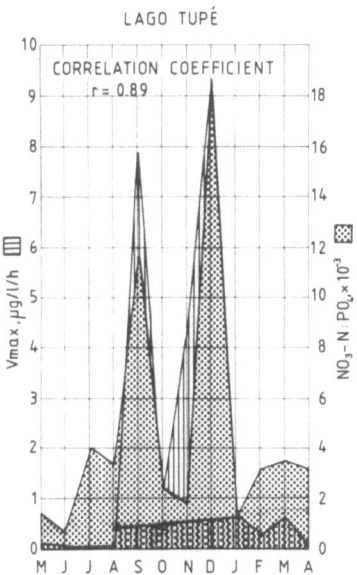

Fig. 4. At the surface of Lago Tupé, V_{max} is related to the nutrient ratio.

References

Andrews, P. & Williams, P. J. L., 1971. Heterotrophic utilization of dissolved organic compounds in the sea. J. mar. biol. Assoc. U.K. 51: 111–125.

Gocke, K., 1977. Heterotrophic activity. In: Rheinheimer G., (ed.) Microbial Ecology of Brackish Water Environment, pp. 198–111. Springer-Verlag, Berlin.

Hill, G. & Rai, H., 1980. A preliminary characterization of the lakes of the Central Amazon by comparisons with Arctic and temperate systems. Proc. Am. Soc. Limnol. & Oceanogr. LA, CA.

Hobbie, J. E. & Crawford, C. C., 1969. Bacterial uptake of organic substrate: new method of study and application to eutrophication. Mitt. int. Verein. Limnol. 17: 725–730.

Overbeck, J., 1975. Distribution pattern of uptake kinetic responses in a stratified eutrophic lake (Pluß-See ecosystem study IV). Verh. int. Verein. Limnol. 19: 2600–2615.

126

Rai, H., 1978. Utilização de glicose por bacterias heteroficas no ecosistema lacustre da Amazônica Central. Acta Amazônica 8: 225–232.

Rai, H., 1979. Microbiology of the Central Amazon lakes. Amazoniana 4: 583–599.

Rai, H. & Hill, G., 1981a. A black water lake – physical and chemical studies of Lago Tupé, a Central Amazonian black water 'Ria lake'. Int. ges. Res. Hydrobiol. 65 (6): 37–82.

Rai, H. & G, Hill., 1981b. Bacterial biodynamics of Lago Typé, a Central Amazonian black water 'Ria lake'. Arch. Hydrobiol. Suppl. 58 (Monograph. Beitr.) 4: 420–468.

Schmidt, G. W., 1973. Primary production in the three types of Amazonian water. III. Primary productivity of phytoplankton in a tropical floodplain lake of Central Amazonia. Lago do Castanho. Amazonas, Brazil. Amazoniana 4: 379–404.

Sioli, H., 1975. Tropical river: the Amazon. In: Whitton, B. A. (ed.) River Ecology. Univ. of California Press, Berkeley.

Wright, R. T. & Hobbie, J. E., 1966. Use of glucose and acetate by bacteria and algae in aquatic ecosystems. Ecology 47: 447–464.

Zsolnay, A., 1975. Total labile carbon in the euphotic zone of the Baltic Sea as measured by BOD. Mar. Biol. 29: 125–128.

Modelling

Nutrient dependence of phytoplankton production in brown-water lakes with special reference to Lake Päijänne

K. Granberg[1] & H. Harjula[2]

[1] *Hydrobiological Research Centre, University of Jyväskylä, Seminaarinkatu 15, SF-40100 Jyväskylä 10, Finland*

[2] *Helsinki Metropolitan Area Water Co., Nuijamiestentie 5 B, SF-00400 Helsinki 40, Finland*

Keywords: aquatic models, phytoplankton primary production, chlorophyll *a*, phosphorus, humic water

Abstract

A method for predicting the mean seasonal chlorophyll *a* concentration, the mean seasonal *in vitro* phytoplankton primary productivity per unit volume, the maximum daily production per unit volume and the seasonal integral production in brown-water lakes is presented. The production values can be calculated when the mean annual concentration of total phosphorus and the mean annual colour of the water are known. This method has been developed especially for practical water pollution studies to permit rapid and inexpensive estimates of major biological consequences of changes in effluent loads. The method can be applied for brown-water lakes where phosphorus is the limiting nutrient for primary production.

Introduction

Water pollution studies often suffer from insufficient budgets and consequently employ sampling programmes incapable of revealing the complex effects of certain effluent loads. There is therefore a need for simple models to predict certain water quality parameters from loading models. In the decade following Vollenweider's (1968) paper increasing attention has been directed towards predicting phosphorus concentration, phytoplankton chlorophyll and hypolimnetic oxygen depletion from nutrient loadings. Equations predicting primary production of phytoplankton have also been presented (see Smith 1979). Unfortunately most of these models have been developed for uncoloured waters, and therefore are not applicable to brown-water lakes without modifications (cf. Smith 1979).

In this study we have aimed to develop a simple model of the nutrient dependence of phytoplankton production in brown-water lakes.

Material and methods

Material for this study was collected from Lake Päijänne and some adjacent lakes belonging to the Kymijoki watercourse (Granberg 1973; Harjula 1979). Systematic studies on physical, chemical and biological water characteristics have been carried out at Lake Päijänne since 1969.

Chlorophyll *a*, primary productivity (measured in a laboratory incubator) and primary production *in situ* were determined as described by Granberg (1973) and Harjula (1979).

The equations derived in this paper are based mainly on data collected from Lake Päijänne unless otherwise indicated.

The data used to verify presented equations are from four sampling sites: Poronselkä, Ristiselkä, Tehinselkä and Asikkalanselkä (see Harjula 1979), of Lake Päijänne in 1972–1979. This material is different from that used to derive the equations, except for that collected in 1978. All these measurements are published in Reports of the Hydrobiological Research Centre, University of Jyväskylä.

Hydrobiologia 86, 129–132 (1982). 0018-8158/82/0862-0129/$00.80.

Results and discussion

In brown-water lakes light conditions and hence phytoplankton production are affected by humic substances, and it is not possible to estimate production in these lakes directly from the phosphorus concentration of the water body (cf. Smith 1979). However we may (1) estimate the maximum production m^{-3} from the chlorophyll a concentration, (2) incorporate the effects of irradiance and its attenuation through the water column. It is then possible to estimate production per unit area.

Estimation of the mean seasonal chlorophyll a concentration from the total phosphorus concentration in a water body is already satisfactory for practical purposes. Published regressions (e.g. Sakamoto 1966; Dillon & Rigler 1974; Jones & Bachmann 1976; Carlson 1977) give closely similar results.

According to Dillon & Rigler (1974) the mean seasonal chlorophyll a concentration in the trophogenic layer is given by:

$$\text{mg Chl } a \text{ m}^{-3} = 0.073 \cdot P_v^{1.449}, \tag{1}$$

where P_v = the total phosphorus concentration at the spring overturn.

Although the phosphorus–chlorophyll relationship of Dillon & Rigler (1974) is calculated mainly from the data of clear water lakes, the regression has been found to give satisfactory results also in humic lakes because the result obtained represents the mean seasonal chlorophyll a concentration in the trophogenic layer (e.g. Table 1).

Chapra & Tarapchak (1976) have presented the following relationship between mean annual and spring overturn total phosphorus concentrations:

$$P = 0.9 \cdot P_v, \tag{2}$$

where P = the mean annual total phosphorus concentration of the water body.

The mean seasonal chlorophyll a concentration and the mean seasonal phytoplankton productivity in the study area are closely correlated (Granberg 1979a). The fitted regression is:

$$y = 11.18 \ x^{1.358}, \ n = 30, \ r^2 = 0.80, \tag{3}$$

where x = chlorophyll a (mg m^{-3})

y = primary productivity (mg C m^{-3} day^{-1}).
Primary productivity and maximum primary production per unit volume are also correlated (Granberg 1979b) with the relationship:

$$y = 1.924 \ x^{0.808}, \ n = 16, \ r^2 = 0.52, \tag{4}$$

where x = primary productivity (mg C m^{-3} day^{-1}).
y = maximum primary production (mg C m^{-3} day^{-1}).

If we know the V/O quotient (Rodhe 1958) it is possible to calculate the production per unit area by dividing the maximum production per unit volume by the V/O quotient.

The V/O quotient is dependent on water colour (Granberg 1973):

$$y = 0.012 \ x -0.015, \ n = 10, \ r^2 = 0.90, \tag{5}$$

where x = water colour (mg Pt l^{-1})
y = V/O quotient.

Transparency and water colour correlate in the following way:

$$y = 142.1 \ x^{-1.255}, \ n = 15, \ r^2 = 0.97, \tag{6}$$

where x = Secchi disk depth (m)
y = water colour (mg Pt l^{-1}).

Equation (6) is based on the data in Kokko *et al.* (1977) and Eloranta (1978) from highly coloured oligotrophic, mesotrophic and polluted lakes, and unpublished data from Lake Päijänne and the very clear Lake Inarinjärvi. Combining equations (5) and (6) and including data from the eutrophic Lake Jyväsjärvi (Granberg 1972) to correct the upper part of the curve yields:

$$y = 1.282 \ x^{-1.172}, \ n \doteq 10, \ r^2 = 0.91, \tag{7}$$

where x = Secchi disk depth (m)
y = V/O quotient.

The V/O quotient can thus be obtained from equation (7), but also from equation (5) with certain limitations (water colour between 20 and 80 mg Pt l^{-1}, oligotrophic or mesotrophic lakes).

Division of the maximum daily primary production per unit volume by the V/O quotient gives the daily integral production. The growing season can be considered to last from the 15th of May until the end of September in the study area. However, in September primary production is only half that calculated due to the decreased intensity of light (see Granberg 1973).

To verify the method presented, we calculated the seasonal *in situ* primary production per m^2 for four sampling sites of Lake Päijänne (Table 1) and compared the calculated values with the observed

Table 1. Calculated and observed results of chlorophyll *a* and integral *in situ* primary production per growing season in the four sampling sites of Lake Päijänne.

Year	Total phosphorus µg l⁻¹ observed	Water colour mg l⁻¹ Pt observed	V/O calculated	Chlorophyll a mg m⁻³ calculated	observed	Primary productivity mg C m⁻³ day⁻¹ calculated	Primary production g C m⁻² growing season⁻¹ calculated	observed
Poronselkä								
1972	32	53	0.62	13.0	–	361	43.2	37.1
1973	30	55	0.65	12.0	–	318	37.2	33.7
1974	27	64	0.75	10.0	–	258	27.3	23.0
1975	24	69	0.81	8.5	–	205	20.9	12.2
1976	30	51	0.60	12.0	12.0	318	40.3	23.7
1977	25	55	0.64	9.0	7.6	222	28.3	19.3
Ristiselkä								
1972	23	49	0.54	8.0	–	188	29.3	24.5
1973	22	43	0.50	7.5	–	173	29.6	29.0
1974	23	48	0.56	8.0	–	188	28.3	27.0
1975	19	65	0.76	6.1	–	129	15.4	13.7
1976	22	46	0.53	7.5	–	172	27.9	34.8
1977	22	50	0.58	6.4	8.4	140	21.6	14.8
1978	19	45	0.53	6.1	5.0	129	22.1	20.4
Tehinselkä								
1972	9	34	0.39	2.1	–	30	9.2	9.2
1973	10	32	0.37	2.4	–	37	11.4	7.5
1974	9	30	0.35	2.1	–	30	10.2	13.7
1975	14	42	0.49	3.9	–	71	14.7	17.8
1976	12	38	0.44	3.1	4.0	52	12.8	20.6
1979	11	25	0.29	2.8	3.3	44	17.2	25.0
Asikkalanselkä								
1972	9	30	0.35	2.1	–	30	10.1	5.2
1973	12	29	0.33	3.1	–	52	17.1	9.1
1974	8	30	0.34	1.7	–	24	8.7	9.0
1975	11	31	0.36	2.7	–	44	9.6	13.6
1976	11	33	0.38	2.8	2.8	44	12.9	11.6
1977	9	34	0.39	2.1	2.4	30	9.2	5.0
1978	10	29	0.33	2.4	3.2	37	12.8	18.9
1979	10	22	0.25	2.4	2.8	37	16.8	10.3

132

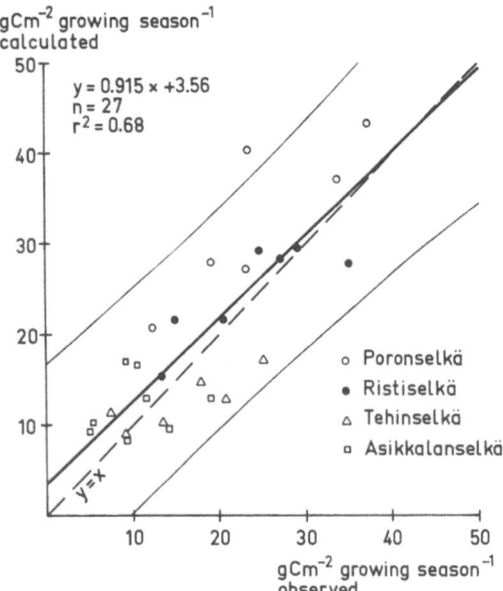

$y = 0.915 x + 3.56$
$n = 27$
$r^2 = 0.68$

o Poronselkä
● Ristiselkä
△ Tehinselkä
□ Asikkalanselkä

gCm^{-2} growing season^{-1}
calculated

gCm^{-2} growing season^{-1}
observed

Fig. 1. Calculated and observed values of integral phytoplankton production per growing season for the four sampling sites of Lake Päijänne. Limits shown are 95% confidence limits for individual points around the regression line.

ones (Fig. 1). The differences between the calculated and the observed values are mainly due to hydrological and climatic conditions which cannot be taken into account. These factors may largely determine the annual primary production, especially in oligotrophic lakes (Ilmavirta 1975). More work is needed to improve the predictions, especially equation (4), which has a particularly low r^2 value. It must also be stated that the regional application of this kind of statistical method is restricted. However, considering that the method presented has been developed for practical water pollution studies to indicate rapidly and cheaply the major biological consequences of changes in effluent loads, we feel that the accuracy of the predictions is adequate.

References

Carlson, R. E., 1977. A trophic state index for lakes. Limnol. Oceanogr. 22: 361–369.

Chapra, S. C. & Tarapchak, S. J., 1976. A chlorophyll model and its relationship to phosphorus loading plots for lakes. Wat. Resour. Res. 12: 1260–1264.

Dillon, P. J. & Rigler, F. H., 1974. The phosphorus–chlorophyll relationship in lakes. Limnol. Oceanogr. 19: 767–773.

Eloranta, P., 1978. Light penetration in different types of lakes in Central Finland. Holarct. Ecol. 1: 362–366.

Granberg, K., 1972. Kasviplankton- ja perustuotantotutkimus Päijänteellä v. 1971. (Summary: Primary production and phytoplankton during 1971 in Lake Päijänne, Central Finland). Jyväskylä Hydrobiol. Res. Inst., Rep. '21: 1–70.

Granberg, K., 1973. The eutrophication and pollution of Lake Päijänne, Central Finland. Ann. bot. fenn. 10: 267–308.

Granberg, K., 1979a. Etelä-Päijänteen rehevöitymisuhan selvitys. (Summary: The eutrophication of Southern Päijänne). University of Jyväskylä, Hydrobiol. Res. Centre, Rep. 105: 1–35.

Granberg, K., 1979b. Äänekoski-Vaajakoski-vesireitin velvoitetarkkailu v. 1978. (Summary: Limnological studies of the watercourse of Äänekoski, Central Finland, in 1978). University of Jyväskylä, Hydrobiol. Res. Centre, Rep. 102: 1–15.

Harjula, H., 1979. Analysis of errors in estimating phytoplankton primary productivity and chlorophyll *a* with special reference to Lake Päijänne. Ann.bot. fenn. 16: 307–337.

Ilmavirta, V., 1975. Dynamics of phytoplankton production in the oligotrophic Lake Pääjärvi, southern Finland. Ann. bot. fenn. 12: 45–54.

Jones, J. R. & Bachmann, R. W., 1976. Prediction of phosphorus and chlorophyll levels in lakes. J. Wat. Pollut. Control Fed. 48: 2176–2182.

Kokko, H., Hakkari, L., Nyrönen, J. & Granberg, K., 1977. Lievestuoreenjärven kalataloudellisista käyttömahdollisuuksista. University of Jyväskylä, Hydrobiol. Res. Centre, Rep. 92: 1–158.

Rodhe, W., 1958. Primärproduktion und Seetypen. Verh. Verein. Limnol. 13: 121–141.

Sakamoto, M., 1966. Primary production by phytoplankton community in some Japanese lakes, and its dependence on lake depth. Arch. Hydrobiol. 62: 1–28.

Smith, V. H., 1979. Nutrient dependence of primary productivity in lakes. Limnol. Oceanogr. 24: 1051–1064.

Vollenweider, R. A., 1968. Water management research. OECD, Paris. DAS/SCI/68.27. Mimeogr. 183 pp.

An oxygen model for Lake Haukivesi

T. Frisk

National Board of Waters, P.O. Box 250, 00101 Helsinki 10, Finland

Keywords: water quality, model, dissolved oxygen, BOD, total phosphorus, phytoplankton biomass

Abstract

A simple water quality model for Lake Haukivesi, heavily loaded by pulp and paper mill effluents, has been developed. The main purpose of the model is to predict the concentration of dissolved oxygen in the hypolimnion. The lake is divided into seven sub-basins, and also into epilimnion and hypolimnion. Transfers between sub-basins are calculated using water balance equations. The state variables of the model are dissolved oxygen concentration, biochemical oxygen demand, phytoplankton biomass, and total phosphorus concentration. The effect of temperature on reaction rate coefficients has been taken into account. Temperature is calculated in the model using a second degree polynomial function. The processes affecting hypolimnetic oxygen consumption are BOD decay, decomposition of phytoplankton, benthic oxygen demand, and decomposition of slowly decaying organic matter.

Introduction

Predictions of water quality are necessary to aid planning and supervision. The development of water quality models for these purposes has been rapid over recent years. A mass balance approach must be used whenever evaluations of the cause and effect relationships between water quality and loading or discharge are to be made. In principle the correct model for describing the mass balances of a water body would consist of a set of three-dimensional second order partial differential equations. The biological and physico-chemical part of the model should include all the most important components of the ecosystem. In practice the models used for predictions are crude simplifications of the ideal three-dimensional model. The hydraulic part of the ideal model can be simplified in many different ways. In Finland, the one-dimensional model of Chen & Orlob (1972), in which vertical differences of water quality are taken into account, has been tested (National Board of Waters, Finland

1978) and further developed (Kinnunen *et al.* 1980). A two-dimensional one-layer model (called VENLA) has been developed by the National Board of Waters based on the models of Virtanen (1977, 1978). Most lake models applied in Finland have been simple phosphorus and oxygen models (e.g. Lappalainen 1974, 1975).

Lake Haukivesi (for general data see e.g. Lappalainen 1974) is heavily loaded by effluents from the wood-processing industry situated in the town of Varkaus. In Lake Haukivesi horizontal differences in water quality must be taken into consideration in predictions made for practical purposes. The lake is thermally stratified, and making predictions of the hypolimnetic oxygen concentration is more important. An oxygen model for Lake Haukivesi was developed in which the lake was divided into seven sub-basins. Both epilimnion and hypolimnion have been considered. The model allows the effects of BOD and phosphorus loading on water quality to be studied.

Hydrobiologia 86, 133–139 (1982).   0018-8158/82/0862-0133/$01.40.

Description of the model

A water quality model can be divided into a hydraulic part and a water quality part. The structure of the hydraulic part of the oxygen model for Lake Haukivesi is based on division of the lake into two layers and into sub-basins (Fig. 1). In the epilimnion the following mass balance equation can be written for the sub-basins:

$$\frac{dc_i^e}{dt} = \frac{Q_{i-1}^e}{V_i^e} c_{i-1}^e - \frac{Q_i^e}{V_i^e} c_i^e - \frac{k_{thi} A_{thi}}{V_i^e z_{thi}} (c_i^e - c_i^h)$$
$$+ S_i^e (c) + \frac{\Delta I_i^e(c)}{V_i^e} \qquad (1)$$

The corresponding equation in the hypolimnion for the sub-basins is

$$\frac{dc_i^h}{dt} = \frac{Q_{i-1}^h}{V_i^h} c_{i-1}^h - \frac{Q_i^h}{V_i^h} c_i^h + \frac{k_{thi} A_{thi}}{V_i^h z_{thi}} (c_i^e - c_i^h)$$
$$+ S_i^h (c) + \frac{\Delta I_i^h(c)}{V_i^h} \qquad (2)$$

where superscript e = epilimnion,
superscript h = hypolimnion,

c_i = concentration of the substance considered in sub-basin i,
Q_i = outflow from sub-basin i,
V_i = average volume of sub-basin i,
$S_i (c)$ = rate of change of concentration c due to biological and physico-chemical (= non-hydraulic) processes inside sub-basin i,

$\Delta I_i (c)$ = additional input of the substance considered into sub-basin i,
A_{thi} = average surface area of the thermocline of sub-basin i,
$\bar{z}_{thi}$ = average thickness of the thermocline of sub-basin i,
k_{thi} = vertical exchange coefficient in sub-basin i.

In the model it has been assumed that there is a hydrological steady state. The volumes of the sub-basins are assumed to be constant. The outflow values, which are calculated from water-balance equations, change simultaneously in the whole system.

The vertical exchange coefficient (k_{th}) can be estimated on the basis of temperatures in the epilimnion and hypolimnion. Application of the method in two-layer oxygen models has been presented by Lappalainen (1978) and Frisk and Kylä-Harakka (1980).

The state variables of the model are dissolved oxygen, biochemical oxygen demand (BOD), total phosphorus concentration, and phytoplankton biomass. In addition, temperature is calculated by the model, but it must be regarded as an external variable rather than a state variable. The structure of the water quality part of the model is presented in Fig. 2.

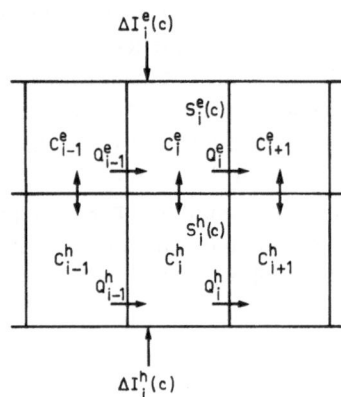

Fig. 1. The structure of the hydraulic part of the model.

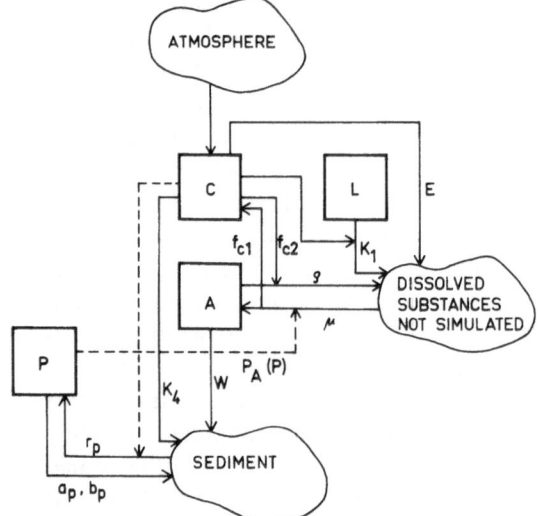

Fig. 2. The structure of the water quality part of the model. C = dissolved oxygen concentration, L = BOD, A = phytoplankton biomass, P = total phosphorus concentration, ⟶ = transfer of a substance from a store to another, - - - → = effect of a substance on a process.

The S(c)-term in equations (1) and (2) describe the rates of change of the different state variables due to biological and physico-chemical processes. Through S(c)-terms the state variables may affect each other. Phytoplankton is simulated in the model since it has a considerable effect on the concentration of dissolved oxygen, even though predictions of phytoplankton dynamics can be inexact if simple structures as in this model are used. The equations for phytoplankton biomass are as follows:

epilimnion

$$S^e(A) = \mu^e \, A^e - \rho^e \, A^e - w^e \, A^e \qquad (3)$$

and

hypolimnion

$$S^h(A) = \frac{V^e \, A_{th}}{V^h \, A_0} \, w^e \, A^e - \rho^h \, A^h - w^h \, A^h, \quad (4)$$

where A = phytoplankton biomass,
μ = growth rate coefficient, dependent on temperature and phosphorus concentration,
ρ = phytoplankton respiration rate coefficient, dependent on temperature,
w = phytoplankton settling coefficient, dependent on temperature.

The second terms on the right hand side of equations (3) and (4) describe the effect of endogenous respiration, predation, and bacterial decomposition of algae. The third terms describe the effect of phytoplankton settling and the first term in equation (4) describes the increase of phytoplankton in hypolimnion due to settling from the epilimnion. Phytoplankton growth is assumed to be possible only in the epilimnion and growth is assumed to be phosphorus-limited. This dependence is described as a Monod function:

$$\mu(P) = \mu_{max} \, \frac{P_A(P)}{KP + P_A(P)}, \qquad (5)$$

where μ_{max} = maximal growth rate coefficient,
KP = half saturation constant of phosphorus,

$P_A(P)$ = theoretical concentration of phosphorus available to phytoplankton, described as a function of total phosphorus concentration.

Orthophosphate does not represent phosphorus available to phytoplankton in most lakes in Finland where much phosphorus is combined with the colloidal and dissolved fractions of humic substances. The interrelation between humic substances and the different forms of phosphorus are not well known, but a chemical equilibrium between them can be assumed to prevail. A simple relation between available phosphorus and total phosphorus was used in the model:

$$P_A(P) = P - (\alpha_P + \beta_P \, P) - f_P \, A \qquad (6)$$

where α_P and β_P =empirical coefficients for calculating the fraction of total phosphorus that is not available to phytoplankton,
f_P = phosphorus content of phytoplankton

For total phosphorus the following equations can be written:

$$S^e(P) = - a_p^e \, (P^e) \, b_p^e + \frac{A_0 - A_{th}}{V^e} r_p^e \qquad (7)$$

and

$$S^h(P) = - a_p^h \, (P^h) b_p^h + \frac{A_{th}}{V^h} r_p^h + \frac{V^e \, A_{th}}{V^h \, A_0} a_p^e \, (P^e) \, b_p^e$$

$$\qquad (8)$$

where P = total phosphorus concentration,
a_p and b_p = empirical phosphorus settling coefficients; a_p is dependent on temperature,
A_0 = surface area at depth z = 0,
A_{th} = surface area at the depth of thermocline,
r_p = the rate of phosphorus release per surface area of the sediment, dependent on temperature and dissolved oxygen concentration.

The first terms on the right hand side of equations (7) and (8) describe the effect of settling of total phosphorus and the second terms the effect of phosphorus release from the sediment. The third term on the right hand side of equation (8) describes the transfer of phophorus from epilimnion to hypolimnion due to settling.

The dependence of phosphorus release on dissolved oxygen concentration has been described in a simple, not very realistic way:

$$r_P(C) = r_{P0}, \text{ when } C \geqslant C_{CRP} \tag{9}$$

$$r_P(C) = r_{P1}, \text{ when } C < C_{CRP}, \tag{10}$$

where C_{CRP} = critical dissolved oxygen concentration for phosphorus release from the sediment,

r_{P0} and r_{P1} = parameters.

The value of r_{P1} is much greater than the value of r_{P0}.

Biochemical oxygen demand (BOD) is assumed to follow first order reaction kinetics:

$$S^e(L) = -K_1^e L^e \tag{11}$$

and

$$S^h(L) = -K_1^h L^h, \tag{12}$$

where L = ultimate BOD,
      K_1 = BOD decay coefficient, dependent on temperature.

In the model the processes affecting the concentration of dissolved oxygen are BOD decay, reaeration, growth and respiration of phytoplankton, benthic oxygen demand, and oxygen consumption due to decomposition of slowly decaying substances:

$$S^e(C) = -K_1^e L^e + K_2^e (C_s^e - C^e) + \mu^e A^e f_{C1} -$$

$$\rho^e A^e f_{C2} - \frac{A_0 - A_{th}}{V^e} K_4^e - E^e \tag{13}$$

and

$$S^h(C) = -K_1^h L^h - \rho^h A^h f_{C2} - \frac{A_{th}}{V^h} K_4^h - E^h, \tag{14}$$

where K_2 = reaeration coefficient, dependent on temperature,
    C_s = saturation coencentration of dissolved oxygen,
    f_{C1} = mg O_2 produced / mg phytoplankton grown,
    f_{C2} = mg O_2 consumed / mg phytoplankton respired,
    K_4 = benthic oxygen demand coefficient (mg O_2 day^{-1}m^{-2}), dependent on temperature,
    E = the rate of change of oxygen concentration due to decomposition of slowly decaying substances, dependent on temperature.

The dependence of the saturation concentration on temperature can be expressed as a second degree polynomial function, for example.

Most coefficients in the model are dependent on temperature. The exponential function of Streeter & Phelps (1925) does not give a successful description of the dependence (e.g. Kinnunen et al. 1981). In the model the temperature correction function developed by Frisk & Nyholm (1980) was used:

$$K(T) = K(T)_s \exp \int_{T_s}^{T} \ln \Theta(T) \, dT, \tag{15}$$

where $\Theta(T) = a_T + b_T T$ (16)

When equation (16) is applied, equation (15) can be integrated analytically:

$$\int_{T_s}^{T} \ln \Theta(T) \, dT = (\frac{a_T}{b_T} + T)(\ln(a_T + b_T T) - 1)$$

$$-(\frac{a_T}{b_T} + T_s)(\ln(a_T + b_T T_s) - 1), \tag{17}$$

where K(T) = value of the reaction rate coefficient at temperature T,
    T_s = standard temperature (20 °),
    a_T and b_T = empirical constants, characteristic of each reaction rate coefficient.

The temperature at different times is needed when calculating the actual values of reaction rate coefficients from the values at 20 °C. The summer-time temporal variation of temperature in Finnish lakes can be approximated by a second degree parabola (Lappalainen 1978):

$$T^e = k_0^e + k_1^e t + k_2^e t^2 \qquad (18)$$

and

$$T^h = k_0^h + k_1^h t + k_2^h t^2 \qquad (19)$$

where t = time from the beginning of simulation, where k_0, k_1, and k_2 = parameters.

Discussion

A water quality model can be either empirical or fundamental (Snodgrass 1979). An empirical model involves a statistical relationship between two or more variables. A fundamental model is based on three principles: the law of conservation of mass, the principle of kinetics, and the principle of stoichiometry. The law of conservation of mass is taken into account in writing the basic equations (= mass balance equations). The principle of kinetics means that the reaction rates are expressed as proportional to concentrations or products of several concentrations. According to the principle of stoichiometry the relationships between different substances are expressed by constant coefficients. The two latter principles are applied in writing the formulae that describe the biological and physico-chemical processes (= S(c) terms). The possibility of building a general model applicable to all lakes does not seem realistic. However, there are some widely applied equations for describing the processes of a lake that can be included in different models. The final choice of the equations of a water quality model must be made on the basis of the characteristic features of the lake or the lakes to which the model will be applied.

The model for Lake Haukivesi resembles in many respects the one developed by Lappalainen (1978) for a project carried out by the National Board of Waters, Finland (1978). The kinetics of BOD decay and reaeration are assumed to follow first order equations as in the classical river model of Streeter & Phelps (1925). Lappalainen (1978) has presented different versions for summer and winter conditions. In the winter version the BOD-DO equations have been solved analytically. In the summer version the horizontal differences of water quality have not been taken into account. The lake has been considered as two continuously-stirred-tank-reactor elements connected by vertical diffusion. In the model of Lappalainen (1978) phytoplankton is not simulated as a state variable. The effects of phytoplankton on oxygen consumption are taken into account using a eutrophication factor based on an empirical relation between phosphorus retention and hypolimnetic oxygen concentration (Lappalainen 1974, 1975). In the model for Lake Haukivesi the equation of phytoplankton biomass resembles the corresponding equation in the river model QUAL II (Norton *et al.* 1974). The main difference is that in QUAL II orthophosphate is used instead of total phosphorus. There are some methodological difficulties in determining the coefficients α_P and β_P that are used in calculating the phosphorus fraction that is available to phytoplankton from total phosphorus values. Bioassays may be applicable if the limitations of the bioassay methods are taken into account and the results can be interpreted in a sufficiently realistic way. A rough estimate of α_P and β_P can be obtained by means of regression analysis if simultaneous measurements of total phosphorus, orthophosphate, and phytoplankton biomass are available. The concentration of the particulate phosphorus in phytoplankton can be estimated by multiplying the phytoplankton biomass by a stoichiometric coefficient.

The model for Lake Haukivesi has been tested using data from 1978. These tests suggest that the basic structure of the model is adequate for making water quality prognoses for practical purposes. The model will be further tested, applied, and developed in the joint project of the National Board of Waters and the national Finnish Fund for Research and Development.

Special attention must be paid to two points. In the model the mass balance of organic matter and phosphorus in the sediment has not been considered. A submodel for sediment will be employed to increase the realism of the model. In the present model BOD is a measure of organic matter from effluents. The oxygen consumption due to slowly decaying organic matter has been taken into account using a separate term that must be calibrated. In the future slowly decaying organic matter will be simulated as a state variable in both water and sediment.

Abbreviations

A	=	phytoplankton biomass
A_0	=	surface area at depth $z = 0$
A_{th}	=	surface area of the thermocline
C	=	dissolved oxygen concentration
C_{CRP}	=	critical dissolved oxygen concentration for phosphorus release from the sediment
C_s	=	saturation concentration of dissolved oxygen
E	=	rate of change of dissolved oxygen concentration due to decomposition of slowly decaying organic substances
K_1	=	BOD decay coefficient
K_2	=	reaeration coefficient
K_4	=	benthic oxygen demand coefficient
KP	=	half saturation constant of phosphorus
L	=	ultimate BOD
P	=	total phosphorus concentration
P_A	=	concentration of phosphorus available to phytoplankton
Q	=	outflow
$S(c)$	=	rate of change of concentration c due to non hydraulic processes
T	=	temperature
T_s	=	standard temperature
V	=	volume
a_P	=	phosphorus settling coefficient
a_T	=	regression constant between Θ and T
b_P	=	phosphorus settling exponent
b_T	=	regression coefficient between Θ and T
c	=	concentration (general symbol)
e	=	as upper index: epilimnion
f_{C1}	=	mg oxygen produced per mg phytoplankton grown
f_{C2}	=	mg oxygen consumed per mg phytoplankton respired
h	=	as upper index: hypolimnion
i	=	as downer index: order number of the hydraulic element
k_{th}	=	vertical exchange coefficient
k_0	=	parameter for calculating temperature
k_1	=	parameter for calculating temperature
k_2	=	parameter for calculating temperature
r_P	=	rate of phosphorus release per surface area of the sediment
r_{P0}	=	r_P when dissolved oxygen concentration is greater than the critical concentration C_{CRP}
r_{P1}	=	r_P when dissolved oxygen concentration is smaller than the critical concentration C_{CRP}
t	=	time
w	=	phytoplankton settling coefficient
z_{th}	=	average thickness of the thermocline
ΔI	=	additional input
Θ	=	temperature correction coefficient
α_P	=	coefficient for calculating the fraction of phosphorus that is not available to phytoplankton
β_P	=	coefficient for calculating the fraction of phosphorus that is not available to phytoplankton
μ	=	growth rate coefficient of phytoplankton
μ_{max}	=	maximal growth rate coefficient of phytoplankton
ρ	=	respiration rate coefficient of phytoplankton.

References

Chen, C. W. & Orlob, G. T., 1972. Ecologic simulation for aquatic environments. Final report for the Office of Water Resources Research, U.S. Department of the Interior. Water Rescources Engineers, Inc., Walnut Creek, California.

Frisk, T. & Kylä-Harakka, T., 1980. Mathematical methods for predicting dissolved oxygen concentration and phytoplankton biomass in lakes loaded by waste waters of the pulp and paper industry. Aqua fenn. 10: 21–31.

Frisk, T. & Nyholm, B., 1980. The effect of temperature on reaction rate coefficients in water quality models. Vesitalous 5: 24–27 (Finnish, English summary).

Kinnunen, K., Nyholm, B., Niemi, J., Frisk, T., Kylä-Harakka, T. & Kauranne, T., 1981. Final report of the joint project between the National Board of Waters and IBM, Finland in the years 1978-1980. Publ. Nat. Bd Wat. Finl. (in press).

Lappalainen, K. M., 1974. Predictions of water quality with different loading alternatives, the Kallavesi watercourse and Lake Haukivesi. Rep. Bd Wat. Finl. 59: 1–84 (Finnish).

Lappalainen, K. M., 1975. Phosphorus loading capacity of lakes and a mathematical model for water quality prognoses. Nordforsk Miljövårdssekreteriatet. Publ. 1975 (1): 425–441.

Lappalainen, K. M., 1978. An oxygen model for lakes. Rep. Nat. Bd Wat. Finl. 149: 1–57 (Finnish).

National Board of Waters, Finland, 1978. Final report for the International Bank for Reconstruction and Development of the research project carried out in 1975-1978 by the National Board of Waters. Publ. Nat. Bd Wat. Finl. 26. 153 p. Helsinki.

Snodgrass, W. J., 1979. Predictive water quality models for the Great Lakes. Some capabilities and limits. In: Scavia, D. & Robertson, A. (eds) Perspectives on Lake Ecosystem Modeling, pp. 171-191. Ann Arbor Science, Michigan.

Streeter, H. W. & Phelps, E. B., 1925. A study of the pollution and natural purification of the Ohio river. Public Health Bull. 146. U.S. Public Health Service. 75 p.

Virtanen, M., 1977. Numerical one-layer models describing the transfer and dilution of waste waters. The National Finnish Fund for Research and Development (SITRA), YVY esitutkimus 28: 1-45 (Finnish).

Virtanen, M., 1978. Possibilities of predicting the physical behaviour of a lake system. In: Nyroos, H. (ed.) Water Quality Models and their Practical Applications, pp. 143-155. Vesi- ja Kalatalousmiehet ry:n koulutuspäivät 1977, Helsinki (Finnish).

PART THREE

Water Management

Limnological research can improve and reduce the cost of monitoring and control of water quality

Curt Forsberg

Institute of Limnology, Box 557, 751 22 Uppsala, Sweden

Keywords: water quality, monitoring, control, limnological research needs

Abstract

An evaluation of the programmes for control and monitoring of water quality in Sweden has demonstrated many short-comings. This paper discusses how improvements can be achieved by using new limnological knowledge.

High-frequency sampling and analysis of mixed samples gives reliable average values with reduced costs. Strong correlations between, for instance, summer average values and summer maximum values, can be used for predicting the worst possible situation, which would be difficult to analyse by low-frequency sampling. Phosphorus load and lake response relationships are valuable tools for lake management and for establishing water quality criteria in physical planning. Future control may be further improved if limnological research gives better information on relationships between structure and function of aquatic ecosystems in relation to the form and size of the water body, water flushing rate, inputs from watershed and air space, etc.

Introduction

During a comparatively short period of time many natural waters in Sweden have received increasing pollution from municipalities, industries and agriculture. Several effects of this pollution are well known. Parallel to the water deterioration, investigations were initiated to map distribution of pollution and to study its effects on water bodies. Programmes for monitoring and control were developed. Depending mainly on the obligation under the Environment Protection Act 1969 to arrange a control programme when permission for water-polluting activity is obtained, the number of these programmes has increased rapidly during the last ten years. However, an evaluation of the control and monitoring activities demonstrated many shortcomings (Ryding 1977).

This paper discusses how the control can be improved immediately using existing limnological knowledge, and also how further improvements can be obtained by limnological research directed at relationships between loading of elements and compounds, and structure and function of aquatic ecosystems.

Evaluation of the present programmes

No central guidelines for monitoring and control of water bodies and water quality exist today. This means that great variations in sampling, analysing and reporting are involved in the control activities. Furthermore, no limnological analysis of the basic prerequisites for this control has been presented. The programmes seem to have been formulated from praxis parallel to great efforts being made to combat different pollution problems. The first discussion evaluating part of the recipient control activities was presented by Forsberg & Ryding (1975), followed by more comprehensive investigations by Ryding (1977, 1979).

Hydrobiologia 86, 143–146 (1982). 0018-8158/82/0862-0143/$00.80.
© Dr W. Junk Publishers, The Hague.

It was found that the programmes as a rule had no defined objective and further, that physical-chemical investigations were often based on a large number of sampling stations, where water sampling was performed only a few times/year, often 2–4 times. As there are great variations in water quality, such low-frequency sampling cannot give reliable information on average values, nor on the worst possible situation. Often analyses were performed following a standard set of parameters which restricted the possibility to study special conditions.

Biological investigations often resulted in comprehensive lists of species, difficult to understand by people outside the small group of experts. In running waters, unsuitable or often no information on water flow limited the value of water quality data. Knowledge of water flow, of transports of elements and the quantitative roles among different pollution sources is an important prerequisite for making the right water protection decisions. Finally it was found that results were often presented and reported only as untreated primary data. Summarized, it can therefore be stated that poor information was obtained at disproportionally high costs.

This evaluation did not include water quality control where water is used for specific purposes, e.g. for drinking water or industrial process water. Nor did it include monitoring programmes for studies of long-term changes due to changes of, e.g., climate, airborne pollution or human customs of different types.

Improvements by using new limnological experience

Concerning the shortcomings listed above, the main interest will here be focused on the problem of obtaining reliable information on water quality and knowledge of relationships between loading and lake response.

Reliable average values

Comprehensive studies on recovery of polluted lakes in Sweden (e.g. Forsberg et al. 1975), have used high-frequency water sampling to demonstrate great variations and often irregular patterns among water quality parameters, especially in shallow lakes and smaller rivers (e.g. Forsberg & Ryding

1975, 1979; Ryding & Forsberg 1976, 1977, 1979; Forsberg 1978).

In the River Fyris, continuous sampling every second hour throughout a three-month period gave the following weekly average values:
Discharge: $0.5–10 \text{ m}^3 \text{ s}^{-1}$
Total-N: $0.84–4.20 \text{ mg l}^{-1}$
Total-P: $25–77 \ \mu\text{g l}^{-1}$
Organic matter (COD): $28–52 \text{ mg l}^{-1}$

Analysis of every single sample will of course enlarge these ranges. In any case it seems impossible to study water quality in this type of water body by using the normal low-frequency sampling.

In order to improve the reliability of the data, Forsberg & Ryding (1975) suggested more frequent sampling but analyses only of mixed samples covering a certain period, e.g. one month, the growing period, or the whole year. Samples from running waters were mixed in proportion to the flow. The reliable average values then obtained were also used for calculations of transports of elements (see e.g. Forsberg & Ryding 1979). This method of using more frequent sampling, at strategically chosen sampling stations, and analyses of mixed samples, where every single sample has been preserved by deep-freezing (see e.g. Forsberg et al. 1975) has been tested for eight years in a number of different lakes and rivers. This approach, giving reliable average values, will also reduce the costs, especially if sampling can be combined with other routines running in a municipality. Thus, in the municipality of Uppsala the cost of an old control programme was reduced from 40 000 to 6 000 Sw. Cr per year when analyses were performed with mixed samples taken from strategically chosen sampling stations (Forsberg 1979).

By statistical evaluation of sampling frequencies, Ryding (1980a) proposed a bi-monthly sampling for achieving a ±10% precision regarding the monthly average situation for both loading and water quality aspects in flowing waters.

Predicting the worst possible situation

Average values are useful for evaluating changes of water quality, e.g. after water protection measures such as introduction of advanced wastewater treatment for phosphorus removal. However, average values do not reveal anything about the worst possible situation, for instance during algal blooms causing the well-known problems for water-works operators. Well founded criticism could therefore easily arise from the use of average proposed by Forsberg & Ryding (1975).

When analysing correlations between eutrophication parameters using data from the high-frequency sampling programme mentioned above, Forsberg & Ryding (1980) found very strong linear correlations between summer average and summer maximum values. For total-N, total-P, chlorophyll *a* and transparency (here minimum value) the correlation coefficients ranged from 0.95 to 0.99. This type of correlation, first reported by Jones *et al.* (1979), can be expected as a certain base level is necessary for having high peak values. These findings seem to open the way for predicting the worst possible situation by use of reliable average figures. Recently Ryding (1980a) complemented these correlation studies with results on other parameters that supported the original finding. For lake water, 16 parameters analysed generally gave correlation coefficients above 0.95. Similar coefficients were also obtained for flowing waters, here based on annual averages.

The results of these correlation studies demonstrate that reliable average values covering a certain period of time can be used for predicting the maximum value, or the worst possible situation during this period, a situation difficult to analyse by low-frequency sampling.

Load–lake response relationships

In order to achieve a better understanding of, e.g., dynamics of transports of elements to lakes, quantitative roles of loadings from different type of sources and relationships between load and lake response, much more information is required on the size and rhythm of discharge. Unfortunately most monitoring programmes of today do not include this information. This situation however must be improved. When better information on loading conditions becomes available new possibilities will appear for control and management of water bodies.

Interesting steps in this direction have been taken during the 1970s. The development of nutrient loading criteria, in particular the phosphorus loading lake water response relationship (Vollenweider 1968, 1975, 1976), has been the base for

comprehensive studies coordinated by OECD with the aim of refining and extending this knowledge. Comprehensive regional reports have been prepared (Rast & Lee 1978; Clasen 1980; Fricker 1980; Ryding 1980b) which will be followed by an overall OECD synthesis report (Vollenweider & Kerekes, in prep.).

The results from these reports demonstrate, among other things, that statistical phosphorus loading lake response models are valuable both for assessing expected effects of changed loading conditions and for establishing phosphorus loading-water quality criteria in physical planning. However, before using a data set for this purpose, some basic prerequisites must be met (see Ryding 1980b, Fig. 51).

Another approach for future control

In the future the methods for health control of water bodies must be based on an extended limnological knowledge of how a water body reacts to different environmental influences. To achieve this point, limnological research has to concentrate on studies giving increased information of structure and function of the aquatic ecosystem in relation to its form, size, water flushing rate, inputs from catchment and air-space, etc.

More attention should also be paid to correlations between different parameters. A composite lake model developed by Ryding (1980b) indicates new openings for monitoring. When basic investigations have demonstrated that a lake conforms the number of analyses can be reduced, as information then can be obtained by existing correlations.

Once these tools have been acquired, qualified limnologists can help to improve control and monitoring at reduced costs. By using limnological experiences during a limited period of time, meaningless random sampling year after year can be abandoned. This will save money. When authorities have emissions under control, resulting in no marked change of pollution load, then the range of water quality variation will be mainly dependent on climatic variation. Unfortunately many control programmes are continuously describing only this variation.

Grossly simplifying the matter, the health control of water bodies can be compared to health control

146

of human bodies. Medical research and experience here demonstrates what the human body can withstand, and how it reacts to, for instance, too much food, toxic compounds or other harmful influences. The conditions of the human body are permitted to vary to a great extent, without repeated controls year after year.

Compared to medicine, limnology is a young science however, which at least partly explains the discrepancies outlined. Limnologists have mostly concentrated on the water body itself, 'a kind of supra-organism with a way of life of its own, governed and expressed by metabolic reactions in response to previous and prevailing conditions' (Rodhe 1979). Therefore much remains to study until this supra-organism can be better considered in relation to its environment. For limnology, it will be important to find principles and general relationships applicable to water management. The time will then be ripe for qualified limnologists to be wholly responsible for the water bodies' health control, to be their physicians.

References

Clasen, J., 1980. OECD cooperative programme for inland waters. Shallow lakes and reservoirs. The Water Research Centre, Medmenham Laboratory, P.O. Box 16, Medmenham, Marlow, Bucks., England, 1–246.

Forsberg, B., 1978. Phytoplankton in Lake Utthan before and after sewage diversion. Nat. sw. Envir. Prot. Bd. P.M. 1029, pp. 1–51.

Forsberg, C., 1979. Monitoring of effluent from municipal waste deposition. Nordforsk Publikation 1979: 2 (in Swedish).

Forsberg, C. & Ryding, S.-O., 1975. Some ideas on municipal recipient control. Vatten 31: 347–358 (in Swedish).

Forsberg, C. & Ryding, S.-O., 1979. Correlations between discharge and transport of elements. Studies from six catchment areas. Vatten 35: 291–300.

Forsberg, C. & Ryding, S.-O., 1980. Eutrophication parameters and trophic state indices in 30 Swedish waste-receiving lakes. Arch. Hydrobiol. 89: 189–207.

Forsberg, C., Ryding, S.-O. & Claeson, A., 1975. Recovery of polluted lakes. A Swedish research programme on the effect of advanced wastewater treatment and sewage diversion. Water Res. 9: 51–59.

Fricker, H., 1980. OECD eutrophication programme. Regional project. Alpine lakes. Swiss Federal Board for Environmental Protection (Bundesamt für Umweltschutz), CH-3003 Bern, Switzerland, 1–234.

Jones, R. A., Rast, W. & Lee, G. F., 1979. Relationship between summer mean and maximum chlorophyll a concentrations in Lakes. Am. chem. Soc. 13: 869–870.

Rast, W. & Lee, G. F., 1978. Summary analysis of the North American (US portion) OECD eutrophication project: Nutrient loading – lake response relationships and trophic state indices. EPA-600/3-78-008, Jan. 1978.

Rodhe, W., 1979. The life of lakes. Arch. Hydrobiol. Beih. Ergebn. Limnol. 13: 5–9.

Ryding, S.-O., 1977. Evaluation of water quality control. Nat. sw. Envir. Prot. Bd., P.M. 868, pp. 1–180 (in Swedish).

Ryding, S.-O., 1979. Shortcomings in present monitoring and control programmes. Possibilities of improvements. Nordforsk Publikation 1979: 2, 103–114 (in Swedish).

Ryding, S.-O., 1980a. Optimization of monitoring programmes by means of general correlations between standard water quality variables. M.S. presented at SIL Symp., Kyoto, Japan, Aug. 1980.

Ryding, S.-O., 1980b. Monitoring of inland waters. OECD eutrophication programme. The Nordic project. Nordforsk Publikation 1980: 2, 1–207.

Ryding, S.-O. & Forsberg, C., 1976. Six polluted lakes: a preliminary evaluation of the treatment and recovery processes. Ambio 5: 151–156.

Ryding, S.-O. & Forsberg, C., 1977. Sediments as a nutrient source in shallow polluted lakes. In: Golterman, H. L. (ed.) Interactions between Sediments and Fresh Water, pp. 227–234. Junk, The Hague.

Ryding, S.-O. & Forsberg, C., 1979. Nitrogen, phosphorus and organic matter in running waters. Studies from six drainage basins. Vatten 35, 46–58.

Vollenweider, R. A., 1968. Scientific fundamentals of the eutrophication of lakes and flowing waters with particular reference to nitrogen and phosphorus as factors in eutrophication. OECD DAS/CSI/68.27. Paris.

Vollenweider, R. A., 1975. Input–output models with special reference to the phosphorus loading concept in limnology. Schweiz. Z. Hydrol. 37: 53–84.

Vollenweider, R. A., 1976. Advances in defining critical loading levels for phosphorus in lake eutrophication. Mem. Ist. Ital. Idrobiol. 33: 53–83.

The limnological basis for planning water quality management

Ingvar Lundqvist

Institute of Limnology, University of Lund, P.O. Box 3060, S-220 03 Lund, Sweden

Keywords: limnology, water quality, planning procedures

Abstract

The planning of water quality management needs to be organized in a rational manner, based on sound limnological and ecological principles. Criteria for management need to be clearly defined. Management plans must be formulated in the contex of activities and interests in the whole catchment area, not just those within the boundaries of water bodies. A procedure is outlined for approaching water quality management decisions.

Introduction

Comprehensive demands for different natural resources will continue for many years to come.

Management of water resources involving an ecological approach must not only produce descriptive documents, but also create criteria that allow estimation of both prevailing and desired water quality. This is even the case in estimating water resource capacity and the consequences of different water requirements.

The main goal that challenges limnologists working in close contact with the planning process is to call attention to ecological principles of importance for water conservation, water restoration and planning water quality management. In the future, water quality management quidelines will be the most important.

Water conservation guidelines

As severe losses in European flora and fauna have been suffered by freshwater ecosystems, water conservation quidelines have been important for many years.

Maximization and optimization of stability, diversity and productivity (as commercial fisheries) are desirable goals in management of water resources, but the eutrophication process may conflict with some of these aims. The degree of stability is of importance in making prognoses about the reaction of ecosystems to environmental and human impacts.

Stability can refer to both environmental stability and community stability (Fig. 1). There are strong reasons for believing that high environmental stability leads to high diversity, and this is in turn an expression of a high degree of specialization in the biota (Pielou 1975). But this diversity can also cause more adaptation problems and less tolerance for stress, so there is also a possible tendency for high diversity to lessen community stability and to have an effect opposite to that of environmental stability.

Water resources can only maintain their existing level of quality if the state of the surrounding catchment is maintained. If the primary goal is conservation, the area included must have a future status affording strict legal protection.

Hydrobiologia 86, 147–151 (1982). 0018-8158/82/0862-0147/$01.00.

148

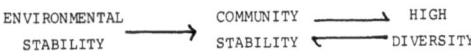

Fig. 1. Determinants of diversity in an ecosystem (Pielou 1975).

Example

Only four lakes in Sweden are colonized by the spring-spawning cisco, *Coregonus trybomi* (Svärdsson 1979). This species may soon remain in only two lakes, Lake Fegen and Lake Ören, however, since the *C. trybomi* stocks of the other two lakes in which it occurs, Lake Hålsjön and Lake Åsunden, are threatened with extinction because of pollution.

Local authorities wanted to build tourist villages around Lake Fegen. It was pointed out that the oxygen budget and nutrient loading would be detrimentally affected by exploition, however, and the regional authority refused to approve the plans (Lundqvist 1975). The spawning area for *C. trybomi* in the deepest part of the lake could easily be destroyed by anaerobic conditions during summer stratification. In the beginning of 1980 Lake Fegen and its surroundings became a nature conservation area.

In describing water resources, it is not suffient to use a scale where the rare and unique are ranked ahead of the widespread and general. Therefore, ecological rules which consider both cultural and scientific nature conservancy planning should protect:
1. Representative as well as unique water resources.
2. Water resources of importance to discrete species of flora and fauna.
3. Water resources that represent an aesthetic value in the landscape.
4. Water resources of importance for scientific reference and educational needs.

Lake restoration guidelines

The stability of lake ecosystems is to a great extent a result of morphological and hydrological conditions. Morphometric factors are dominant in determining trophic type and productivity in larger lakes. Under natural circumstances, ordered energy flow through the entire system produces low

entropy. Increasing nutrient loading raises the entropy level and checks succession in the lake ecosystem with destroyed niches, extinction of organisms and rapid nutrient circulation as a result. When organic loading to the sediment increases to such an extent that the redox potential is depressed, then the lake receives an internal loading during anaerobic conditions.

Example

The need for more arable land in Sweden, especially during the last century, caused very drastic changes in many lakes due to drainage and lowering of the surface level. The drained lakes became more sensitive to later nutrient loading from urban sewage etc.

In the Stockholm region, where nearly all lakes have been lowered and polluted, the sewerage system today mostly consists of a rock tunnel system. Five treatment plants were built with their outlets in Lake Mälaren and the Baltic Sea. The former lake recipients in the urban area have been freed from most of the sewage, but there are still problems to solve before the lakes are again in good conditions (Ripl & Lundqvist 1977; Lundqvist & Gavelin 1980).

Lake Trekanten in Stockholm has a totally urbanized catchment. The lake is used for recreational purposes. Some external nutrient loading still remains from storm water outlets. During the former recipient period the sediment redox potential decreased, and the sediment acted as a nutrient source. The internal phosphorus loading was four times greater than the external laoding during 1976 (Fig. 2). During the summer period, blue-green algae were the dominant phytoplankton. There was no bottom fauna, and the fish population was influenced by earlier fish kills. Eutrophication directed the natural succession in the lake backwards. Environmental stability, and hence community stability, was adversely influenced, and the diversity of phytoplankton, bottom fauna and fish was drastically reduced.

Man must and can help damaged lake ecosystems in the purification process (Björk 1968, 1980; Ripl & Lundqvist 1977). Lake restoration guidelines should include an overall assessment of the needs and conditions for restoration measures in urbanized areas. This can be done in two stages,

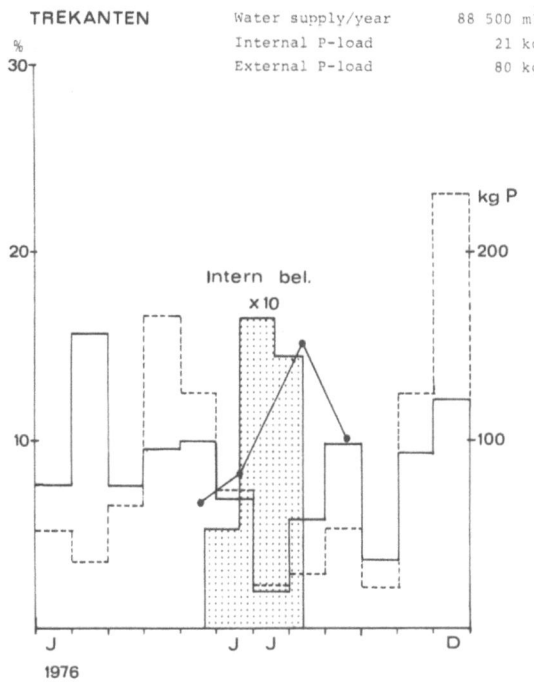

TREKANTEN
Water supply/year   88 500 m³
Internal P-load   21 kg
External P-load   80 kg

Fig. 2. Comparison of external and internal P loading and water supply to Lake Trekanten (Stockholm) during 1976. Monthly distribution (%) of annual contribution of water (---) and phosphorus (—). Net release from the sediment (change in the total P content of the lake minus external loading) is expressed in % of the external P loading. ●—● designates the total P content of the lake in kg. (Ripl & Lundqvist 1977).

where the first stage is comprised of three steps:
1. Determine the value of the lake for recreation.
2. Determine the needs for areas and facilities for recreation around the lake.
3. Determine the limnological conditions for the restoration work.

On the basis of these three factors, the lakes studied can be divided into two main groups:
1. Lakes where detailed limnological investigations can begin with the object of determining suitable restoration methods.
2. Lakes where further studies must be made to monitor ecosystem development (water quality and biota).

The second stage consists of detailed limnological investigations in accordance with the goals of the restoration work.

Water quality management guidelines

The historical development of water resources planning procedures in Sweden can roughly be separated into two periods: (1) up to the end of the 1960s there have been survey-type plans, mainly to insure water supply to urban areas and industries, and (2) from the beginning of the 1970s guidelines have focused on water conservation plans. Problems of sewage water treatment and recipient usage have been dealt with in such plans. However, by the mid-1970s human impact in some areas had risen to such a level that water allocation problems became both quantitative and qualitative. In such areas there is often a need for comprehensive water resources planning, i.e. a multifunctional planning that integrates all aspects of water quantity, water quality, water use, and related land and urban planning into a well-balanced whole for the benefit of all legitimate interests. In other areas water quality management may be enough where specific human requirements have to be approved in relation to their water quality impact, the background quality for the region and future water quality needs. In the future, therefore, more efforts must be devoted to planning guidelines instead of to only ranking lakes according to those conservation items and restoration guidelines mentioned above.

Incorporation of water resources into land-use plans is the first step in both water conservation and planning water quality management. Human water requirements in different land-use categories, such as urban areas, rural areas and areas for outdoor recreation, have their own specific water quantity and quality requirements. Once the planning area has been delineated into distinct functional (urban, agricultural, forest, etc.) areas, it is possible to begin an assessment of the existing condition of the total land and water environment. Land use changes among the three categories can also be considered in estimation of human water requirements in the future as well as in the past.

In Figure 3 some stages in a model for planning water quality management are arranged. The stages are logically grouped in a flow-chart.

STAGE A: *Resource characteristics*
Integrate existing data on the planning area:
1. Land and water use, population, industries and pollution.

150

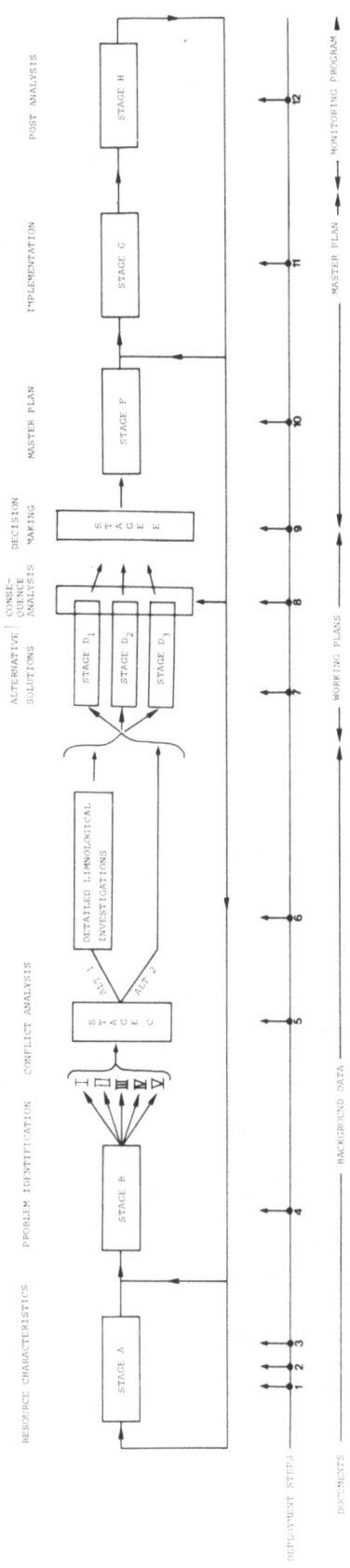

Fig. 3. A model for planning water quality management.

151

2. Geological and ecological investigations in the terrestrial surroundings, flora, fauna organism communities, etc.
3. Limnological investigations (structure and function of inland water ecosystems).

STAGE B: *Problem identification*

The actual planning for a discrete waterbody in the area starts with identification of problems. This requires a description of the system, beginning with requirements for water quality and with the aims of water users as important ingredients. The different water quality requirements must be understood in relation to the limnological classification of the water resources. The planners have to design a program to insure the necessary future water quality.

STAGE C: *Conflict analysis*

This stage shows the planners if more detailed limnological investigations are required for prognoses. Conflict analysis must take into account both administrative problems and the scientific problems described in stage B. For example, in the planning process the administrative and legal boundaries are not the same as the hydrological. Detailed limnological investigations are tailor-made and give a problem-oriented basis for the continued planning process. Prognoses are made for the different types of influence.

STAGE D: *Planning document (s)*

The planning document consists of three parts:
1. System design or identification of the problems.
2. Problem solution.
3. Consequence analysis.
For each problem there are often one or more possible alternative solutions, and for each a consequence analysis must be made.

STAGE E: *Decision making*

The weighing of alternatives is a political process based on the documentation of the analysis. The choice is directed by water quality criteria written to correspond to established goals. Documentation of

the analysis should include all relevant factors in a way that is comprehensible to the decision makers.

STAGE F: *Master plan*

STAGE G: *Implementation*

STAGE H: *Post analysis*

Analysis of effects is made so that consequences of changes for water users and for water quality can be taken into consideration in the future planning process. It is of general value to obtain a record of how the planning model used corresponded in reality to the goals expressed.

Concluding remarks

It must be emphasized that human use of water resources must be done in a way that gives future generations freedom of choice in their water requirements. Physical planning has to direct the development of society within the limits of the sensitivity of natural resources.

References

Björk, S., 1969. Methods and scientific problems in restoration of lakes. Vatten 24: 57–71 (in Swedish).
Björk, S., 1980. Restoration of degraded lake ecosystems. In: Duncan, N. & Rzoska, J. (eds.) Land Use Impacts on Lake and Reservoir Ecosystems, pp. 196–219. MAB 5 Workshop held at Warsaw, Poland, May 26–June 2, 1978.
Lundqvist, I., 1975. Limnological investigations of lakes Fegen and Kalvsjön. Institute of Limnology, Lund. Memeogr. 240 pp. (in Swedish).
Lundqvist, I., 1980. Water quality management planning and lake restoration in Stockholm, Sweden. Vatten 36: 219–230 (in Swedish).
Pielou, E. C., 1975. Ecological Diversity. John Wiley, London. 165 pp.
Ripl, W. & Lundqvist, I., 1977. Restoration program for lakes in Stockholm, Sweden. Institute of Limnology, Lund. 267 pp (in Swedish).
Svärdsson, G., 1979. Speciation of Scandinavian Coregonus. Report, Institute of Freshwater Research, Drottningholm 57: 1–95.

The phosphorus economy of a hypertrophic seepage lake in Scania, south Sweden groundwater influence

Magnus Enell

Institute of Limnology, University of Lund, P.O.Box 3060, S-220 03 Lund, Sweden

Keywords: lake water, phosphorus, sedimenting material, groundwater, water and nutrient budget, hypertrophy

Abstract

The phosphorus dynamics and economy of Lake Bysjön, a hypertrophic seepage lake in Scania, southern Sweden, were investigated during 1973–1977. The mean dissolved inorganic phosphorus concentration (1973–1977) was 580 $\mu g \cdot l^{-1}$. There were no correlations between dissolved inorganic P, total organic P, dissolved organic P, particulate P and phytoplankton biomass. Groundwater inflow and lake water outflow through the ground are the most important factors for maintaining a constant water volume. Groundwater seepage is also important for water quality. Groundwater inflow, together with planktonic activity, keeps the P concentration high in the lake water.

Introduction

Groundwater input to lakes is poorly understood, and usually referred to as a quantitatively and qualitatively unknown factor. Recently some authors (Trew *et al.* 1978; Schaffer & Oglesby 1978; McColl 1978) have suggested that groundwater may represent a significant source of nutrients in certain watersheds. Few studies have attempted to quantify groundwater inflow (Lee 1977; Lock & John 1978). Nitrogen and phosphorus loading calculations for Lake Tahoe made by Vollenweider (1968), Dugan & McGauhey (1974) and Goldman (1974) were shown by Loeb & Goldman (1979) to be invalid because of the omission of groundwater sources. Lee *et al.* (1980) presented a detailed documentation of diffuse groundwater inflow through a lakebed, and their study revealed several aspects of surface-groundwater interactions. Advection and dispersion can be important processes in lakebed environments. Winter (1978) has presented a theoretical work that should lead to a clearer understanding of the interrelationship of lakes and groundwater.

Diffuse seepage into lakes is an important source of water, and knowledge of bulk groundwater inflow is essential for predicting regeneration and recycling of materials from the sediment outside the minerogenic littoral zone.

The main purpose of the present work was to clarify the phosphorus economy in Lake Bysjön, a hypertrophic seepage lake in southern Sweden. The large and rapid variations in chemical composition and physical characteristics of the water were investigated together with *in situ* relationships between biomass, productivity and species distribution of bacteria, phyto- and zooplankton (Coveney *et al.* 1977; Coveney *et al.* 1979).

Materials and methods

Phosphorus concentrations were determined in the lake itself (water, interstitial water and sediment), in sedimenting material and in the groundwater entering the lake. Laboratory experiments, made under aerobic and anaerobic conditions with undisturbed sediment–water systems showed the P

Hydrobiologia 86, 153–158 (1982). 0018-8158/82/0862-0153/$01.20.

release–adsorption properties under different environmental conditions. The results obtained were used to explain the high phosphorus concentration of lake water and to clarify the present phosphorus economy of the lake ecosystem.

Methods for phosphorus analyses, interstitial water separation from whole sediment and sampling procedures for sedimenting material are published elsewhere (Enell 1980). Groundwater level measurements have been made monthly in the area since 1948 by the Malmö waterworks. This is done in drilled holes S and SE of Lake Bysjön, as well as at one point 1.2 km N of the lake. Groundwater for phosphorus analyses was taken in wells 200–500 m from the lake.

The groundwater investigation by waterworks personnel in the Lake Bysjön area was started because of infiltration of Lake Vombsjön water to the drinking water supply for the cities of Malmö and Lund.

Results and discussion

Geology

Lake Bysjön (N 55°4′, E 13°33′) is one of the smallest lakes in the so-called Vomb depression. The bedrock is dominated by cretaceous rocks from the Senon period, and the quaternary deposits overlying the bedrock consist mainly of gravel and sand (Fig. 1). Clay layers occur locally. The lakes in

the Vomb depression are all considered to be of the kettle type. Among these, Lake Bysjön is a typical example with steep slopes above and below the water line and a nearly circular shoreline (shoreline development = 1.08). Lake Bysjön is comparable to the pot-hole lakes in the Ericson Area in Canada (Barica, pers. com.).

Lake Bysjön – a seepage lake

There is no surface drainage to the lake, and due to a lowering of the water table at the end of the 19th century, the outlet functions only during short periods (1–2 weeks) of high water.

Some morphometric and hydrological data are as follows:

Surface area	0.12 km²	Altitude	22.0	m a.s.l.
Water volume	0.42 Mm³	Drainage area	0.8	km²
Epilimnion	0.35 Mm³	Village	18	%
Hypolimnion	0.07 Mm³	Farmland	23	%
Maximum depth	8.3 m	Lake	17	%
Mean depth	3.6 m	Shoreline development	1.08	

Lake water

There is usually a stable temperature stratification in the lake during summer, and from the middle of May to the middle of September the hypolimnion (4–8 m) is anoxic.

The mean value for dissolved inorganic phosphorus (DIP) in the epilimnion was 580 $\mu g \cdot l^{-1}$ (range 260–1 000 $\mu g \cdot l^{-1}$) for the 5 investigated years. There are no sewage discharges from private houses, farms, industries or other sources into the lake. The surrounding cultivated land is fertilized with super-phosphate.

Mean ($\bar{x}$) and range (R, minimum and maximum values) of selected phosphorus fractions in the epi- and hypolimnion are shown in Table 1.

During the spring diatom outburst there was a 40 to 65% reduction in the DIP concentration of the epilimnion from the winter maximum. Subsequent mineralization of the algae involved a liberation of DIP. Zooplankton development, succeeding the diatom bloom, seemed to be unaffected by DIP fluctuations and vice versa. A minimum in DIP concentration was reached simultaneously with or some days after the summer maximum of bluegreen algae, which was always ended by a collapse.

High phytoplankton photosynthesis induced a

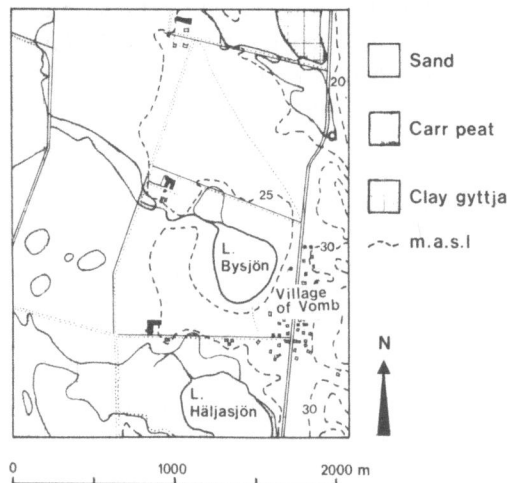

Fig. 1. Soil deposits in the area surrounding Lake Bysjön.

Table 1. Lake Bysjön 1973–1977. Mean (x̄) and rage (R, minimum and maximum values) of selected phosphorus fractions in the epi- en hypolimnion. DIP = dissolved inorganic P, TOP = total organic P, DOP = dissolved organic P, TP = total P, TDP = total dissolved P, PP = particulate P.

Factor			1973	1974	1975	1976	1977
DIP	µg/l						
Epilimnion		x̄	775	579	537	430	586
Hypolimnion		x̄	1 024 .	812	837	714	743
Epilimnion		R	759–1 000	304– 891	261– 694	305– 594	284– 632
Hypolimnion		R	757–1 213	510–1 512	441–2 046	333–1 393	306– 931
TOP	µg/l						
Epilimnion		x̄	–	191	209	152	60
Hypolimnion		x̄	–	182	368	140	100
Epilimnion		R	–	8– 601	83– 453	42– 228	30– 148
Hypolimnion		R	–	15–1 310	59–1 524	11– 296	61– 457
DOP	µg/l						
Epilimnion		x̄	–	127	132	97	14
Hypolimnion		x̄	–	37	58	74	28
Epilimnion		R	–	2– 272	33– 441	11– 145	14– 86
Hypolimnion		R	–	0– 285	0– 515	0– 144	0– 196
TP	µg/l						
Epilimnion		x̄	–	770	746	582	646
Hypolimnion		x̄	–	994	1 205	854	843
Epilimnion		R	818– 950	509–1 219	499–1 013	455– 744	350– 730
Hypolimnion		R	837–1 364	653–2 358	500–3 181	461–1 672	376–1 334
TDP	µg/l						
Epilimnion		x̄	–	706	669	527	600
Hypolimnion		x̄	–	849	895	788	771
Epilimnion		R	–	394– 891	467– 961	371– 696	327– 713
Hypolimnion		R	–	569–1 565	451–2 247	333–1 393	306– 999
PP	µg/l						
Epilimnion		x̄	–	64	77	55	46
Hypolimnion		x̄	–	145	310	66	72
Epilimnion		R	–	1– 427	12– 193	10– 138	16– 89
Hypolimnion		R	–	2–1 260	4–1 009	17– 332	12– 397

pH increase, which sporadically included a biogenic Ca-carbonate precipitation. However, the DIP fraction and Ca did not decline simultaneously. Decrease in DIP concentration during bluegreen algal blooms was apparently more dependent on assimilation by the algae than on Ca-carbonate coprecipitation.

Annual variations in dissolved organic phosphorus (DOP) concentrations were characterized by irregular fluctuations without distinct maxima or minima. The same pattern was found in all the investigated years and has also been recognized by Barica (pers. com.). Hypolimnetic DIP represented 90–100% of total dissolved phosphorus (TDP) during the whole year, except for a short period in August–September when there was an increase in the DOP concentration. This was caused by mineralization of phytoplankton settling to the hypo-

156

limnion. Otherwise mineralization of algae occurred chiefly in the epilimnion.

Interstitial water, sediment and sedimenting material

Interstitial DIP is the sediment P fraction that is most sensitive to changes in environmental conditions and has the highest degree of mobility. Olsen (1958, 1964) concluded that bottom deposits must be regarded as PO_4 reservoirs, i.e. a kind of savings bank at which PO_4 will be withdrawn or deposited whenever the equilibrium is disturbed.

There was an increase in phosphorus in the hypolimnion during the anaerobic stratification period, but no significant correlation existed between P in sedimenting material and the increase in P in the hypolimnion and interstitial water.

In Table 2 some characteristic data on hypolimnetic oxygen demand, increase of DIP in interstitial water, aerobic and anaerobic nutrient exchange and accumulation via sedimenting material are given.

Accumulation of P in the hypolimnion was caused by a release from the sediment of about 27 mg DIP $\cdot$ m^{-2} $\cdot$ day^{-1} during summer stratification and an input from the epilimnion via sedimenting material of about 9.5 mg P $\cdot$ m^{-2} $\cdot$ day^{-1}. The dominant part of sedimenting material is incorporated into the sediment before regeneration of P occurs, which means that the hypolimnetic P increase is a result of release from the sediment (Enell 1978).

Table 2. Oxygen demand, increase in interstitial water P concentration, P exchange, accumulation of P and hypolimnetic P increase in Lake Bysjön for 1973–1977.

Oxygen demand 4 °C	0.40	g O$_2$/m^2 $\cdot$ day
8 °C	0.52	g O$_2$/m^2 $\cdot$ day
Decrease of oxygen in the hypolimnion immediately after stratification	0.31	g O$_2$/m^2 $\cdot$ day
DIP increase in the interstitial water during anaerobic stratification	52	μg/l $\cdot$ day
Nutrient exchange, anaerobic release	27	mg DIP/m^2 $\cdot$ day
aerobic release	7	mg DIP/m^2 $\cdot$ day
mean during the year	17	mg DIP/m^2 $\cdot$ day
Accumulation via the sedimenting material	9.5	mg P/m^2 $\cdot$ day
P increase in the hypolimnion during the anaerobic summer stratification 1974–1977	20	mg P/m^2 $\cdot$ day

Input of P to the hypolimnion via sedimenting material was comparable to a sedimentation of about 4.1 cm $\cdot$ yr^{-1}. This value is not corrected for resuspension, decomposition processes, zooplankton and macrobenthos grazing and consolidation (compaction). Corresponding calculations for dry matter and nitrogen in sedimenting material gave sedimentation rates of 2.7 and 2.6 cm $\cdot$ yr^{-1}, respectively. Preliminary results from a ^{137}Cs dating gave a sedimentation rate of 0.9 cm $\cdot$ yr^{-1}.

Groundwater and water budget

Groundwater levels and movements are well documented in the Vomb area. The direction of groundwater movement around Lake Bysjön is SE to NW through the lake. When the groundwater enters the lake, the flow turns to the north. During the 5-year period investigated the direction of groundwater has been the same, with minor alterations.

The uppermost layer of the soil deposits consists of sand, underlayered by clay. It is groundwater in the superficial sandy soil layer (thickness 10–25 m) that seeps into Lake Bysjön. Groundwater inflow into the lake occurs in the minerogenic littoral zone down to a depth of 2–2.5 m (Albing *et al.* 1975).

The water level in Lake Bysjön fluctuated around 21.5 m a.s.l. during 1973–1977. Fluctuation was dependent on the groundwater level in the area and changes in precipitation and evaporation. 95–97% of water added by precipitation on the lake surface was evaporated. The groundwater table seemed to be fairly constant during the study, with a variation of around 0.5–1 m.

Evaporation during June–August, 1973–1977, was significantly higher than the lake water volume decrease during the corresponding period, indicating the importance of groundwater inflow. Groundwater inflow and lake water outflow can be expected to be of about the same quantity from one year to another, if the groundwater level in the catchment area is constant.

During 1973, 1975 and 1976 outflow of groundwater was 61 290 m^3 more than inflow. In 1974 and 1977, however, inflow of groundwater exceeded the outflow by 27 230 m^3 (Table 3). The difference, calculated for a 5-year period, was 34 000 m^3, corresponding to 8.2% of the total lake volume and a water level decrease of 290 mm.

Table 3. Water budget (m³) for Lake Bysjön during 1973–1977, compared with some data for 1931–1960. ( )$^+$ means corresponding volume in mm.

	1973	1974	1975	1976	1977	1973–77	1931–60
Precipitation on the lake surface	70 630	74 550	60 350	62 660	76 970	69 010	68 090
Surfacewater input	0	0	0	0	0	0	0
Groundwater input	X_{1973}	X_{1974}	X_{1975}	X_{1976}	X_{1977}	X_{73-77}	X_{31-60}
Evaporation from the lake surface	65 090	63 120	70 630	63 120	62 550	64 850	67 280
Surfacewater output	0	0	0	0	0	0	0
Groundwater output	Y_{1973}	Y_{1974}	Y_{1975}	Y_{1976}	Y_{1977}	Y_{73-77}	Y_{31-60}
Change in water volume	-19 620	+17 310	- 5 770	-31 160	+35 770	- 3 460	?
Net precipitation June–August	-27 120 $(235)^+$	-18 460 $(160)^+$	-29 430 $(255)^+$	-23 540 $(204)^+$	-16 270 $(141)^+$	-22 960 $(199)^+$	-38 540 $(334)^+$
Net precipitation January–December	+ 5 540 $(+48)^+$	+11 430 $(+99)^+$	-10 280 $(-89)^+$	- 460 $(-4)^+$	+14 420 $(+125)^+$	+ 4 160 $(+36)^+$	+ 810 $(+7)^+$
Groundwater input minus Groundwater output	-25 160	+ 5 880	- 4 510	-31 620	+21 350	- 7 620	?

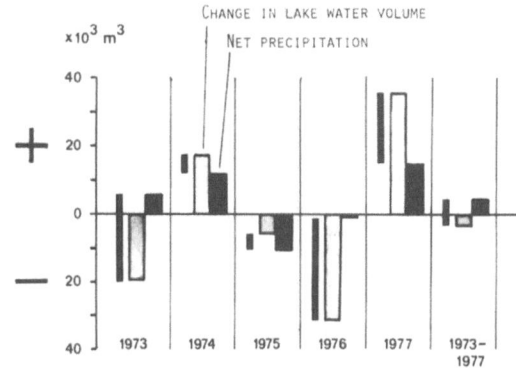

Fig. 2. Changes in lake water volume and net precipitation in Lake Bysjön 1973–1977. Solid black bars indicate differences between groundwater input and lake water outflow.

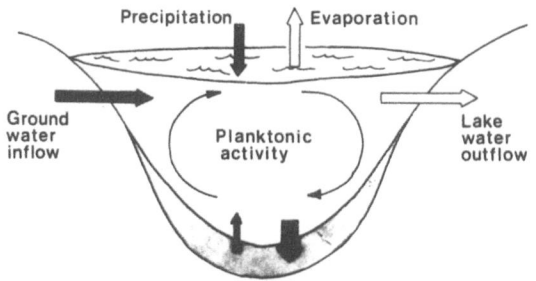

Fig. 3. Schematic function of Lake Bysjön as a groundwater-dependent lake as well as a nutrient trap, especially for phosphorus.

Groundwater inflow and lake water outflow through the ground are the most important factors maintaining a constant water volume in Lake Bysjön. The quantity of water flowing in and out was not measured, but the difference between inflow and outflow was calculated based on precipitation, evaporation and water level data. There was a marked variation during 1973–1977 concerning net precipitation and changes in lake water volume (Fig. 2). No connection exists between net precipitation and the change in lake water volume. Differences between groundwater inflow and outflow were linearly correlated with the change in lake water volume ($r^2 = 0.97$), indicating the importance of groundwater seepage.

In the present investigation, it is suggested that the lake water ecosystem functioned as a nutrient trap, especially for P. Groundwater inflow, together with planktonic activity kept the P concentration high in lake water (Fig. 3). This function as a nutrient trap is represented by a concentration P_1 of phosphorus in the groundwater inflow and a concentration P_2 in lake water seeping out of the lake. The decrease in P concentration ($P_1 > P_2$), when groundwater is 'passing through' the lake water body, is caused by planktonic assimilation and adsorption to particulate material. Long periods with anoxic hypolimnetic conditions keep the sediment reduced and, consequently, the phosphate in solution.

158

References

Albing, P., Sandkvist, J. & Tilly, L., 1975. The hydrogeology of the Vomb depression. An estimate of balance and attempt to evaluation. Dept. Geology, Inst. Tech., Univ. Lund. 35 pp., mimeo. (In Swedish).

Coveney, M. F., Cronberg, G., Enell, M., Larsson, K. & Olofsson, L., 1977. Phytoplankton, zooplankton and bacteria – standing crop and production relationships in a eutrophic lake. Oikos 29: 5–21.

Coveney, M. F., Cronberg, G. & Enell, M., 1979. Relationships between bacteria, phyto- and zooplankton. Research report. Mimeo. Inst. Limnol., Univ. Lund.

Dugan, G. L. & McGauhey, P. H., 1974. Enrichment of surface waters. J. Wat. Pollut. Control Fed. 46: 2261–2280.

Enell, M., 1978. Sedimentation of phosphorus during summer stratification in a eutrophic lake in Scania, Sweden. In: Hongve, D., Skogheim, O., Faafeng, B. & Laake, M. Interactions between Sediment and Water, pp. 30–33. 6th Nordic Symposium on Sediments, 9–12 March, 1978.

Enell, M., 1980. The phosphorus economy of a hypertrophic seepage lake in Scania, South Sweden. Diss. Inst. Limnol., Univ. Lund.

Goldman, C. R., 1974. Eutrophication of Lake Tahoe emphasizing water quality. EPA-660/3-74-034.

Lee, D. R., 1977. A device for measuring seepage flux in lakes and ertuaries. Limnol. Oceanogr. 22: 140–147.

Lee, D. R., Cherry, J. A. & Pickens, J. F., 1980. Groundwater transport of a salt tracer through a sandy lakebed. Limnol. Oceanogr. 25: 45–61.

Lock, M. A. & John, P. H., 1978. The measurement of groundwater discharge into a lake by a direct method. Int. Revul ges. Hydrobiol. 63: 271–275.

Loeb, S. L. & Goldman, C. R., 1979. Water and nutrient transport via groundwater from Ward Valley into Lake Tahoe. Limnol. Oceanogr. 24: 1146–1154.

McColl, R. H. S., 1978. Lake Tutira: the use of phosphorus loadings in a management study. N.Z. J. mar. Freshwater Res. 12: 251–256.

Olsen, S., 1958. The balance of phosphate between sediment and water in Lake Furesø. Experiments with radioactive phosphorus. Folia Limnol. Scand. 10: 39–96. (In Danish).

Olsen, S., 1964. Phosphate equilibrium between reduced sediments and water. Laboratory experiments with radioactive P. Verh. int. Ver. limnol. 15: 333–341.

Schaffer, W. R. & Oglesby, R. T., 1978. Phosphorus loadings to lakes and some of their responses. Part 1. A new calculation of phosphorus loading and its application to 13 New York lakes. Limnol. Oceanogr. 23: 120–134.

Trew, D. O., Beliveau, D. J. & Yonge, E. J., 1978. The Baptiste lake study. Summary report. Alberta Environment Pollution Control Division. Water Quality Control Branch. 105 pp.

Vollenweider, R. A., 1968. Scientific fundamentals of the eutrophication of lakes and flowing waters with particular reference to nitrogen and phosphorus as a factor in eutrophication. OECD. Report DAS/CIS/68.27.

Winter, T. C., 1978. Ground-water component of lake water and nutrient budgets. Verh. int. Ver. Limnol. 20: 438–444.

Changes in quality of Finnish inland waters revealed by grouping analysis

Reino Laaksonen

National Board of Waters, P.O. Box 250, SF-00101 Helsinki 10, Finland

Keywords: water quality, monitoring, lakes .

Abstract

Grouping analysis offers a fresh approach to monitoring the quality of inland water bodies. In this study it has been applied to certain parameters from a network of stations on lake deeps. General trends were sought by examining the distribution of the means for 1971–1977 among five classes created by grouping the means for 1965–1970. Marked changes in water quality are seen to have taken place, mainly in lakes whose condition was almost natural (loss of purity). These changes are indicated by differences in the distributions of the values for conductivity, alkalinity, total nitrogen and oxygen. The results suggest an increase in the importance of loading from non-point sources.

Introduction

Since the early 1960s, the Finnish Board of Waters has been responsible for extensive monitoring of inland waters in Finland, and has made continuous observations of their water quality, using a network of stations on rivers and lake deeps. The parameters monitored have been chiefly physico-chemical. After initial use, the results are stored in registers, and the data collected in this way are occasionally scrutinized for possible trends, chiefly with the aid of regression analysis and sliding means (e.g. Laaksonen & Wartiovaara 1973; Laaksonen & Malin 1980). An alternative approach is offered by grouping analysis.

Methods

Grouping analysis is used to delimit groups of observations in such a way that the variance within the groups is as small as possible, and the variance between the groups is as great as possible. An iterative procedure is used, i.e. the optimal solution is searched for mechanically. For the present study, grouping analysis was applied to the mean results of measurements made at the observation stations on lake deeps in 1965–1970. The final number of groups was five.

Results and discussion

Tables 1 and 2 show the groups or classes obtained and the percentage distribution of the observation stations among these groups. The general impression is that the observation stations are concentrated at the 'clean end', i.e. in groups 1 and 2, where the concentrations of dissolved substances are small (Table 1). There is no doubt that the majority of the, generally large, lake basins included in study are still comparatively unaffected. Group 5, seen in Table 2 to have a mean share of 3–4% of the observation stations, may be considered to represent waters that, for many practical purposes, may be classified as polluted. Among the many trends revealed by the results of the grouping is a general increase in water purity from west to

Table 1. Classes formed by grouping analysis based on means of observations made on lake deeps in 1965–1970.

		Groups				
		1	2	3	4	5
O$_2$ sat.	%	>81	81–73	72–61	60–33	<33
γ_{25}	mS m^{-1}	<3.3	3.3–4.9	5.0–8.1	8.2–13.7	>13.7
Alkal.	mval. l^{-1}	<0.10	0.10–0.16	0.17–0.29	0.30–0.62	>0.62
KMnO$_4$-cons.	mg l^{-1}	<9.4	9.4–16.0	16.1–31.5	31.6–65.4	>65.4
Tot.N	mg l^{-1}	<0.2	0.2–0.4	0.5–0.7	0.8–1.5	>1.5
Tot.P	μg l^{-1}	<11	11–24	25–54	55–137	>137
Tot.S	mg l^{-1}	<1.7	1.7–3.1	3.2–6.4	6.5–12.8	>12.8

Table 2. Percentage distribution of the observation stations among the groups in 1960s and 1970s.

		Groups				
		1	2	3	4	5
		%				
O$_2$	1965–70	38	23	17	15	7
	1971–77	32	25	21	16	6
γ_{25}	1965–70	20	41	24	10	5
	1971–77	11	39	35	9	6
Alkal.	1965–70	19	42	31	6	2
	1971–77	16	59	17	6	2
KMnO$_4$-cons.	1965–70	55	35	6	2	2
	1971–77	55	33	9	1	2
Tot.N	1965–70	15	34	40	9	2
	1971–77	1	45	39	13	2
Tot.P	1965–70	51	26	12	7	4
	1971–77	52	27	16	4	1
Tot.S	1965–70	19	44	26	7	4
	1971–77	17	42	31	7	3

east and from south to north.

The analysis was used to identify changes in water quality by examining the distribution of the corresponding mean values for 1971–1977 among the groups obtained for 1965–1970.

Comparison of the proportions of the observation stations belonging to the different groups in the 1960s and the 1970s reveals changes in water quality that may be regarded as important, in view of the fact that these waters are generally the central basins in large lakes (Table 2). The changes, which mainly consist of an increase in dissolved substances, are found in previously undisturbed waters, chiefly from groups 1 and 2. In contrast, few changes are apparent in the polluted and disturbed waters of groups 4 and 5, with the exception of some high values for total phosphorus. The same observation has been made earlier in other circumstances, and has been termed 'quality blurring' (e.g. Laaksonen 1974). Such changes in the quality of Finnish waters in the last decade represent a loss of purity rather than an increase in pollution, and indicate loading from non-point sources.

Evidence of change was found in shifts in the distribution of the values for electrolytic conductivity, alkalinity, total nitrogen and oxygen. An increase in dissolved salts has proved to be a general index of change in various connections (e.g. Laaksonen & Wartiovaara 1973). The grouping of the stations based on the values for alkalinity also deviates considerably from the corresponding grouping in the sixties: a clear shift is evident from the middle group to adjoining group with lower values. The marked increase in the values for total nitrogen and the smaller changes in the oxygen concentration mainly relate to the waters in a natural condition.

References

Laaksonen, R. & Wartiovaara, J., 1973. Vesistöjen veden laadun muutoksista 1960-luvulla. Summary: Changes in water quality in water courses in the 1960s. Publ. Wat. Res. Inst. 6.

Laaksonen, R., 1974. Veden laadun rakenteesta. Summary: On the factor structure of water quality. Publ. Wat. Res. Inst. 9.

Laaksonen, R. & Malin, V., 1980. Vesistöjen veden laadun muutoksista vuosina 1962–1977. Summary: Changes in water quality in Finnish lakes and rivers 1962–1977. Publ. Wat. Res. Inst. 36.

Muddy odour: a problem associated with extreme eutrophication

Per-Edvin Persson
University of Helsinki, Department of Limnology, Viikki, SF-00710 Helsinki 71, Finland

Keywords: flavour, geosmin, 2-methylisoborneol, *Oscillatoria agardhii,* actinomycetes, hypereutrophy

Abstract

The results of a 4-year study in a hypereutrophic bay, Kaupunginselkä Bay at Porvoo on the south coast of Finland, indicated a significant correlation between muddy odour in bream *(Abramis brama)* and the amount of the blue-green alga *Oscillatoria agardhii* in the phytoplankton. This algal strain has previously been shown to produce the muddy-smelling compound geosmin. The numbers of muddy-smelling actinomycetes in the water and sediments of the study area were not clearly related to muddy odour in fish, nor to phytoplankton biomass. In the hypereutrophic L. Tuusulanjärvi, muddy odour in bream and pikeperch *(Stizostedion lucioperca)* was also related to the amounts of blue-green algae in the phytoplankton.

Introduction

Muddy or earthy odours in drinking water supplies and fish for human consumption are a world-wide problem affecting the utilization of aquatic resources. Some species of blue-green algae and actinomycetes (mainly streptomycetes) have been cited as a source of muddy odour in natural waters (Silvey & Roach 1975; Persson 1977; Gerber 1979). These organisms are capable of producing the muddy-smelling compounds geosmin and 2-methylisoborneol. Geosmin and methylisoborneol have been detected in natural waters (see Persson 1979) and in muddy-smelling fish (Yurkowski & Tabachek 1980). Muddy odour has generally been reported from eutrophic or hypereutrophic environments (Cees *et al.* 1974; Tabachek & Yurkowski 1976; Persson 1979).

This paper discusses the causes of muddy odour in some species of fish in two extremely eutrophic waters in southern Finland.

Material and methods

The primary study area during 1976–1979 was an extremely eutrophic brackishwater bay, Kaupunginselkä Bay at Porvoo, on the south coast of Finland. The area and the methods used have been described elsewhere (Persson 1979, 1980a). In 1978, muddy odour in bream, pike *(Esox lucius)* and pikeperch from the hypereutrophic L. Tuusulanjärvi was studied. For a synopsis of the lake, see Ryding (1980) and Vakkuri (1980). Phytoplankton data for the lake were obtained from the Helsinki City Water Works.

Results and discussion

In Kaupunginselkä Bay, there was a significant $(P < 0.05)$ correlation between muddy odour in bream and the amount of *Oscillatoria agardhii* in the phytoplankton (Fig. 1). In previous work, geosmin production by this strain has been proved (Persson 1979). Other muddy-smelling strains of this alga are also known (Persson 1980b). Muddy

Hydrobiologia 86, 161–164 (1982).   0018–8158/82/0862–0161/$00.80.

162

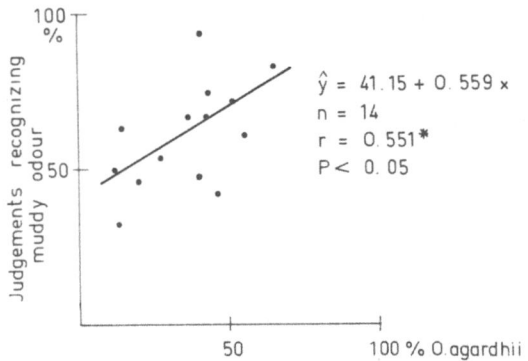

$\hat{y} = 41.15 + 0.559 \, x$
$n = 14$
$r = 0.551^{*}$
$P < 0.05$

Fig. 1. Relationship between adjudged muddy odour in bream and the percentage *O. agardhii* in the total phytoplankton biomass in Kaupunginselkä Bay at Porvoo. Data from 1976–1979.

odour in bream from the study area did not seem to depend on the absolute biomass of *O. agardhii*, but rather on the proportion of this alga in the phytoplankton. This may have been a result of several factors affecting the uptake of muddy odour by the fish, e.g. sedimentation and adsorption effects (Lovell 1973). Since uptake of muddy odour by fish is facilitated when the fish are feeding (Persson & York 1978), factors affecting the ingestion of algae by the fish would also be important. The results may also reflect a variability in the odour production by *O. agardhii*, as reported for some other species of blue-green algae (Henley 1970; Leventer & Eren 1970), and discussed by

Persson (1980b). The phytoplankton data for 1977–1979 (Fig. 2) from the study area indicated that muddy odour in bream occurred during the decline of the biomass peak of *O. agardhii*. However, it should be noted that the fish supply was irregular, and that fish were not always obtained during peak biomasses.

In 1977 and 1978, the numbers of muddy-smelling actinomycetes in the water of the study area were not related to muddy odour in bream (Fig. 2). In 1979, actinomycetes might have contributed to the muddy odour in fish, but it should be noted that during all three years a similar early summer maximum of actinomycetes occurred. Persson (1979, 1980b) interpreted this phenomenon as a result of the spring flood, i.e. wash out from soils. Streptomycetes are soil inhabitants (Lechevalier 1974), and are considered relatively inactive in lakes (Johnston & Cross 1976). In 1977 and 1978, the early summer maxima of actinomycetes probably did not reflect an active growth, as the prevailing water temperatures were low: in 1977, < 12 °C, and in 1978, < 15 °C. At such temperatures, streptomycetes would probably not be actively growing (Silvey & Roach 1975). In 1979, the water temperature in early summer was 17–19 °C, with possible active growth of actinomycetes.

In 1977 and 1978, the number of actinomycetes in the study area was not related to phytoplankton biomass (Fig. 2). In 1979, the concurrence of maximal phytoplankton biomass and actinomycetes may have been coincidental, the number of acti-

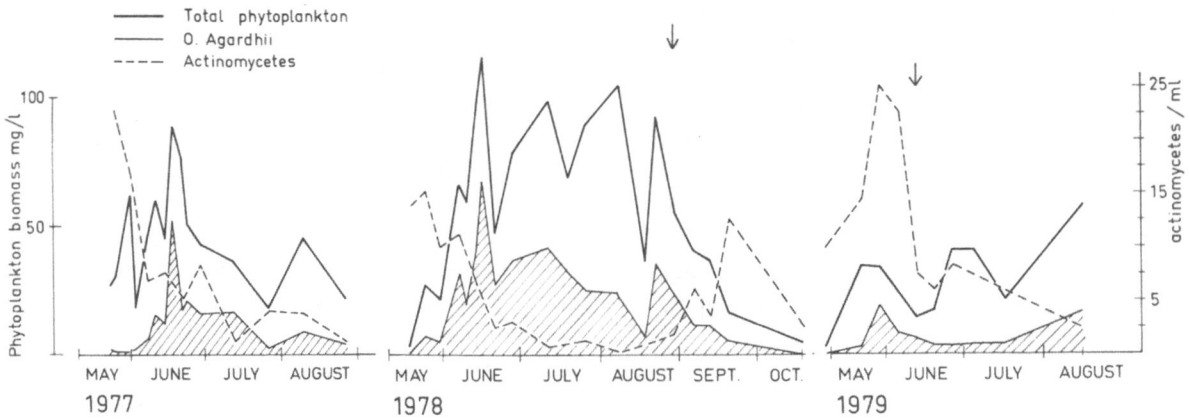

Fig. 2. Biomass of total phytoplankton and *O. agardhii*, and the concentration of actinomycetes in Kaupunginselkä Bay at Porvoo during the summers of 1977, 1978 and 1979. The arrows indicate occurrence of muddy odour in bream. Data for 1977 compiled from Persson (1979), data for 1978 partly from Persson (1980b).

nomycetes reflecting the spring flood as discussed above. However, a causal relationship between algae and actinomycetes in this year cannot be excluded (cf. Silvey & Roach 1964). The number òf actinomycetes in the shore sediments of the study area bore no relationship to muddy odour in fish, nor to phytoplankton biomass (Fig. 2).

The concentrations of actinomycetes in the study area were low compared to the concentrations associated with muddy odour episodes elsewhere. Several studies have indicated that odour production by actinomycetes is inhibited in unsterilized natural water, but proceeds in sterilized water (Seppänen & Jokinen 1969; Leventer & Eren 1970; Persson & Sivonen 1979). Much of the evidence brought up in favour of actinomycetes as a source of muddy odour in natural waters is based on work with laboratory cultures (Gerber 1979). Mere isolation of a muddy-smelling actinomycete from an area does not constitute evidence for actual odour production *in situ* (Persson 1979). The data from Kaupunginselkä Bay indicate that the role of actinomycetes as a source of muddy odour in this area is insignificant compared to that of *O. agardhii*. However, actinomycetes can still be a major source of muddy odour in natural waters in warmer climates (Raschke *et al.* 1975; Silvey & Roach 1975; Weete *et al.* 1977).

In L. Tuusulanjärvi, the muddy odour in bream and pikeperch was related to the amount of blue-green algae in the phytoplankton (table 1), but for pike no such relationship could be established. However, different susceptibilities to muddy odour should be expected for different species of fish. As the intestinal tract seems to be a major route for

uptake of muddy odour by fish (Persson & York 1978), the trophic positions of the fish are important, and so is the habitat of the fish compared to the odorous algae. Vakkuri (1980) indicated a relationship between blue-green algae *(Anabaena* sp.) and the odour of the lake water in L. Tuusulanjärvi. Persson (1980b) found a correlation between muddy odour in pikeperch and the amount of *O. agardhii* in the phytoplankton of the lake. Thus, several species of blue-green algae may contribute to the muddy odour in the lake, and moreover, different species may assume importance in different compartments of the ecosystem.

Acknowledgements

Financial support by the following institutions during different periods of the study is gratefully acknowledged: University of Helsinki, The Academy of Finland, The Finnish Culture Foundation, The Maj and Tor Nessling Foundation, The Finnish Research Foundation for Natural Resources, and the Environmental Protection Board of the Rural Municipality of Borgå.

References

Cees, B., Zoeteman, J., & Piet, G. J., 1974. Cause and identification of taste and odour compounds in water. Sci. Tot. Environ. 3: 103–115.
Gerber, N. N., 1979. Volatile substances from actinomycetes: their role in the odor pollution of water. C.R.C. Crit. Rev. Microbiol. 9: 191–214.
Henley, D. E., 1970. Odorous metabolite and other selected studies of Cyanophyta. Ph. D. Diss., North Texas State Univ., Denton, Texas.
Johnston, D. W. & Cross, T., 1976. Actinomycetes in lake muds: dormant spores or metabolically active mycelium? Freshwat. Biol. 6: 465–470.
Lechevalier, H. A., 1974. Distribution et role des actinomycètes dans les eaux. Bull. Inst. Past. 72: 159–172.
Leventer, H. & Eren, J., 1970. Taste and odor in the reservoirs of the Israel national water system. In: Shuval, H. (ed.) Developments in Water Quality Research, pp. 19–37. Ann Arbor-Humphrey Sci. Publ., Ann Arbor & London.
Lovell, R. T., 1973. Environment-related off-flavours in intensively cultured catfish. F.A.O. Tech. Conf. Fish Prod., Tokyo, Dec. 1973, F II: FP/73/E-46.
Persson, P.-E., 1977. Muddy/earthy off-flavours in fish. Ympäristö ja terveys 8: 515–521 (Finnish, English summary).
Persson, P.-E., 1979. The source of muddy odor in bream (Abramis brama) from the Porvoo sea area (Gulf of Finland). J. Fish. Res. Bd Can. 36: 883–890.

Table 1. Correlations between muddy odour in fish and the amount of blue-green algae in the phytoplankton of L. Tuusulanjärvi in 1978 (y = % of judgements recognizing muddy odour, x = biomass of blue-green algae, mg/l).

Regression line	r	Significance level
Pike		
y = −2.82x + 64.95	−0.348	n.s.
Pikeperch		
y = 4.45x + 7.95	0.569	P < 0.10
Bream		
y = 3.74x + 38.11	0.607	P < 0.05

164

Persson, P.-E., 1980a. Sensory properties and analysis of two muddy odour compounds, geosmin and 2-methylisoborneol, in water and fish. Water Res. 14: 1113–1118.

Persson, P.-E., 1980b. Muddy odour in fish from hypertrophic waters. Develop. Hydrobiol. 2: 203–208.

Persson, P.-E. & Sivonen, K., 1979. Notes on muddy odour. V. Actinomycetes as contributors to muddy odour in water. Aqua fenn. 9: 57–61.

Persson, P.-E. & York, R. K., 1978. Notes on muddy odour. II. Uptake of 2-methylisoborneol by rainbow trout (Salmo gairdneri) in continuous flow aquaria. Aqua fenn. 8: 89–90.

Raschke, R. L., Carroll, B. & Tebo, L. B., 1975. The relationship between substrate content, water quality, actinomycetes and musty odours in the Broad River basin. J. appl. Ecol. 12: 535–560.

Ryding, S. O. (ed.), 1980. Monitoring of inland waters. Nordforsk, Secretariat of Environmental Sciences, Publ. 1980:2.

Seppänen, P. & Jokinen, S., 1969. On the actinomycetes causing odours and tastes in blue-green algae blooms. Suom. Limnol. Yhd. Limnologisymposion 1968, pp. 69–87. Helsinki (Finnish, English summary)

Silvey, J. K. G. & Roach, A. W., 1964. Studies on microbiotic cycles in surface waters. J. Am. Wat. Works Ass. 56: 60–72.

Silvey, J. K. G. & Roach A. W., 1975. The taste and odor producing actinomycetes. C.R.C. Crit. Rev. Environ. Control 5: 233–273.

Tabachek, J. L. & Yurkowski, M., 1976. Isolation and identification of blue-green algae producing muddy odor metabolites geosmin and 2-methylisoborneol in saline lakes in Manitoba. J. Fish. Res. Bd. Can. 33: 25–35.

Vakkuri, T., 1980. Observations on the connection between blue-green algae and the odour of water leaving Lake Tuusulanjärvi. Vesitalous 21 (2): 30–34 (Finnish, English abstract).

Weete, J. D., Blevins, W. T., Wilt, G. R. & Durham, D., 1977. Chemical, biological and environmental factors responsible for the earthy odor in the Auburn City water supply. Bull. Agr. Exp. Sta. Ala. 490: 1–46.

Yurkowski, M. & Tabachek, J. L., 1980. Geosmin and 2-methylisoborneol implicated as a cause of muddy odor and flavor in commercial fish from Cedar Lake, Manitoba. Can J. Fish. Aquat. Sci. 37: 1449–1450.

Factors influencing odour production by actinomycetes

Kaarina Sivonen
Department of Limnology, University of Helsinki, Viikki, 00710 Helsinki 71, Finland

Keywords: actinomycetes, streptomycetes, odour, flavour, water

Abstract

Odour production by actinomycete *(Streptomyces spp.)* strains isolated from hypereutrophic natural waters in which muddy odours in fish have occurred, were studied by the ISP (International Streptomyces Project) carbon utilization method. The streptomycete strains were isolated from water, bottom mud and aquatic plants. Nine different carbon sources were used. Odour character was determined by sniffing the cultures. Odour production varied depending on the strain and the carbon source used. Some of the strains produced similar odours in all media regardless of the carbon source. In other strains, the odour varied depending on the carbohydrate used. The total colony counts of actinomycetes may not necessarily indicate the role of actinomycetes in odour problems in the aquatic environment because the odour production by actinomycetes depends on environmental factors.

Introduction

Actinomycetes, particularly streptomycetes (Fig. 1) and some species of blue-green algae produce geosmin and 2-methylisoborneol which cause water supplies and fish to smell or to taste earthy, muddy or musty (Silvey & Roach 1975; Persson 1977; Gerber 1979). As these odours are difficult to distinguish from each other the present study will describe them as muddy or earthy. Muddy odour or flavour renders surface water and fish unpalatable and is therefore of economic significance.

Odour production by actinomycetes depends on the culture medium, culture conditions and on the strain used (Roach & Silvey 1958; Rosen *et al.* 1970; Gerber 1977, 1979). The studies of Ferramola (1949), Lewis (1966) and Weete *et al.* (1977) indicate that the carbon source may influence odour production. In these experiments only one or two strains were tested.

The aim of the present study was to investigate, under laboratory conditions, the significance of the

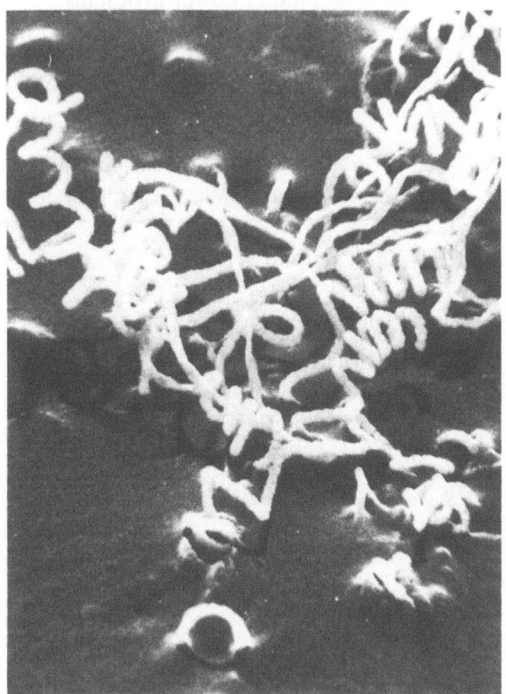

Fig. 1. Streptomyces spheroides (P-K-116) × 1200.

Hydrobiologia 86, 165–170 (1982).   0018-8158/82/0862-0165/$01.20.

carbon source in growth and odour production by strains of actinomycetes isolated from natural waters in Finland.

Material and methods

Actinomycete strains were obtained from the water, bottom mud and aquatic plants *(Nymphaea, Scirpus* and *Typha)* during the summers of 1976 –1978 from the Baltic coast at Oulu (Persson 1974) and Porvoo (Persson 1979) and from Lake Tuusulanjärvi (Vakkuri 1980). The strains were isolated by membrane filtration and incubation on sodium caseinate agar with actidione at 20 °C for 14 days (Vajdic 1968). Actinomycete isolation agar (Difco 1968) and chitin agar (Hsu & Lockwood 1975) were also used to isolate some strains. The isolates were identified by ISP (International Streptomyces Project) methods (Shirling & Gottlieb 1966) and by comparison with reference samples (Sivonen in prep). Earthy smelling strains were selected because the study focused on muddy odour problems.

For the carbon utilization test, Pridham and Gottlieb's basal salts medium was used (Shirling &

Gottlieb 1966). Dry carbon sources: D(+) glucose (BDH), L(+) arabinose (Merck), sucrose (BDH), D(+) xylose (Merck), inositol (Merck), D(-) mannitol (Merck), D(-) fructose (Merck), L(+) rhamnose (Fluka) and raffinose (Merck), were sterilized with ethyl ether and added to the basal medium to a final concentration of 1%. Before inoculation the strains were grown in tryptone-yeast-extract broth (Pridham & Gottlieb 1948). Two streaks of washed inoculum were applied across each petri dish (Shirling & Gottlieb 1966). A total of 17 strains, replicated for each carbon source, were incubated.

Following incubation for two weeks in the dark and at room temperature, each replicate was sniffed by two persons thoroughly familiar with muddy odour. The odour was described and the growth recorded.

Results

The growth of *Streptomyces* strains on different carbon sources varied markedly (Table 1). Glucose, fructose and mannitol were most widely used.

Variable odours were recorded depending on the

Table 1. Growth of *Streptomyces* strains on different carbon sources.
Notation: ++ = good growth, + = moderate growth, - = no growth or only traces of growth.

Strain	No C	Glu	Ara	Suc	Xyl	Ino	Man	Fru	Rha	Raf
Streptomyces flavogriseus (P-V-032)	-	++	++	-	+	++	++	++	-	
S. flavogriseus (P-K-148)	-	++	++	-	+	-	+	++	++	-
S. nigrescens (P-M-066)	-	++	-	++	+	++	++	+	-	++
S. galbus (P-M-068)	-	++	+	-	+	+	+	++	+	-
S. prasinus (P-V-040)	-	+	+	++	++	++	+	+	+	-
S. globisporus (P-M-005)	-	++	+	-	+	-	++	++	+	-
S. globisporus (P-V-041)	-	++	++	-	+	-	++	++	++	-
S. spheroides (P-M-023)	-	++	+	+	+	-	++	++	++	-
S. spheroides (P-K-116)	-	++	+	++	+	-	++	++	+	-
S. levoris (P-V-154)	-	++	++	-	+	-	++	++	-	-
S. coelicolor (P-K-175)	-	++	++	-	+	-	++	+	-	-
S. lipmanii (O-V-00X)	-	++	-	-	+	-	++	++	++	-
Streptomyces sp.										
(lavendulae group) (P-V-153)	-	++	-	-	-	-	-	+	-	-
(lavendulae group) (P-V-022)	-	++	-	-	-	-	-	+	-	-
(lavendulae group) (P-V-011)	-	++	-	++	-	-	++	+	-	-
(lavendulae group) (P-V-012)	-	++	-	-	-	-	++	+	-	-
(lavendulae group) (P-V-161)	-	++	-	-	-	-	++	+	-	-

* Abbreviations of the substrates: No C = no carbon, Glu = glucose, Ara = arabinose, Suc = sucrose, Xyl = xylose, Ino = inositol, Man = mannitol, Fru = fructose, Rha = rhamnose, Raf = raffinose.

Table 2. Recorded odours of *Streptomyces* strains on different carbon sources. Notation: (earth) = a note of earthy odour. - = no odour.

Strain	No C	Glu	Ara	Suc	Xyl	Ino	Man	Fru	Rha	Raf
					Substrate*					
Streptomyces flavogriseus (P-V-032)	-	sweet/ straw- berry	dry hay	-	(earth)/ sea shore	-	(earth)/ sweet	yeast/ fruit	-	-
S. flavogriseus (P-K-148)	-	(earth)/ sweet/ fruit	mild fruit	-	(earth)	-	yeast	fruit- flower	fruit- flower	-
S. nigrescens (P-M-066)	-	yeast/ sour milk	-	yeast/ sour milk	-	yeast/ sour milk	yeast/ sour milk	-	-	sour milk
S. galbus (P-M-068)	-	earth/ medicine	malt/ dry hay	-	-	-	-	earth/ mentol	earth/ mentol/ cheese	-
S. prasinus (P-V-040)	-	yeast	sweet	sweet	sweet	sweet/ yeast	mild sweet	mild sweet	mild sweet	-
S. globisporus (P-M-005)	-	earth/ mild nut	-	-	mild earth	-	earth/ medicine	earth/ fruit	mild earth	-
S. globisporus (P-V-041)	-	earth	earth/ rush	-	toffee	-	earth/ pepper	earth/ soft	(earth)	-
S. spheroides (P-M-023)	-	(earth)/ yeast	malt/ dry hay	mild earth	-	-	(earth)/ bad	earth/ brisk	toffee	-
S. spheroides (P-K-116)	-	pepper/ fruit	malt/ dry hay	pepper	pepper	-	fruit- flower	pepper	pepper/ cheece	-
S. levoris (P-V-154)	-	(earth)/ stale	(earth)/ yeast/ sour milk	-	-	-	(earth)/ stale	(earth)/ yeast/ sour milk	-	-
S. coelicolor (P-K-175)	-	earth	mild earth	-	-	-	earth/ yeast	yeast/ ground toast	-	-
S. lipmanii (0-V-00X)	-	(earth)/ pepper	-	-	earth/ rush	-	(earth)/ pepper	(earth)/ pepper	(earth)/ pepper	-
Streptomyces sp. (lavendulae group) (P-V-153)	-	(earth) sour milk	-	-	-	-	-	-	-	-
Streptomyces sp. (lavendulae group) (P-V-022)	-	(earth) sweet/ spice/ medicine	-	-	-	-	-	-	-	-
Streptomyces sp. (lavendulae group) (P-V-011)	-	(earth)/ mild nut	-	nut	-	-	(earth)/ nut	-	-	(earth)/ nut
Streptomyces sp. (lavendulae group) (P-V-012)	-	earth/ damp cellar	-	-	-	-	earth +?	earth +?	-	-
Streptomyces sp. (lavendulae group) (T-V-161)	-	(earth)/ yeast	-	-	-	-	earth	-	-	-

* Abbreviations of the substrates: No C = no carbon, Glu = glucose, Ara = arabinose, Suc = sucrose, Xyl = xylose, Ino = inositol, Man = mannitol, Fru = fructose, Rha = rhamnose, Raf = raffinose.

168

Table 3. Actinomycetes producing earthy-smelling metabolites (geosmin and methylisoborneol) reported in the literature.

Strain	Reference
GEOSMIN	
Microbispora rosea 3748	Gerber (1967)
Nocardia sp. SS1/1	Gerber (1967)
Nocardia sp. W-68	Gerber (1967)
Nocardia sp. 1–15	Gerber (1967)
Streptomyces alboniger 12464	Gerber (1967)
S. albosporeus (Biwako-C)	Kikuchi *et al.* (1973)
S. antibioticus IMRU 3720	Gerber & Lechevalier (1965)
S. antibioticus IMRU 3491	Gerber & Lechevalier (1965)
S. filipinensis (Biwako-D)	Kikuchi *et al.* (1973)
S. fradiae IMRU 3535	Gerber & Lechevalier (1965)
S. fradiae IMRU 3535-R7	Gerber & Lechevalier (1965)
S. griseoluteus IMRU 3718*	Rosen *et al.* (1968)
S. griseus LP-16	Gerber & Lechevalier (1965)
S. lavendulae 3440 1-Y	Gerber (1967)
S. lavendulae 3440-14	Rosen *et al.* (1970)
S. lavendulae CBS 16245	Piet *et al.* (1972)
S. odorifer IMRU 3334	Gerber & Lechevalier (1965)
S. odorifer ATCC 6246	Piet *et al.* (1972)
S. resistomycificus (Biwako-B)	Kikuchi *et al.* (1973)
S. viridochromogenes 94	Gerber (1967)
Streptomyces sp. 27–20	Gerber (1967)
Streptomyces sp. B-5a	Gerber (1967)
Streptomyces sp. B-7	Gerber (1967)
Streptomyces sp. 37–12	Gerber (1967)
Streptomyces sp. (lavendulae group) (P-V-011)	Persson (1979)**
METHYLISOBORNEOL	
Actinomadura sp. 1–15	Gerber (1969)
Nocardiopsis (Actinomadura) dassonvillei IMRU 1324	Gerber (1979)
S. albosporeus (Biwako-C)	Kikuchi *et al.* (1973)
S. antibioticus Nr. 5234	Medsker *et al.* (1969)
S. filipinensis (Biwako-D)	Kikuchi *et al.* (1973)
S. griseus ATCC 10137	Medsker *et al.* (1969)
S. lavendulae IMRU 3440-1 Y	Gerber (1969)
S. lavendulae CWW3	Gerber (1979)
S. lavendulae 3440-14	Rosen *et al.* (1970)
S. lavendulae CBS 16245	Piet *et al.* (1972)
S. odorifer	Collins *et al.* (1970)
S. odorifer ATCC 6246	Piet *et al.*, (1972)
S. paraecox ATCC 3374	Medsker *et al.* (1969)
S. resistomycificus (Biwako-B)	Kikuchi *et al.* (1973)
Streptomyces sp 100-1	Gerber (1969)
Streptomyces sp. AY-19	Gerber (1969)
Streptomyces sp. (lavendulae group) (P-V-011)	Persson (1979)**

* Gerber & Lechevalier (1965) were not able to detect the presence of geosmin in the products of this strain.
** Identified later (Sivonen, in prep.)

carbon source and the strain used (Table 2). Three strains of the 17 studied did not produce a muddy odour in any media. *Streptomyces nigrescens* (P-M-066) had a yeasty, *S. prasinus* (P-V-040) a sweet and one of the two *S. spheroides* strains (P-K-116) mainly a peppery aroma.

S. globisporus (P-M-005), *S. levoris* (P-V-154), *S. lipmanii* (O-V-00X) and all *Streptomyces* sp. strains, except one designated as (P-V-011), produced an earthy odour on all carbohydrates whenever odour production was recorded. The odour character showed minor variation. For example, *S. globisporus* (P-M-005) produced a nutty note in glucose, medicinal note in mannitol and a fruity note in fructose.

S. flavogriseus (P-V-032), (P-K-148), *S. galbus* (P-M-068), *S. coelicolor* (P-K-175) and *Streptomyces* sp. (P-V-011) strains produced a muddy odour only on certain carbon sources.

Discussion

Several species and strains of actinomycetes have been shown to produce the muddy smelling geosmin and methylisoborneol (Table 3). Some of the *Streptomyces* sp. (lavendulae group) species used in the present study are closely related to the strains of the *S. lavendulae* (Table 3). One of the strains used in the present study has been shown to produce geosmin and 2-methylisoborneol (Table 3). An earthy smelling component produced by *S. coelicolor* was identical to that produced by *Oscillatoria splendida* (Telitchenko *et al.* 1972).

Odour production by actinomycetes may be influenced by several ecological factors. It is known that odour production has been associated with the active growth of actinomycetes (Ferramola 1949) and is generally considered to occur at temperatures above 15 °C (Silvey *et al.* 1950; Weete *et al.* 1977) and in aerobic environments (Higgins & Silvey 1966).

In the primary study area, Kaupunginselkä Bay at Porvoo, the numbers of actinomycetes (mainly streptomycetes) have been rather low, probably because of the discharge of fresh water into the bay. No correlation between the numbers of actinomycetes in water and muddy odour in fish was found (Persson 1979, 1980). Based on a limited literature survey, Bays *et al.* (1970) concluded that

the evaluation of actinomycete counts appeared to bear little, if any, relationship to the intensity of earthy odours in water supply. However, Morris (1962) noticed a relationship between threshold odour levels of the water and numbers of actinomycetes. Other reports (Silvey *et al.* 1950; Silvey & Roach 1953, 1964) have also indicated actinomycetes to be a source of muddy odour in water.

Streptomycetes isolated from lake muds have been considered inactive and to occur mainly as spores (Johnston & Cross 1976). Furthermore, odour production seems to be strain specific (Gerber & Lechevalier 1965); Gerber 1967; present study). The count of actinomycetes must therefore be an overestimation of numbers, since it does not distinguish the odour producing strains.

The factors discussed above indicate that colony counts of actinomycetes do not necessarily indicate their role in odour problems in the aquatic environment, but may reflect growth patterns and therefore be useful in the examination of odour problems associated with actinomycetes.

Acknowledgements

The present study was based in the University of Helsinki, Finland. I thank Prof. R. Ryhänen for providing me with working facilities in the Department of Limnology. I also thank Prof. I. M. Szabó, Eötvös L. University, Budapest, Hungary, for providing with working facilities and guidance in the identification of actinomycetes. I wish to thank Lic. P.-E. Persson, University of Helsinki, for his kind help, advice and valuable criticism of the manuscript. I am grateful to Prof. E. Eklund, University of Helsinki, for her kind advice and to Mr. Darrell Sequeira for checking the language.

References

Bays, L. R., Burman, N. P. & Lewis, W. M., 1970. Taste and odour in water supplies in Great Britain: a survey of the present position and problems for the future. Wat. Treatm. Exam. 19: 136–153.

Collins, R. P., Knaak, L. E. & Soboslai, J. W., 1970. Production of geosmin and 2-exo-hydroxy-2-methylbornane by *Streptomyces odorifer*. Lloydia 33: 199–200.

Difco, 1968. Supplementary literature. Difco Laboratories, Detroit, Michigan, U.S.A.

170

Ferramola, R., 1949. Summary: Earthy odor produced by Streptomyces in water. Asoc. Interam. ingenieria sanitàra 2: 371–380.

Gerber, N. N., 1967. Geosmin, an earthy-smelling substance isolated from actinomycetes. Biotechnol. Bioeng. 9: 321–327.

Gerber, N. N., 1969. A volatile metabolite of actinomycetes, 2-methylisoborneol. J. Antibiotics (Tokyo) 22: 508–509.

Gerber, N. N. 1977. Three highly odorous metabolites from an actinomycete: 2-isopropyl-3-methoxy-pyrazine, methylisoborneol and geosmin. J. chem. Ecol. 3: 475–482.

Gerber, N. N., 1979. Volatile substances from actinomycetes: their role in the odor pollution of water. CRC Crit. Rev. Envir. Control. 9: 191–214.

Gerber, N. N. & Lechevalier, H. A., 1965. Geosmin an earthy-smelling substance isolated from actinomycetes. Appl. Microbiol. 13: 935–938.

Higgins, M. L. & Silvey, J. K. G., 1966. Slide culture observations of two freshwater actinomycetes. Trans. Am. Microsc. Soc. 85: 390–398.

Hsu, S. C. & Lockwood, J. L., 1975. Powdered chitin agar as a selective medium for enumeration of actinomycetes in water and soil. Appl. Microbiol. 29: 422–426.

Johnston, D. W. & Cross, T., 1976. Actinomycetes in lake muds: dormant spores or metabolically active mycelium? Freshwat. Biol. 6: 465–470.

Kikuchi, T., Mimura, T., Itoh, Y., Moriwaki, Y., Negoro, K.-I., Masada, Y., & Inoue, T., 1973. Odorous metabolites of actinomyces Biwako-C and -D strain isolated from the bottom deposits of Lake Biwa. Identification of geosmin, 2-methylisoborneol and furfural. Chem. Pharm. Bull. 21: 2339–2341.

Lewis, W. M., 1966. Odours and tastes in water derived from the River Severn. Wat. Treatm. Exam. 15: 50–74.

Medsker, L. L., Jenkins, D., Thomas, J. F., & Koch, C., 1969. Odorous compounds in natural waters. 2-exo-hydroxy-2-methylbornane, the major odorous compound produced by several actinomycetes. Envir. Sci. Technol. 3: 476–477.

Morris, R., 1962. Actinomycetes studied as taste and odor cause. Water Sewage Works 109: 138, 140.

Persson, P.-E., 1974. On flavour tainting of fish, with special reference to the Oulu sea area (Bothnian Bay). Rep. Nat. Bd Wat. Finl. 65: 1–262 (Swedish, English summary).

Persson, P.-E., 1977. Muddy/earthy off-flavours in fish. Ympäristö ja terveys 8: 515–521. (Finnish, English summary).

Persson, P.-E., 1979. The source of muddy odor in bream (Abramis brama) from the Porvoo sea area (Gulf of Finland). J. Fish. Res. Bd Can. 36: 883–890.

Persson, P.-E., 1980. Muddy odour in fish from hypertrophic waters. Dev. Hydrobiol. 2: 203–208.

Piet, G. J., Zoeteman, B. C. J., & Kraayeveld, A. J. A., 1972. Earthy smelling substances in surface waters of the Netherlands. Wat. Treatm. Exam. 21: 281–286.

Pridham, T. G. & Gottlieb, D., 1948. The utilization of carbon compounds by some Actinomycetales as an aid for species determination. J. Bacteriol. 56: 107–114.

Roach, A. W. & Silvey, J. K. G., 1958. The morphology and life cycle of fresh-water actinomycetes. Trans. Am. microsc. Soc. 77: 36–47.

Rosen, A. A., Mashni, C. I. & Safferman, R. S., 1970. Recent developments in the chemistry of odour in water: The cause of earthy/musty odour. Wat. Treatm. Exam. 19: 106–114.

Rosen, A. A., Safferman, R. S., Mashni, C. I. & Romano, A. H., 1968. Identity of odorous substance produced by Streptomyces griseoluteus. Appl. Microbiol. 16: 178–179.

Shirling, E. B. & Gottlieb, D., 1966. Methods for characterization of Streptomyces species. Int. J. Syst. Bact. 16: 313–340.

Silvey, J. K. G. & Roach, A. W., 1953. Actinomycetes in Oklahoma City water supply. J. Am. Wat. Works Ass. 45: 409–416.

Silvey, J. K. G. & Roach, A. W., 1964. Studies on microbiotic cycles in surface waters. J. Am. Wat. Works Ass. 56: 60–71.

Silvey, J. K. G. & Roach, A. W., 1975. The taste and odor producing aquatic actinomycetes. CRC Crit. Rev. Envir. Control 5: 233–273.

Silvey, J. K. G., Russel, J. C., Redden, D. R., & McCormick, W. C., 1950. Actinomycetes and common tastes and odors. J. Am. Wat. Works Ass. 42: 1018–1026.

Telitchenko, M. M. Tambiev, A. K., & Ermolovich, L. P., 1972. Gaschromagraphic methods for studying condensates of odors, of a biogenic nature in drinking water. Metod. Biol. Issled. Vod. Toksikol., Stroganoff, N. S. (ed.), Nauka, Moscow, 1971, 93. Chem. Abstr. 77, 57220r.

Vajdic, A. H., 1968. The isolation and enumeration of actinomycetes from water samples. Ontario Wat. Resources Comm., Div. Research RP 2016, 20 pp.

Vakkuri, T., 1980. Observations on the connection between blue-green algae and odour of water leaving Lake Tuusulanjärvi. Vesitalous 21:30–34 (Finnish, English abstract).

Weete, J. D., Blevins, W. T., Wilt, G. R. & Durham, D., 1977. Chemical, biological and environmental factors responsible for the earthy odor in the Auburn City water supply. Bull. Agr. Exp. Sta. Ala. 490: 1–46.

Faecal indicator bacteria at fish farms

Maarit Niemi[1] & Irmeli Taipalinen[2]
[1] *National Board of Waters, P.O. Box 250, SF-00101 Helsinki 10, Finland*
[2] *Kuopio Water District Office, P.O. Box 49, SF-70101 Kuopio 10, Finland*

Keywords: total coliforms, faecal coliforms, faecal streptococci, fish farms, rainbow trout

Abstract

The observed concentrations of bacteria at two large fish farms were not high, but due to the great volume of the discharge the total amount of bacteria was large. Total coliform (TC) bacteria identified belonged mainly to the genera *Enterobacter, Citrobacter* and *Aeromonas*. The majority of faecal coliform (FC) strains were *Escherichia coli*. *E. coli* was absent, or occurred at very low concentrations, in the influent water, but was present in the effluent water, in the sediment, and at one fish farm also in fish faeces. FC bacteria were not observed in the fish feed. The concentrations of faecal streptococci (FS) in the influent water were low, but strains isolated were identified as group D streptococci. The concentrations of FS were low in the feed and sediment samples but were elevated in the fish faeces and also in the effluent.

Introduction

The farming of aquatic animals is becoming an increasingly important source of food. Aquaculture is rapidly developing from a traditional craft into a sophisticated technology with profound ecological consequences (Ackefors & Rosén 1979).

In Finland *ca.* 130 professional farms produced *ca.* 3300 tons of rainbow trout (*Salmo gairdneri*) in 1979, which is less than the production in other European countries (Moisala 1979). However, the number of fish farms producing trout is increasing rapidly.

The majority, 80%, of the farms are situated by inland waters. Due to the necessity for high quality raw water, the farms utilize otherwise relatively unpolluted water bodies.

Investigations by the water authorities have revealed elevated concentrations of coliform bacteria and FS in the effluents from fish farms in Finland (Haavisto 1974; Myllymaa, personal communication).

TC, FC and FS have been isolated from the intestines of free-living fish (Geldreich & Clarke 1966; Trust & Sparrow 1974; Souter *et al.* 1976) and coliform bacteria from cultured fish and from fish pond water (Watanabe *et al.* 1971; Aoki 1975; Ruane *et al.* 1977).

Routine investigations at the relatively unpolluted Rautalampi watercourse indicated elevated counts of TC, FC and FS in the water below two trout-producing farms. The counts of these bacteria in water, feed, sediment and fish faeces were therefore investigated at both farms. In November the counts for all these bacteria were very low, but the results for August when the water temperature was 17 °C, indicated contamination. The August results are presented in this paper.

Materials and methods

The fish farms

Nilakkalohi Ltd. is situated by the rapids of Äyskoski in the Rautalampi watercourse. The

Hydrobiologia 86, 171–175 (1982). 0018-8158/82/0862-0171/$01.00.

172

maximum annual production capacity of the farm is 300 tons of rainbow trout. During 1979, 463 tons of pelleted dry feed and 35 tons of fresh fish material was used as feed. The majority of feed was used during May and June. About 500 m³ of silt deposit was removed in Autumn 1979 from the mud bottomed ponds. The average discharge in August was 2.6 m³ s⁻¹. The effluent from the fish ponds flows to the rapids of Äyskoski after a sedimentation pond. A separate sewage system is provided for domestic wastes.

Savon Taimen Ltd. is situated by the River Tyyrinvirta further down the Rautalampi watercourse. The maximum annual production capacity of this farm is 400 tons of rainbow trout. During the summer, 150 to 200 tons of feed per month is used, mainly pelleted dry feed. About 775 m³ of silt deposit was removed from the mud bottomed ponds during the summer of 1979. The volume of the discharge was 5.7 m³ s⁻¹ in August. During the period of this study the domestic wastes drained into the effluent of the cultivation channel, which flows untreated to the river.

The samples

Samples were taken from influent and effluent waters and from sediments into sterile borosilicate bottles. Faecal material was collected from five fish for a composite sample. At Nilakkalohi Ltd. fish were harvested from the area enclosed from the river with nets and at Savon Taimen Ltd. from the rearing channel. Samples were taken from Nilakkalohi Ltd. on 27 August and from Savon Taimen Ltd. on 28 August 1979.

The determination of bacteria was carried out mainly with the filtration technique using Sartorius 11406 AC membranes, but solid samples were analysed with the MPN technique when the sample size was 0.01 g or more.

TC bacteria

In the filtration technique LES Endo agar (Difco) was used. The plates were incubated at 35 °C for 24 h. In the MPN technique lactose broth (3 times 5 tubes) and incubation at 35 °C for 48 h were used for the primary fermentation. For the confirmation, brilliant green bile broth and incubation at 35 °C for 48 h were used. Eosin methylene blue agar was used for the isolation of bacteria.

FC bacteria

In the filtration technique mFC agar, modified by addition of 0.25 g Water Blue instead of Aniline Blue (Niemelä, personal communication) and incubation at 44 °C in the water bath for 24 h, was used. In the MPN technique the same procedure was used as for TC bacteria but in the confirmation test the tubes were incubated at 44 °C in the water bath.

Identification

The coliform bacteria were purified twice on tryptone yeast extract glucose agar before identification with API-20E identification kits (Analytab. Products Inc.).

FS bacteria

Both the KF Streptococcus and m Enterococcus agars (Difco) were used in the filtration technique. The plates were incubated at 35 °C for 48 h. In the MPN technique dextrose azide broth and incubation at 35 °C for 48 + 24 h was used for the primary fermentation. For the confirmation, samples from positive tubes were inoculated on PSE agar (Pfizer). Those strains hydrolyzing esculin within 24 h were recorded as FS. For the serotyping, bacteria were purified twice on BHI agar (Difco) before testing for the presence of Lancefield group D antigens with a Streptex kit (Wellcome Laboratories).

Results

The concentrations of FS were low in the influent waters, but elevated concentrations were observed in the effluents (Table 1). Low concentrations of FS occurred in the sediment samples and in the feed. High concentrations of FS were observed in the fish faeces.

The colony counts on KF Streptococcus and m Enterococcus agars did not differ significantly. Of the FS strains isolated as typical from KF Streptococcus agar, from m Enterococcus agar and using the MPN technique, only 77, 60 and 56%, respectively, belonged to group D (Table 1).

TC bacteria were common in the influent waters, but their concentrations were still higher in the

Table 1. FS and group D streptococci at fish farms.

Sample	Method	Nilakkalohi Ltd.		Savon Taimen Ltd.	
		FS ml⁻¹, g⁻¹	D+/FS	FS ml⁻¹, g⁻¹	D+/FS
Influent	KF	0.00		0.01	2/2
	SB	0.01	2/2	0.02	4/5
Effluent					
rearing pond or	KF			0.37	9/10
channel	SB			0.25	6/10
middle channel	KF			0.31	6/10
	SB			0.18	6/10
sedimentation pond	KF	1.41	7/10		
	SB	0.59	7/10		
Fish faeces	KF	143 000	9/10	6 500	7/10
	SB	168 000	6/8	26 500	5/10
Sediment					
rearing pond or	KF	0		0	
channel	SB	(5 500)	3/10	0	
	MPN	104	1/20	5	9/13
middle channel	KF			0	
	SB			0	
	MPN			5	14/20
sedimentation pond	KF	0			
	SB	0			
	MPN	3	5/13		
Feed	MPN	5	15/18	5	14/20
Total at the				%	
two farms	KF		40/52	77	
	SB		39/65	60	
	MPN		58/104	56	
	total		134/221*	62	

* Eleven strains lost before serotyping.

effluent waters (Table 2). The concentrations of TC bacteria were very high in the sediment and in the fish faeces, although only one of the 223 TC isolates was *Escherichia coli* (Table 3). A very high proportion of the strains isolated from typical TC colonies did not grow long enough to enable identification (Table 3). Twenty-six per cent of the TC bacteria tested could not be identified. The majority of the strains identified belonged to the genera *Enterobacter* and *Citrobacter*. *Aeromonas hydrophila* was quite common among the isolates.

The counts of FC bacteria at the fish farms were far lower than those of TC bacteria (Table 2). The influent waters were of good quality, but the effluent waters were somewhat contaminated. FC bacteria were not observed in the fish faeces harvested from the river area enclosed with the nets (Table 2). A low concentration of FC bacteria was, however, observed in the faeces from fish fed with fresh fish material and harvested from the pond. The concentrations of FC in the sediment were far lower than those of TC.

The majority of FC bacteria isolated were *E. coli* (Tables 2 and 3). *E. coli* was present in the effluent waters, sediment samples and faeces from fish harvested from the pond. A few *Enterobacter*, *Citrobacter* and *Klebsiella* strains were isolated as FC bacteria (Table 3). *Aeromonas hydrophila* and *A. shigelloides* strains were isolated from the effluent waters. Of the eight *Aeromonas* strains, seven were isolated from plates with bad colour production because of dampness.

Discussion

Geldreich & Clarke (1966) observed that the minimum temperature for the multiplication of FC

Table 2. TC, FC and *E. coli* at fish farms.

Sample	Nilakkalohi Ltd.			Savon Taimen Ltd.		
	TC ml^{-1}, g^{-1}	FC ml^{-1}, g^{-1}	*E. coli*/FC	TC ml^{-1}, g^{-1}	FC ml^{-1}, g^{-1}	*E. coli*/FC
Influent	0.46	0.005	1/1	2.55	0.00	
Effluent						
rearing pond or channel				19.5	0.11	18/20
middle channel				8.0	1.2	4/20
sedimentation pond	5.2	0.48	10/19			
Fish faeces	56 000[+]	0.00*[+]		490 000[§]	1[§]	7/7
Sediment						
rearing pond or channel	235 000	5*	13/13	10 500	3*	11/13
middle channel				19 000	8*	8/13
sedimentation pond	7 500	84*	20/21			
Feed	0.04*	0.00*		2.3*	0.00*	

*Results from MPN tubes; others are from membrane filtration technique.
[+] Fish harvested from the enclosed river area.
[§] Fish harvested from the pond.

and *Streptococcus faecalis* in the fish intestines was between 10 and 20 °C. These bacteria were commonly encountered in fish intestines when the ambient water temperature was between 13 and 18 °C. *Enterobacter, Aeromonas* and *Acinetobacter* were the predominant genera, while *E. coli* and FS were also present in the intestines of free-living salmonid fish when the water temperature was between 14 and 22 °C (Trust & Sparrow 1974). Aoki (1975) isolated antibiotic resistant strains of *Aeromonas,* Enterobacteriaceae (including *E. coli*), *Vibrio* and *Pseudomonas* from the intestines of cultured fish and pond water when the water temperature was between 16 and 21 °C. High levels of FC bacteria were present in high density catfish cultures when the water temperature ranged from

24 to 29 °C, the concentrations increasing with increase in temperature (Ruane *et al.* 1977).

During this study, the temperature was 17 °C. However, during the summer the water temperature can rise up to 26 °C (Haverinen 1979). The concentrations of TC, FC and FS were clearly elevated in the effluents, although the concentrations were not very high. However, an increase in water temperature could increase the counts of bacteria, since the multiplication rate is probably regulated by the water temperature when decomposing feed in the fish intestines and feed and fish faeces in the sediment provide an adequate nutrient supply. It seems evident that during the growing season fish ponds can act as multiplication sites of Enterobacteriaceae – *E. coli, Enterobacter* and *Citrobacter* – *Aeromonas* and group D streptococci.

FS entered the farms via influent water and feed. The major multiplication site seemed to be fish intestines.

TC bacteria were present in the influent water and at low concentrations in the feed. They seemed to multiply in the fish intestines and possibly in the sediment. It is not possible, however, to exclude the possibility that the high concentrations of TC bacteria in the sediment originated from fish faeces without multiplication in the sediment itself, although multiplication in the sediment seems probable.

Table 3. Identity of isolated coliform bacteria.

	TC		FC	
	No.	%	No.	%
Isolates	223		129	
Lost	62	28	11	8.5
Escherichia coli	1	0.5	92	71
Enterobacter	54	24	6	4.7
Citrobacter	27	12	2	1.5
Klebsiella pneumoniae	4	1.8	2	1.5
Aeromonas hydrophila	17	7.6	6	4.7
Aeromonas shigelloides	0	0.0	2	1.5
Unidentified	58	26	8	6.2

The source of FC bacteria at the fish farms is not clear. According to the official control, no feed containing FC is accepted. However, even if the raw water was satisfactory for drinking water, occasional FC bacteria due to nonpoint sources, could not be excluded. Wild animals, e.g. sea-gulls, might also act as vectors. At Savon Taimen Ltd., domestic wastes might have contaminated the rearing channel before providing a separate sewer system in October 1979. The probable sites of multiplication of FC bacteria and *E. coli* seem to be sediment and fish intestines.

The large discharge volume at fish farms increases the total amount of bacteria washed into the watercourse. At the time of sampling, the total number of FS discharged was *ca.* 10^8 s^{-1}, of TC between 10^7 and 10^9 s^{-1} and of FC between 10^6 and 10^8 s^{-1}. The effluents from Nilakkalohi Ltd. constituted 18% of the discharge volume of the rapids of Äyskoski and the effluents of Savon Taimen Ltd. 17% of the discharge volume of the River Tyyrinvirta in August 1979.

The total number of indicator bacteria in the effluents from fish farms was high enough to be detected in the receiving water body. The multiplication at fish farms of indicator bacteria of faecal contamination of homoiothermic animals disturbs the routine control of water hygiene. Additional determinations of bacteria pathogenic to man, domestic animals and free-living fish are needed. The choice of bacteria to be determined is, however, difficult, and due to the already great dilution of bacteria in the effluent water the methods to be used must be very sensitive. In addition, the large discharge volume makes efficient control difficult.

Japanese investigations (Watanabe *et al.* 1971; Aoki 1975) have revealed the presence of antibiotic-resistant bacteria carrying R factors at fish farms. The propagation of these bacteria has been proposed to be due to the treatment of diseased fish with antimicrobial drugs. R factors have also been detected in bacteria isolated from fish caught from rivers (Lavoie & Mathieu 1974; Aoki 1975). The possibility that fish can act as vectors of bacteria carrying R factors adds to the concern over effluents from fish farms.

Abbreviations: TC = total coliforms, FC = faecal coliforms, FS = faecal streptococci, MPN = most probable number, KF = KF Streptococcus agar, SB = m Enterococcus agar.

Acknowledgements

A grant from the Maj and Tor Nessling Foundation enabled the serotyping of FS at the Department of Microbiology, University of Helsinki. Our thanks are due to the personnel of Kuopio Water District Office and of the Water Research Institute for their technical assistance. We would especially like to thank Mrs. Tuovi Vartio and Miss Helena Ruutianen for their skilled assistance. Finally, our thanks are due to Mr. Michael Bailey for revising the English.

References

Ackefors, H. & Rosén, C.-G., 1979. Farming aquatic animals. The emergence of a world-wide industry with profound ecological consequences. Ambio 8: 132–143.

Aoki, T., 1975. Effects of chemotherapeutics on bacterial ecology in the ponds and the intestinal tracts of cultured fish, ayu (Plecoglossus altivelis). Jap. J. Microbiol. 19: 7–12.

Geldreich, E. E. & Clarke, N. A., 1966. Bacterial pollution indicators in the intestinal tract of freshwater fish. Appl. Microbiol. 14: 429–437.

Haavisto, P., 1974. The magnitude of loading from natural fish ponds and from fish farms. Rep. Nat. Bd Wat. Finl. 74: 1–67 (in Finnish).

Haverinen, L., 1979. The loading from fish farms. Rep. Nat. Bd Wat., Finl. 170: 1–76 (in Finnish).

Lavoie, M. & Mathieu, L. G., 1974. Incidence of hemolytic and R factors in lactose-fermenting enteric bacteria isolated from the digestive tract of fish. Rev. Can. Biol. 33: 185–191.

Moisala, E., 1979. The state of the art of fish farming and its pursuits. Ympäristö ja terveys 10: 257–259 (in Finnish).

Ruane, R. J., Chu, T.-Y. J. & Vandergriff, V. E., 1977. Characterization and treatment of waste discharged from high-density catfish cultures. Water Res. 11: 789–800.

Souter, B. W., Sonstegard, R. A. & McDermott, L. A., 1976. Enteric bacteria in carp (Cyprinus carpio) and white suckers (Catostomus commersoni). J. Fish.Res. Bd Can. 33: 1401–1403.

Trust, T. J. & Sparrow, R. A. H., 1974. The bacterial flora in the alimentary tract of freshwater salmonid fishes. Can. J. Microbiol. 20: 1219–1228.

Watanabe, T., Aoki, T., Ogata, Y. & Egusa, S., 1971. R factors related to fish culturing. Ann. N.Y. Acad. Sci. 182: 383–410.

Goals, methods and possibilities for directing development of limnic ecosystems

Sven Björk

University of Lund, Institute of Limnology, Box 3060, S-220 03 Lund, Sweden

Keywords: aeration, biomanipulation, restoration, sediment removal, wetlands

Abstract

Case studies illustrating lake and wetland ecosystem problems as well as restoration methods are given. Among these methods, aeration of deep lakes, sediment removal from shallow, polluted lakes, sediment manipulation in polluted and acidified lakes, biomanipulation and wetland management methods are considered. A treatment program for directing ecosystem development is designed in each individual case according to lake type, degradation problems and goals of restoration. The most common goal in treatment of lakes is to meiotrophicate (oligotrophicate) hypertrophicated ecosystems. In the case of wetlands, conservation and restoration aspects are combined with a growing interest focused on biomass production by emergent macrophytes. Within practical frames of applied limnology, basic limnological research and training of doctorands have been organized as team-work for ecosystem-oriented investigations.

Introduction

As the characteristics of lakes are dependent on qualities and activities in their catchment areas, the active management of waters must start on land. The direct and firm bonds between watershed and lake make this a postulate (Likens & Bormann 1979). It should not be forgotten that airborne contamination is also of terrestrial origin.

In Sweden there is a legitimate, undisguised pride in having efficient plants for sewage treatment (Fig. 1). The external loading of nutrients in surface waters has decreased dramatically, and with some exceptions polluted lakes recover. However, although water degradation due to classical eutrophication has now been brought under control, tens of thousands of Swedish lakes are hit by acidification, mainly caused by airborne sulphuric compounds. Irreversibly eutrophicated, hypertrophic lakes, drained lakes and acidified waters need help. In most of the world, water management for preservation of diversified natural systems is still an unknown or not understood concept. Limnologists are first called only when lake ecosystems collapse.

By individually-designed measures, lake ecosystems can be directed towards a normal function. In case land-based preventive measures fail, ecosystems can be managed to help them function properly and to save that which can be saved.

Oxygen deficiency is a classical problem in lake ecosystems. Removal of hypolimnetic water (Olszewski's siphon), artificial total circulation and hypolimnetic oxygenation are well-known methods to tackle such problems, especially in summer at our latitudes. In order to illustrate advantages and disadvantages of aeration methods, an example will be given from a man-made lake in Colombia (Björk & Gelin 1980).

Lake water management

The El Penol reservoir is situated slightly north of the equator and was created by damming the

Hydrobiologia 86, 177–183 (1982). 0018-8158/82/0862-0177/$01.40.

178

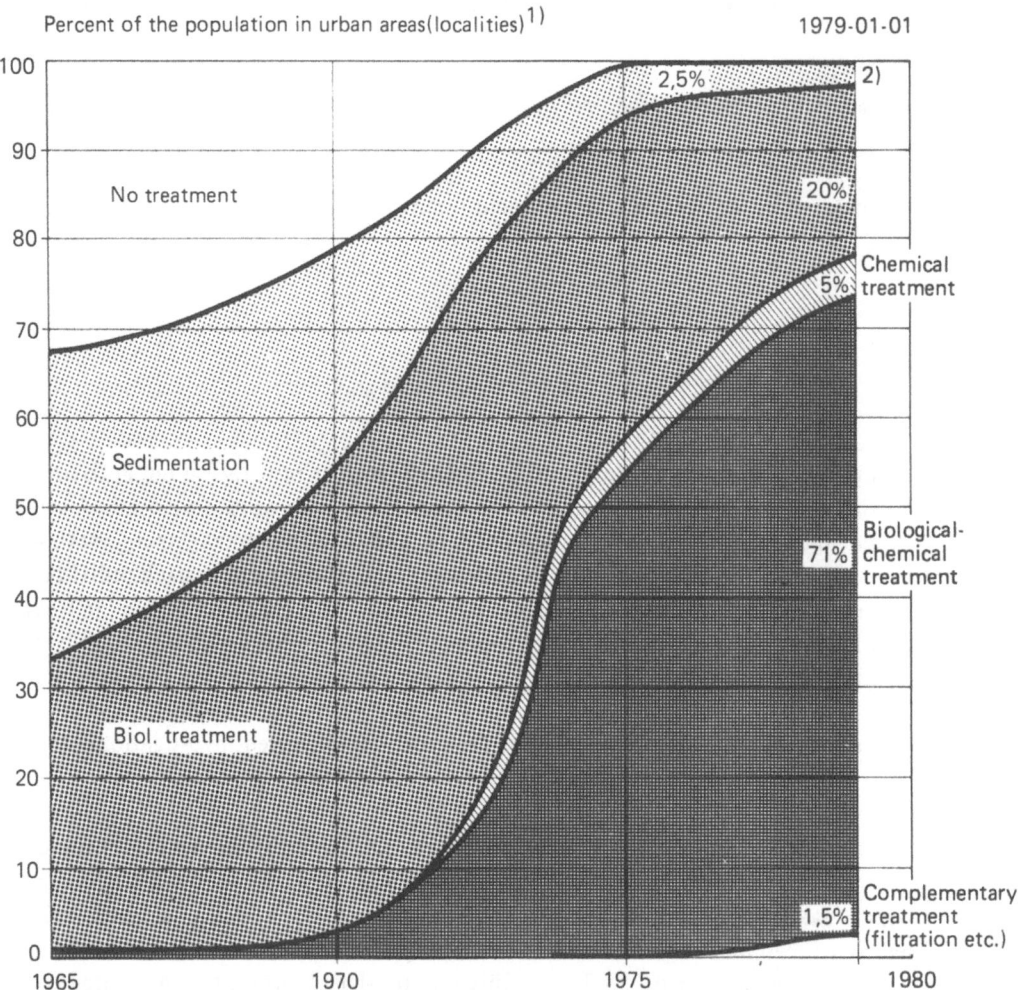

Percent of the population in urban areas(localities)[1]   1979-01-01

100
90   No treatment
80
70
60   Sedimentation
50
40
30   Biol. treatment
20
10
0
     1965   1970   1975   1980

2,5%   2)
20%
Chemical treatment
5%
Biological-chemical treatment
71%
Complementary treatment (filtration etc.)
1,5%

Type of sewage treatment	Number of treatment plants	Number of persons served
No treatment		7 000 (− 1 000)
Sedimentation	156 (−34)	181 000 (− 21 000)
Biological treatment	380 (−37)	1 398 000 (−110 000)
Chemical treatment	141 (+ 3)	324 000 (+ 22 000)
Bio. chemical treatment	625 (+18)	4 833 000 (+ 106 000)
Complementary treatment	18 (+ 2)	107 000 (+ 7 000)
	1 320 (−48)	6 850 000

The figures in brackets refer to the change since January 1 st, 1978

[1] Population clusters with at least 200 inhabitants are called localities if the distance between buildings as a rule does not exceed 200 meters. Localities account for approximately 83% of the total population.

[2] Half of which are without connection to municipal sewer system.

Fig. 1. Sewage treatment in Sweden 1965-1979. From The National Environment Protection Board, Sweden.

Rio Negro river. As in other tropical lakes, El Penol (*ca.* 60 km², 1870 m a.s.l., depth *ca.* 25–40 m depending on water level) has a thermal stratification resulting in oxygen deficiency below 6–8 m. The soils of the catchment area are of the poor, leached lateritic type, and the water is therefore poor in nutrients and macroconstituents.

The only outlet from the reservoir is a tunnel to the hydroelectric power plant situated in bedrock 810 m below the lake. The intake to the tunnel is about 2 m above the bottom.

The temperature of Rio Negro water is lower than that of water at any depth in the lake. Therefore, the river water flows in a submerged stream along the deepest part of the impounded river, i.e. the old river channel, towards the tunnel intake. Rio Negro is heavily loaded by minerogenic erosional material but not yet with organic matter or chemicals. As the oxygen concentration of river water is high and oxygen demand low, the river causes a peculiar oxygen stratification along its path through the lake. The river flows as a tongue-like protrusion towards the tunnel outlet, but oxygen delivered with the river water is totally consumed over about two-thirds of the distance. Conditions for development of hydrogen sulphide are not yet favourable in the reservoir, but a weak smell could be noticed from hypolimnetic water as well as in the hydroelectric power plant. Bubbling systems were installed in the reservoir outside the tunnel intake to eliminate this smell.

As oxygen disappears in the lake, iron goes into solution. Therefore the iron concentration increases along the submerged current through the lake, and in laterally situated backwaters hypolimnetic iron concentrations are considerably higher.

Bubbling breaks the thermal and chemical stratification whereby at least part of the iron is oxidized. However, it does not precipitate in the lake but is dispersed throughout the whole water column. Instead, precipitation takes place in the cooling system of the hydroelectric power plant. The precipitate accumulates, which necessitates stopping the turbines.

In this case bubbling should be replaced by hypolimnetic aeration, treating the layers along which water is transported to the tunnel intake. The aeration front should be located as far away from the outlet as possible so precipitation of oxidized iron can take place.

In man-made lakes like El Penol the minerogenic bottom makes the otherwise intriguing and challenging internal loading of nutrients negligible. Even the external loading has in this ecosystem a highly personal profile which is difficult to fit into formulas and models, as tributaries supplying the water body with nutrients disappear from the trophogenic zone to a considerable extent.

Lago Paranoa at Brasilia, another man-made South American lake, is also void of organic sediment (Björk 1979). The epilimnion is supplied with sewage causing a permanent, heavy water-bloom. Diversion of sewage is the simple remedy to make the lake suitable for the recreational purposes for which it was created. Unpolluted tributaries furnish the lake with water with a specific conductivity of less than 20 μS and phosphate phosphorus concentrations mostly less than 10 μg l^{-1}. After sewage diversion the lake would recover rapidly as no permanent damage has been done during the recipient period.

Sediment manipulation

Lakes at our latitudes have been compared to refrigerators – allochthonous and autochthonous material is preserved and accumulated, which delays recovery when too much has been supplied and produced. Lake Trummen provides an example of restoration by removal of a considerable part of the ecosystem's working and bound capital of nutrients.

The heavily polluted Lake Trummen was oligotrophicated (or meiotrophicated) by suction dredging the top sediment layer. This sediment was deposited during the recipient period and during the 12 years of nutrient recirculation following sewage diversion, when it appeared that the lake was damaged irreversibly.

Changes in environmental conditions are shown in Fig. 2, overall changes are schematically illustrated in Fig. 3, and Dr. Gertrud Cronberg describes the development of the ecosystem as reflected in the phytoplankton in this volume (Cronberg 1981). In summary, the restoration of Lake Trummen replaced an environmental problem with an environmental asset, water for recreation in an urbanized area.

Removal of nutrients in huge volumes of sedi-

180

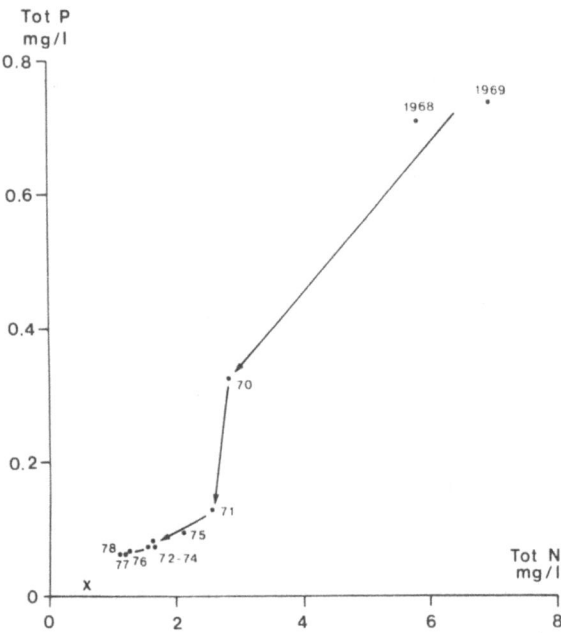

Fig. 2. Concentrations of total phosphorus and nitrogen in Lake Trummen following restoration in 1970–1971. Weighted mean values for the summer period (15 June–15 September). x = mean value of 412 samples from 250 lakes in the surrounding area (Kronoberg county). Raw data from Wilhelm Ripl and Länsstyrelsen i Kronobergs län. Calculations and figure design by Gunnar Andersson 1979a (compl.).

ment was necessary in the shallow Lake Trummen. However, if it is not necessary to deepen a lake loaded with highly oxygen-demanding sediments rich in nutrients, the sediment can be oxidized and phosphorus firmly bound directly in the lake. The designer of this Riplox procedure was Professor Wilhelm Ripl (Ripl 1976, 1978a, 1980).

A sediment harrow operated with compressed air is the tool by which chemicals are injected into the top sediment layer (Fig. 4). For oxygenation calcium nitrate solution is used. Through denitrification nitrogen is removed to the atmosphere, organic matter is oxidized and phosphate is bound to iron in the sediment. In order to make the environmental conditions as suitable as possible for these processes, pretreatment of the sediment might be necessary. Thus, in Lake Lillesjön (central south Sweden), where the method was first tested on a full scale, ferric chloride was added to alleviate the lack of iron in the sediment and pH was adjusted by injecting a slaked lime solution. Lake Lillesjön, treated in 1975, was converted from an irreversibly damaged sewage recipient to a lake suitable for swimming in.

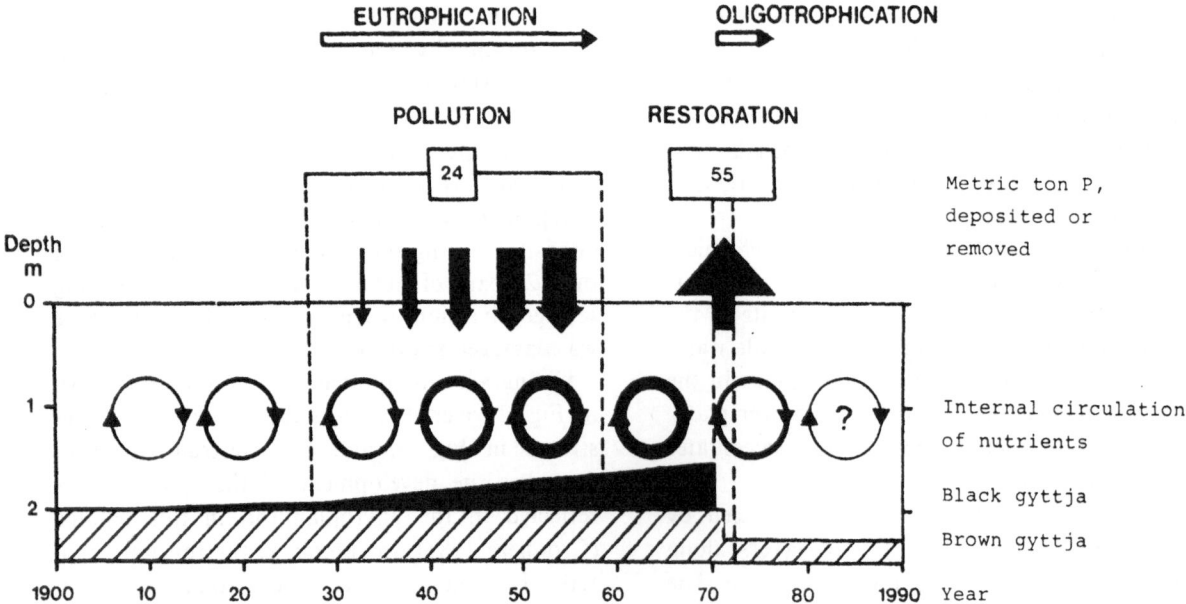

Fig. 3. A schematic illustration of the development of Lake Trummen during the 20th century. The question mark at the right denotes uncertainty concerning future loading of the lake from the urbanized watershed. From Gunnar Andersson 1975.

Fig. 4. Sediment manipulation according to Riplox and Contracid methods. 1. Field laboratory. 2. Supply of chemicals. 3. Portable compressor. 4. Tank for dilution of chemicals and mixing through bubbling. 5. Harrow-like device to loosen the sediment with compressed air and to inject chemicals. Riplox method uses addition of calcium nitrate and Contracid addition of sodium hydroxide or sodium carbonate solutions. 6–7. Air feed lines. 8. Air-driven pump. 9. Supply line for chemicals. 10. Guide line. 11. Air-driven dilution pump. 12. Dilution water intake. 13. Pneumatic winch arrangement. Drawing by Atlas Copco Airpower N.V.

Costs for sediment manipulation are, of course, considerably lower than for sediment removal. Furthermore, after thorough preinvestigations and tests, the method is quick. Only troublesome sediments, i.e. those with high oxygen demand and nutrient leaching, need to be treated. Epilimnetic sediments do not need treatment, and usually only 20–30% of the lake bottom requires manipulation. Preinvestigations must include a careful mapping of horizontal and vertical variation in sediment characteristics and an evaluation of the role of the sediment in the functioning of the lake ecosystem.

Regulation of pH

The most serious environmental problem in Sweden as a whole at present is acidification of surface and ground waters. It is estimated that

Sweden imports about 70% of the airborne sulphur compounds making her waters acid. Preventive measures at the emission sources and international agreements are the only definite solutions to this problem. However, in order to try to save valuable waters (with rare species, valuable fisheries, etc.) efforts are now being made to raise pH and keep it at a tolerable level. So far liming has been the only method used. After laboratory and small-scale field tests a new method – again designed by Ripl (1978b, 1980) – is now being applied on a full scale in Lake Lilla Galtsjön (southeastern Sweden). The technical device is a sediment harrow by means of which sodium carbonate solution (Na_2CO_3) is injected into the superficial sediment (Fig. 4). In this way the top sediment layer is used as an ion-exchanger. A long-term effect is expected as hydrogen ions from acid precipitation are successively exchanged with sodium ions from the prepared sediment. As in other restoration and management projects at the Institute, the study is carried out as ecosystem-oriented team-work for training limnologists. This Contracid method is principally intended for humic waters where liming is ineffective.

Water level lowering

Among the large group of lakes damaged due to water level lowering, Lake Hornborga (*ca.* 30 km²) is the best known. The Swedish government has decided to restore it as a waterfowl lake, which will be the biggest restoration project so far in Sweden, *ca.* 13 million US$. Before the water level is raised, 11 km² overgrown by *Phragmites* will be prepared to ensure its future preservation as open water with submerged vegetation. Several different types of amphibious and pontoon machines for wetland management (cutting of vegetation and preparation of the bottom to remove the root felt of *Phragmites, Schoenoplectus, Carex,* etc.) have been constructed within this project (Björk 1975, Björk & Granéli 1978b).

Wetland as energy fields

Due to drainage, Sweden has lost most of its wetlands, and remaining areas are still threatened. However, as the production of biomass is very high in wetlands – in comparison to terrestrial ecosytems – there is a growing interest in utilizing wetlands as valuable natural resources. A new project at the Institute of Limnology deals with productivity studies on *Phragmites australis,* a species characterized by wide variation among genetically different types (Björk & Granéli 1978a, b; Granéli 1980a, b). In good combinations of environmental conditions and *Phragmites* types, a biomass of up to 10 tons ha^{-1} can be harvested yearly in winter when the stems have a water content of only about 15%. The energy content of 2 kg *Phragmites* dry matter corresponds to 1 l of oil. As the potential areas for *Phragmites* cultivation are large, fuels made from reed could in the future make a contribution to Sweden's energy production in combination with other kinds of biomass, e.g. straw and wood chips.

The energy-reed project includes investigations on environmental conditions and *Phragmites* types. The demonstration *Phragmites* plantations are designed in consideration of nature conservancy and wildlife management in order to get the greatest possible benefit from the increase in wetland area.

Thus, while the Lake Hornborga project means removal of reed, the energy reed project means improvement of *Phragmites* growth. Together the projects justify preservation, increase, and management of wetland areas.

Biomanipulation

The restoration and management projects treated above have in common environmental operations to direct the development of organisms. After nutrient concentrations were lowered in Lake Trummen, efforts have been concentrated on governing the ecosystem by means of biomanipulation (Andersson 1979b; Andersson *et al.* 1978). The development of phyto- and zooplankton has been directed via the fish population. Even the exchange of nutrients between sediment and water can be controlled by this means. This field of research is thrilling and promising. As does other ecosystem-oriented restoration and management research, it improves our knowledge about relationships in ecoystems.

Conclusion

Restoration and management limnologists are called for environmental service when ecosystems have been brought to man-made collapse. Even if it is possible to solve several types of problems, waters should absolutely not be allowed to spoil with the implicit knowledge that they are reparable. The combination of present limnological knowledge and common sense makes it possible to plan soundly for protection, use and management of land and water resources. Case studies from Sweden on the limnological basis for water quality management planning are reported by Ingvar Lundqvist (1981, this volume).

References

Andersson, G., 1975. Results and experiences from restoration of Lake Trummen. (Resultat och erfarenheter från restaureringen av sjön Trummen.) Nordforsk, Miljövårdssekretariatet, publ. 1975 (1): 495–505 (in Swedish).

Andersson, G., 1979a. Internal loading of phosphorus and how to reduce it. Some examples from Lake Trummen. In: Björk, S. et al. Lake Management. Studies and results at the Institute of Limnology in Lund. Arch. Hydrobiol. Beih. Ergebn. Limnol. 13: 31–55.

Andersson, G., 1979b. The effect of fish on trophic conditions in eutrophic lakes. (Fiskens inverkan på trofiförhållandena i eutrofa sjöar.) Mimeogr. Institute of Limnology, Lund. LUNBDS/(NBLI-3024)/1–22 (in Swedish).

Andersson, G., Berggren, H., Cronberg, G. & Gelin, C., 1978. Effects of planktivorous and benthivorous fish on organisms and water chemistry in eutrophic lakes. Hydrobiologia 59: 9–15.

Björk, S., 1975. The degradation and restoration of Lake Hornborga. Inadvertent effects of man on the hydrological cycle. A Nordic case book. Nordic IHD Report 8: 111–119. Oslo.

Björk, S., 1979. The Lago Paranoa restoration project, Brasilia, Brazil. Mimeogr. Institute of Limnology, Lund. WHO project BRA-2341, final report: 1–45, append. I–V.

Björk, S. & Gelin, C., 1980. Limnological function and management of the El Penol Reservoir, Colombia. Mimeogr. Institute of Limnology, Lund. LUNBDS/(NBLI-3038)/1–58.

Björk, S. & Granéli, W., 1978a. Energy reeds I. (Energivass I.) Mimeogr. Institute of Limnology, Lund. Reprinted 1980 by Nämnden för Energiproduktionsforskning. NE 1980: 12, 1–77 (Swedish, English summary).

Björk, S. & Granéli, W., 1978b. Energy reeds and the environment. Ambio 7: 150–156.

Cronberg, G., 1981. Phytoplankton changes in Lake Trummen induced by restoration. Developments in Hydrobiology (this volume).

Granéli, W., 1980a. Energy reeds II. (Energivass II.) Mimeogr. Institute of Limnology, Lund. LUNBDS/(NBLI-3028)/1–37 (in Swedish).

Granéli, W., 1980b. Energy reeds III. (Energivass III.) Mimeogr. Institute of Limnology, Lund. LUNBDS/(NBLI-3029)/1–39 (in Swedish).

Likens, G. E. & Bormann, F. H., 1979. The role of watershed and airshed in lake metabolism. Arch. Hydrobiol. Beih. Ergebn. Limnol. 13: 195–211.

Lundqvist, I., 1981. The limnological basis of water quality management planning. Developments in Hydrobiology (this volume).

Ripl, W., 1976. Biochemical oxidation of polluted lake sediment whith nitrate – A new lake restoration method. Ambio 5: 132–135.

Ripl, W., 1978a. Oxidation of lake sediments with nitrate – a restoration method for former recipients. Mimeogr. Institute of Limnology, Lund. LUNBDS/(NBLI-1001)1–151.

Ripl, W., 1978b. Restoration of acidified lakes by the Contracid method. (Restaurering av försurade sjöar enligt Contracidmetoden.) Mimeogr. Institute of Limnology, Lund: 1–13.

Ripl, W., 1980. Lake restoration methods developed and used in Sweden. Mimeogr. Communications Department, Atlas Copco Airpower, Wilrijk, Belgium: 1–10.

Changes in the phytoplankton of Lake Trummen induced by restoration

Gertrud Cronberg

Institute of Limnology, University of Lund, Box 3060, 220 03 Lund, Sweden

Keywords: lake restoration, phytoplankton, community changes

Abstract

Lake Trummen, previously an oligotrophic lake, was heavily polluted during 1930–1957. The lake was restored by suction dredging of the top sediment layer. After restoration nutrient concentrations and phytoplankton biomass were drastically reduced. Eutrophic species disappeared and more oligotrophic species returned to the lake. The investigation covers 12 years-monthly-bimonthly countings of algae together with electronmicroscopical studies of taxonomically difficult taxa. The investigation was made at the species level (300 species identified and 80 quantitatively recorded).

Introduction

Lake Trummen is situated in the city of Växjö in the south Swedish uplands. The drainage area is 13 km^2 and consists of granites and till poor in lime and basic minerals (Fig. 1).

Originally Lake Trummen was an oligotrophic lake typical of the region (Digerfeldt 1972), but during this century it recieved sewage and waste water from a flax factory. The worst pollution period was 1936–1957, and although sewage was diverted from the lake in 1958, the lake did not recover.

During the 1960s Lake Trummen showed all the signs of an overexploited recipient with dense waterblooms during summer, steadily expanding areas of *Phragmites* and *Typha* and plaur formations along the shores and sometimes fish kills during winter. There was no submerged vegetation.

Annual sediment growth increased from 0.4 mm to 8 mm during the pollution period. Finally about 0.5 m FeS-coloured black nutrient-rich sediment had been deposited on top of the brown nutrient-poor layers.

At the end of the 1960s the city authorities of Växjö planned to build a new university on the shore of Lake Trummen, and conditions around the lake had to be improved. A restoration plan was designed (Björk 1966) and the following time-table for research and restoration measures was applied:
1968–1969 Pre-investigation
1970–1971 Restoration
1972–1973 Post-investigation
1974–1980 Follow-up investigation (Fig. 2).

The restoration was carried out during the summers of 1970 and 1971. About 0.5 m of FeS-enriched sediment was removed through suction-dredging, and the sediments were deposited in settling ponds beside the lake (Fig. 2). Run-off water from the ponds was treated with aluminium sulphate in order to reduce the phosphorus concentration before it returned to the lake. In 1971 the shores were cleaned, and the bulk of the macrophyte vegetation was taken away. Some completely overgrown bays were diked and used for sediment deposition.

Hydrobiologia 86, 185–193 (1982).   0018-8158/82/0862-0185/$01.80.
© Dr W. Junk Publishers, The Hague.

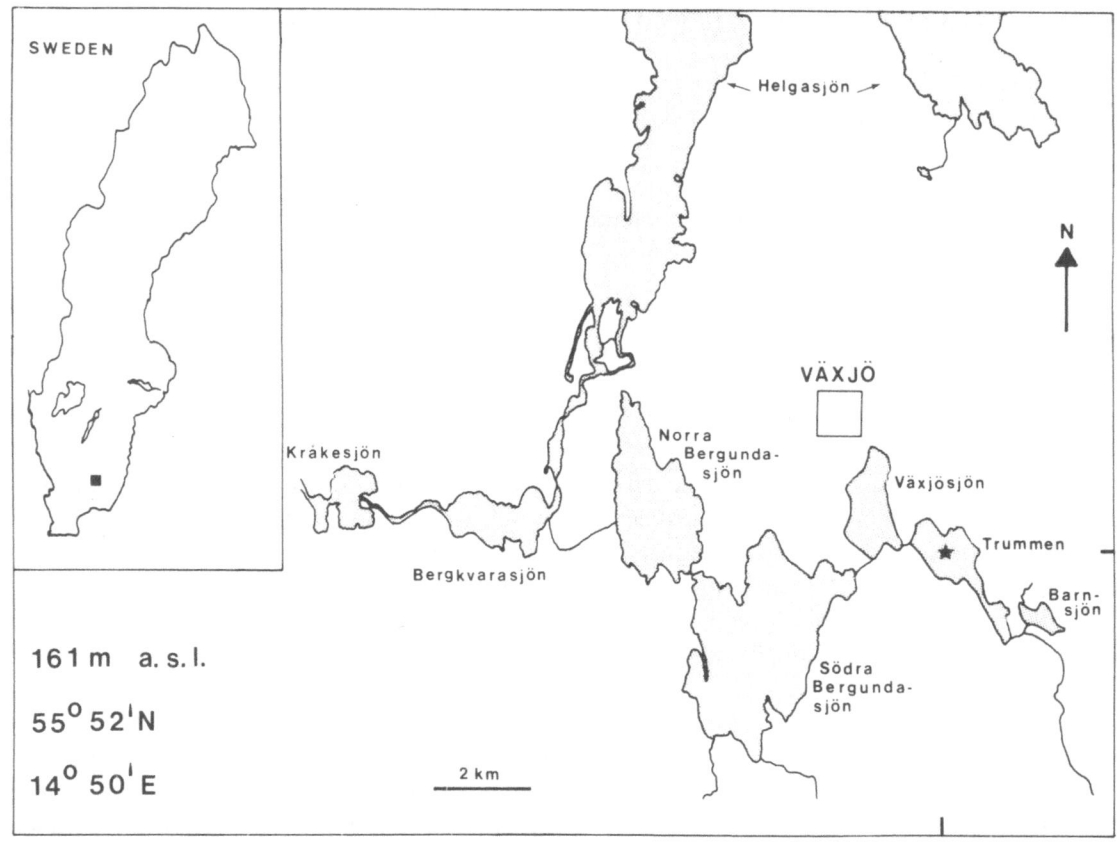

Fig. 1. The position of Lake Trummen.

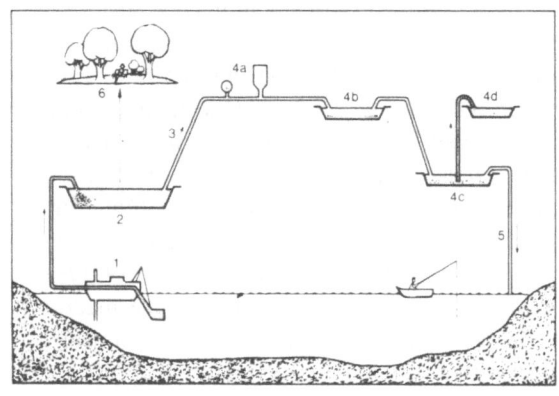

Fig. 2. Lake Trummen's tailor-made treatment: (1) suction dredger designed to operate with minimal turbidity and mixing; (2) settling pond; (3) runoff water; (4) precipitation of phosphorus and suspended matter with aluminium sulphate: (4a) automatic dosage, (4b) aeration, (4c) sedimentation, (4d) sludge pond/; and (5) clarified run-off water. (6) The dried sediment is used as fertilizer for parks and lawns. From Björk (1972).

Materials and methods

Samples for phytoplankton and chemical analyses were taken monthly at the middle of the lake at 0.2 m depth. For quantitative phytoplankton analyses, samples were taken with a Ruttner sampler and immediately fixed with Lugol's solution (Willén 1962). Net samples (45 μm) were fixed to a 4% solution of formalin.

Phytoplankton were counted in sedimentation chambers with an inverted microscope (Utermöhl 1958). Single cells were counted and filamentous algae measured (Cronberg 1980).

To obtain acceptable reliability, 60–100 cells of each species were counted. At least 20 cells of each species or group were measured for calculation of mean volume on every occasion. Phytoplankton density was taken as 1.0 for calculation of fresh weight.

Samples rich in blue-greens were first sonicated

with a Rapidis 50 Ultrasonic Disintegrator at 20 kHz for 15-60 s to split bundles of *Aphanizomenon* into filaments, *Anabaena* into small chains and *Microcystis* into cells. Preparations for SEM and TEM were made according to Cronberg (1980).

For calculating species diversity, Brillouin's index (1956) was used and the index calculated on a cellular basis.

Results

Phytoplankton 1968-1978

During the period 1968-1978 the phytoplankton showed similar seasonal variations. In winter very few algae were found in the lake, mainly Chrysophyceae, Cryptophyceae, and small green algae, sometimes forming blooms for short periods. Blue-green algae started to develop in the beginning of June. Filamentous blue-greens dominated initially, but in July-August they were succeeded by *Microcystis* spp. In autumn a new maximum of diatoms appeared, consisting of *Melosira* and/or *Synedra*. Most algae disappeared when the ice cover developed. In the summers 1968 and 1969, i.e. before restoration, the total biomass of phytoplankton was high (Fig. 3). However, there was a drastic decrease from the start of restoration in 1970 and onwards.

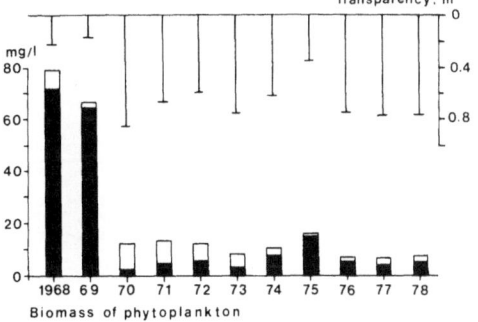

Fig. 3. Changes in phytoplankton biomass and transparency in Lake Trummen 1968-1978 following the restoration in 1970-1971. Weighted mean values of total biomass of phytoplankton (whole column) and biomass of blue-green algae (black column) for the summer period (15 June-15 Sept.).

Cyanophyta

Before restoration blue-green algae were most important in the Lake Trummen plankton community. In 1970 the biomass of blue-greens was reduced to 5% of the prerestoration values. *Microcystis*, especially *M. aeruginosa*, formed heavy blooms in 1968 and 1969, but was much reduced after restoration. It was succeeded by small blue-green algae such as *Aphanothece*, *Synechococcus* and *Cyanodictyon* (Fig. 4).

Of the filamentous blue-green algae, *Aphanizomenon flos-aquae*, *Oscillatoria agardhii*, *Raphidiopsis mediterrannea* and *Anabaena solitaria* f. *smithii*, all disappeared after resoration, but *Aphanizomenon gracile* and *Anabaena lemmermannii* increased (Fig. 4).

Chlorophyta

Before restoration green algae developed large biomass peaks in spring and autumn. During restoration and especially in 1971, the biomass of green algae increased, and the maximum appeared during summer. Many different species were represented at that time, but the most abundant genera were *Pediastrum* and *Scenedesmus* (Figs. 5 and 6).

Pediastrum boryanum was the most dominant *Pediastrum* species before restoration, but its biomass decreased thereafter. *P. duplex* and *P. gracillimum* appeared in great numbers during the restoration period. The highest peak recorded was in August 1979 when most macrophyte vegetation was removed from the littoral zone. In 1972-1975 *P. biradiatum* was frequent, and during the last few years *P. angulosum* and *P. tetras* increased in numbers.

After restoration the biomass of the genus *Scenedesmus* decreased, and the species composition changed. During the whole observation period the genus was represented with 21 species. Before restoration the most common species were *S. abundans*, *S. acuminatus*, *S. arcuatus*, *S. armatus*, *S. oahuensis* and *S. quadricauda*. After restoration they all decreased. During the course of restoration the number of large *Scenedesmus* species decreased while the number of small species, such as *S. subspicatus*, increased. After restoration the size of *Scenedesmus* coenobia also decreased.

Other green algae that were reduced after restoration were *Chlamydomonas* spp., *Micractinium pusillum* and *Chorella* sp.

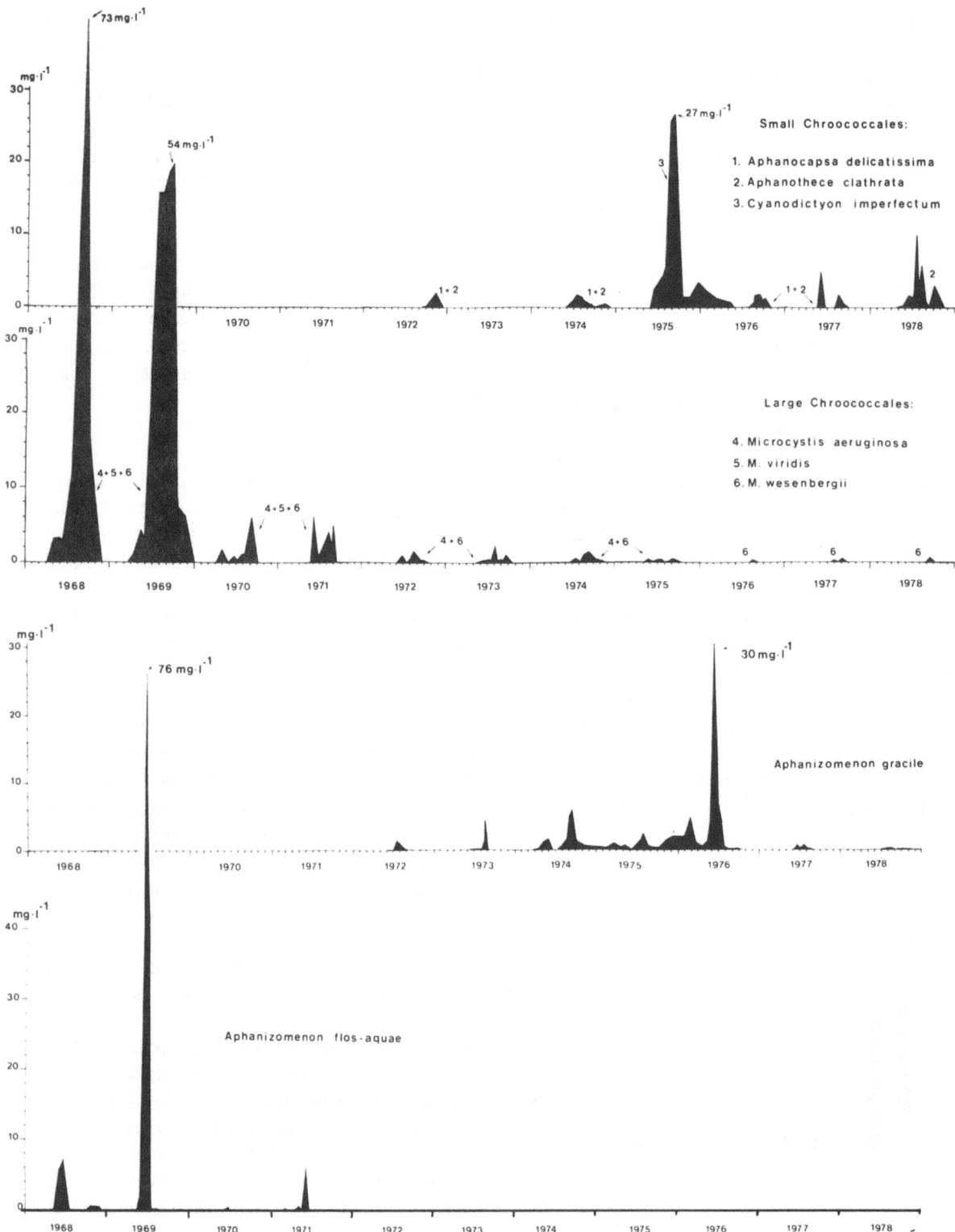

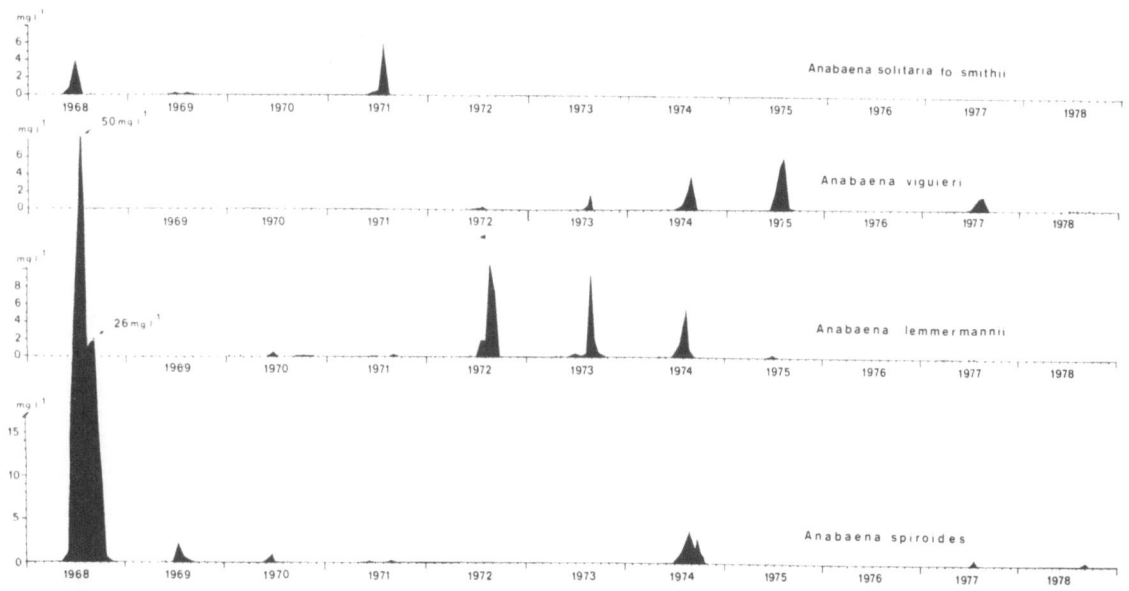

Fig. 4. Biomass of cyanophyte species in Lake Trummen 1968–1978.

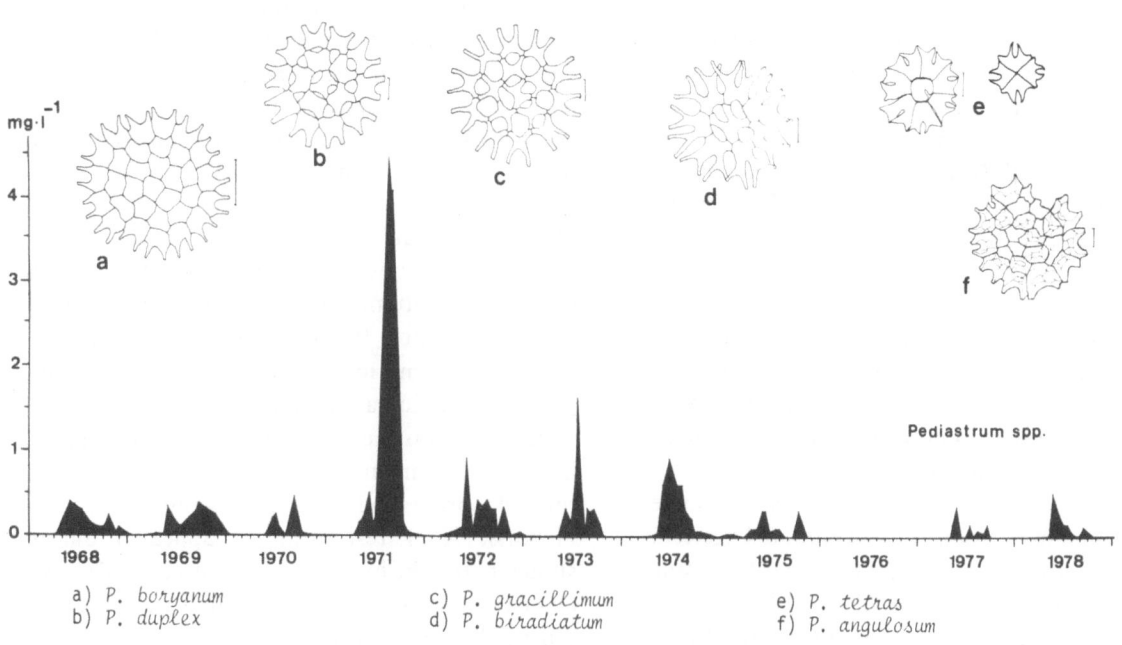

a) P. boryanum   c) P. gracillimum   e) P. tetras
b) P. duplex   d) P. biradiatum   f) P. angulosum

Fig. 5. Biomass of *Pediastrum* in Lake Trummen 1968–1978.

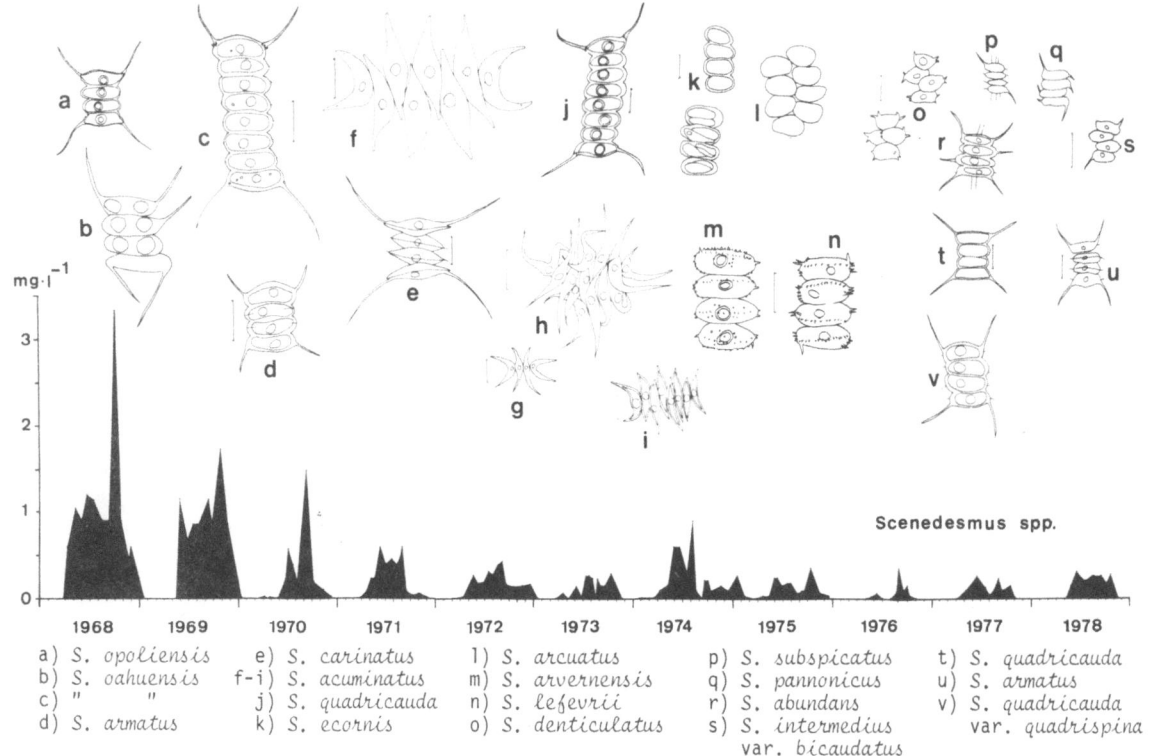

a) *S. opoliensis* e) *S. carinatus* l) *S. arcuatus* p) *S. subspicatus* t) *S. quadricauda*
b) *S. oahuensis* f-i) *S. acuminatus* m) *S. arvernensis* q) *S. pannonicus* u) *S. armatus*
c) " " j) *S. quadricauda* n) *S. lefevrii* r) *S. abundans* v) *S. quadricauda*
d) *S. armatus* k) *S. ecornis* o) *S. denticulatus* s) *S. intermedius* var. *quadrispina*
  var. *bicaudatus*

Fig. 6. Biomass of *Scenedesmus* in Lake Trummen 1968–1978.

Chrysophyceae

Before restoration a few species of Chrysophyceae occurred at low densities in Lake Trummen. After restoration the number increased from 11 species to 50. The most important genera were *Dinobryon, Mallomonas* and *Synura.* They increased immediately after restoration. The highest peak of Chrysophyceae was recorded in spring 1971 (Fig. 7a), after which a spring maximum appeared every year, but with reduced size.

Mallomonas eoa was the dominant *Mallomonas* species and formed cysts when the water temperature rose above +6 °C (Cronberg 1973). Altogether 15 species of *Mallomonas* were identified from Lake Trummen.

During the 1960s *Synura petersenii* was sometimes found in great quantities (Björk & Digerfeldt 1965). *Synura* was, however, not recorded during the preinvestigation period 1968–1970. In February, 1971, *S. petersenii* formed a bloom under the ice, and a high biomass was recorded. During 1972 few *S. petersenii* colonies were found, but instead a

great number of *S. spinosa.* In April, 1974, the highest biomass of *Synura* was recorded. On this occasion *S. echinulata* dominated, and the other *Synura* species appeared in low numbers only. After 1974 *S. echinulata* was the most common *Synura* in Lake Trummen.

Diatomophyceae

Before restoration diatoms appeared with both spring and autumn maxima and high biomass. After restoration the seasonal pattern changed. Diatoms became more frequent during summer, but the total biomass decreased (Fig. 7b).

Melosira was the most important diatom genus in Lake Trummen. Before restoration *Melosira* formed spring and autumn maxima with high biomass. After restoration it appeared during summer, and the biomass was reduced by 57%. No changes in *Melosira* species composition was seen during the investigation period. The same species dominated, viz. *Melosira ambigua, M. granulata* var. *augustissima* and *M. italica* subsp. *subarctica.*

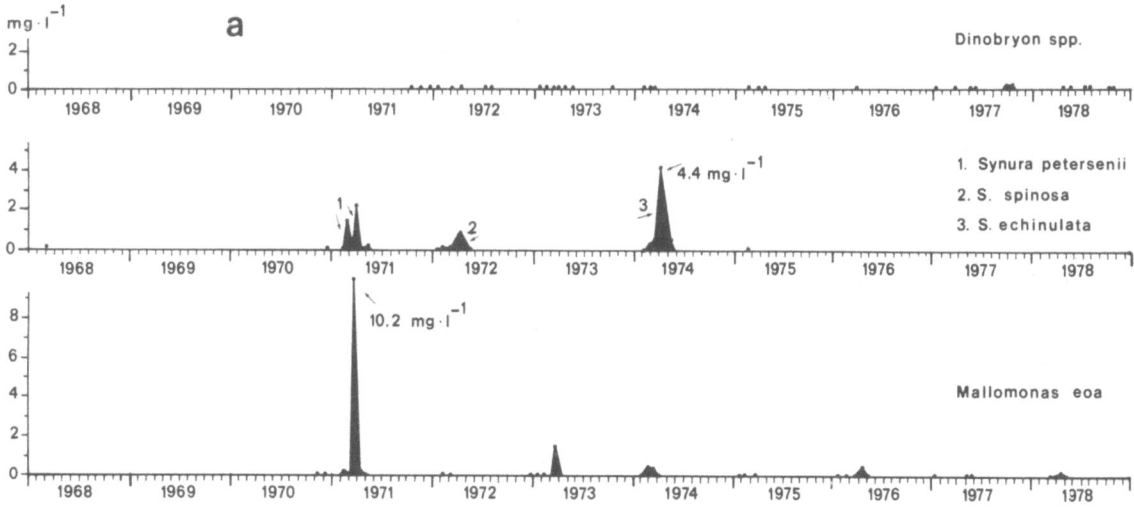

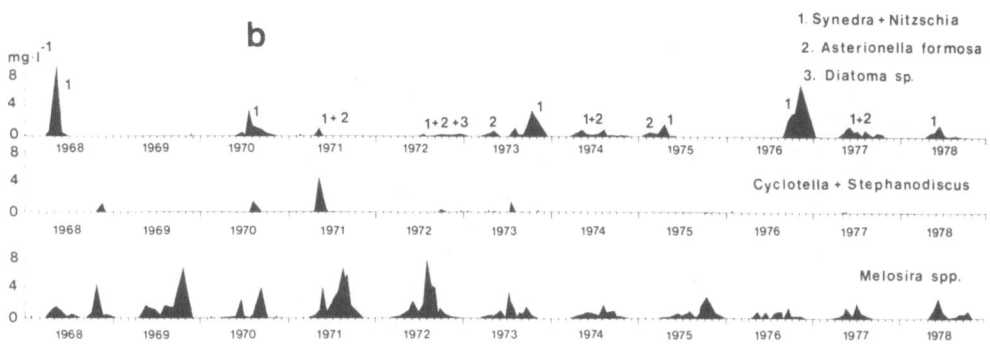

Fig. 7. Biomass of a) *Dinobryon, Synura* and *Mallomonas eoa* and b) different diatoms in Lake Trummen 1968–1978.

Stephanodiscus hantzschii was reduced after restoration, but *Cyclotella* spp. seemed to increase.

Cryptophyceae and Dinophyceae

Before and during restoration both groups were frequent and reached high numbers for short periods. After restoration they decreased drastically.

Undetermined µm-algae

This is an artificial group where undetermined small algae were counted in size groups, 2–12 µm. Small cryptomonads, chrysomonads, green and blue-green algae were assigned to this group.

Before restoration the µm-algae appeared distinctly in spring and autumn peaks. During the restoration µm-algae increased and became frequent in summer. After 1974 the µm-algae decreased in number and did not show regular seasonal variation.

Community structure

More species of blue-green and green algae were represented in Lake Trummen before than after restoration. The decrease in connection with restoration was mainly in the orders Chroococcales and Chlorococcales. After restoration, on the other hand, the number of Chrysophyceae species increased threefold.

The number of species of Pyrrhophyta, Diatomophyceae and Xanthophyceae remained more or less the same during the whole investigation period. It was evident that the number of periphytic and benthic species was reduced (Chlorococcales), while true planktonic species increased (Chrysophyceae) in connection with the restoration.

The restoration of Lake Trummen induced an immediate and drastic reduction in phytoplankton biomass. Species diversity increased during the restoration period (Cronberg 1980) and a few years after, but has now stabilized at nearly the same level as before restoration. However, species composition changed completely as a result of the restoration.

Discussion

From the results obtained in this phytoplankton investigation of Lake Trummen it is obvious that the restoration induced large changes in the phytoplankton community. The external loading was eliminated in 1958, but no real improvements were observed. When sewage water from the polluted Lake Norrviken outside Stockholm (Ahlgren 1978; Ahlgren et al. 1979) was diverted in 1969–1970, the conditions in the lake improved and the lake successively recovered. The nutrient concentrations decreased while the species diversity of phytoplankton increased (Tinnberg 1979). The recovery of Lake Norrviken was initially caused by a dilution and later by decreased release of nutrients from the sediments.

Lake Trummen, however, had developed up to 0.5 m thick nutrient-rich sediment and equally rich macrophyte vegetation. The nutrients deposited in the sediments and macrophytes recirculated within the ecosystem. But the restoration of Lake Trummen caused an extremely rapid decrease in nutrients and phytoplankton biomass. Altogether 50 Mg phosphorus and 450 Mg nitrogen were removed from the lake through the sediment dredging (Andersson 1979). At the same time as the top sediment layer was removed cysts, spores, auxospores, zygotes, hormogonia etc., resting cells

of algae and other organisms belonging to the actual plankton community also disappeared. The sediment removal caused a reduction mainly in the blue-green algal community.

In the polluted Lake Trummen the sedimentation rate was 8 mm yr^{-1} (Digerfeldt 1972) and cells were rapidly covered in the sediment. Once submerged they may have been inhibited from leaving the sediments. However, in sediments algae with visible chloroplasts can be viable for many years. Stockner & Lund (1970) investigated the viability of algae in sediment cores from different lakes in the English Lake District. They tested the viability through culturing algae from cores and found viable diatom cells from samles with an age of 175–275 yr.

Livingstone and Jaworski (1980) succeeded in growing *Aphanizomenon* and *Anabaena* akinetes from sediment with an age of up to 18 and 64 years, respectively. They showed that akinetes of blue-green algae have not only a temporary or overwintering function but may also ensure long-term survival. It is assumed that the new bottom uncovered in the restored Lake Trummen did not contain many resting cells of eutrophic species. After restoration another blue-green algal community was established with *Aphanocapsa delicatissima*, *Aphanothece clathrata*, *Anabaena lemmermannii* and *A. viguieri* as dominating species. *Aphanothece clathrata* and *Anabaena lemmermannii* are especially common in adjacent oligotrophic lakes. Thus the new blue-green algae community that appeared after restoration may refer to hatching of old spores from the 'new' sediment. The great maximum of *Mallomonas eoa* the first spring after restoration was certainly a result of cells developing from cysts at least 60 years old.

Before restoration the sediments in Lake Trummen were periodically anaerobic, during blue-green algal blooms in summer and in winter when the lake was ice-covered. After restoration the oxygen conditions improved considerably (Andersson 1979). Some blue-green algae (*Microcystis* spp.) are reported to survive better when anaerobic conditions prevail (Sirenkov et al. 1969). Thus the disappearance of the heavy blooms of blue-green algae in connection with the restoration may be a result of nutrient reduction, improved oxygen conditions and elimination of spores.

It is clear that the restoration of Lake Trummen improved the conditions in the lake drastically and started a meiotrophication, which is still going on. The phytoplankton community changed and has now much in common with adjacent oligotrophic lakes.

References

Ahlgren, I., 1978. Response of Lake Norrviken to reduced nutrient loading. Verh. int. Verein. Limnol. 20: 846–850.

Ahlgren, I. *et al.*, 1979. Lake metabolism studies and the results at the Institute of Limnology in Uppsala. Arch. Hydrobiol. Beih. Ergebn. Limnol. 13: 10–30.

Andersson, G., 1979. Internal loading of phosphorus and how to reduce it. Some examples from Lake Trummen. In: Björk, S. *et al.* Lake Management. Arch. Hydrobiol. Beih. Ergebn. Limnol. 13: 31–55.

Björk, S., 1966. Lake restoration. Skånes Natur 53: 1–5 (in Swedish).

Björk, S., 1972. Swedish lake restoration program gets results. Ambio 1: 156–165.

Björk, S. & Digerfeldt, G., 1965. Notes on the limnology and post-glacial development of Lake Trummen. Bot. Notiser 118: 305–325.

Brillouin, L., 1956. Science and Information Theory. Academic Press, New York.

Cronberg, G., 1973. Development of cysts in Mallomonas eoa examined by scanning electron microscopy. Hydrobiologia 43: 29–38.

Cronberg, G., 1980. Phytoplankton changes in Lake Trummen induced by restoration. Long-term whole-lake studies and experimental biomanipulation. Inst. Limnol. Univ. Lund, Sweden: 1–182.

Digerfeldt, G., 1972. The post-glacial development of Lake Trummen. Regional vegetation history, water level changes and palaeolimnology. Folia limnol. scand. 16: 1–96.

Livingstone, D. & Jaworski, G. H. M., 1980. The viability of akinetes of blue-green algae recovered from sediments of Rostherne mere. Br. Phycol. J. 15: 357–364.

Stockner, J. G. & Lund, J. W. G., 1970. Live algae in post-glacial lake deposits. Limnol. Oceanogr. 15: 41–58.

Tinnberg, L., 1979. Phytoplankton diversity in Lake Norrviken 1961–1975. Holarct. Ecol. 2: 150–159.

Utermöhl, H., 1958. Zur Vervollkommnung der quantitativen Phytoplanktonmethodik. Mitt. int. Verein. Limnol. 9: 1–39.

Willén, T., 1962. The Utål lake chain, central Sweden and its phytoplankton. Oikos, Suppl. 5: 1–156.

The recovery of L. Vesijärvi following sewage diversion

Juha Keto

The Municipal Laboratory of Lahti, P.O. Box 108, SF-15141 Lahti 14, Finland

Keywords: lake covery, blue-green algae, diatoms, algal blooms, sewage

Abstract

For 60 years the sewage of the City of Lahti was discharged into L. Vesijärvi. The eutrophication of the lake was observed as early as the 1920s but in the 1960s the pollution became obvious. The sewage was completely diverted in 1976. As a result of the diversion the bacteriological defects were eliminated within one year, restoring the recreational value of the lake. The recovery of water quality was rapid during the first two years. The oxygen content increased markedly but hypolimnetic oxygen depletion persisted. The phosphorus content decreased about 60% and the nitrogen content about 30%. After that period the recovery slowed down. The biomass of phytoplankton decreased but the abundance of heterocystous blue green algae increased causing blooms in July and August. The primary production has lately been reduced despite the lack of improvement in chemical water quality. Hypolimnetic aeration was started in order to accelerate the recovery. The results during winter stagnation have been encouraging.

Introduction

L. Vesijärvi is part of the Kymijoki watercourse and is located in Päijät-Häme, Southern Finland, between the Salpausselkä Ridges. The lake is divided into five basins Enonselkä, Paimelanlahti, Komonselkä, Laitialanselkä and Kajaanselkä (Fig. 1), of which Enonselkä and Paimelanlahti are the most eutrophic.

The aim of this study was to consider the recovery of the polluted L. Vesijärvi, concentrating upon the main basins, Enonselkä and Kajaanselkä. Hydrological characteristics of the study area are as follows:

	Whole lake	Enonselkä	Kajaanselkä
Drainage area, km²	515	84	138
Surface area, km²	110	26	44
Volume at MSL, ×10⁶m³	663	176	300
Maximum depth, m	40	33	40
Mean depth, m	6.0	6.8	6.8
Theoretical retention time, yr	5.4	5.6	2.4

Originally, L. Vesijärvi was a clear-water oligo-humic lake due to the rich ground water and spring drainage, but for 60 years the mechanically treated sewage of the City of Lahti was discharged into the southern part of the lake. Since 1915 the population of the city has increased from less than 5 000 to 95 000 in 1975, and the sewage load attained a maximum of 60 000 inhabitants in the early 1970s. The eutrophication of Enonselkä was first observed

Hydrobiologia 86, 195–199 (1982). 0018-8158/82/0862-0195/$01.00.

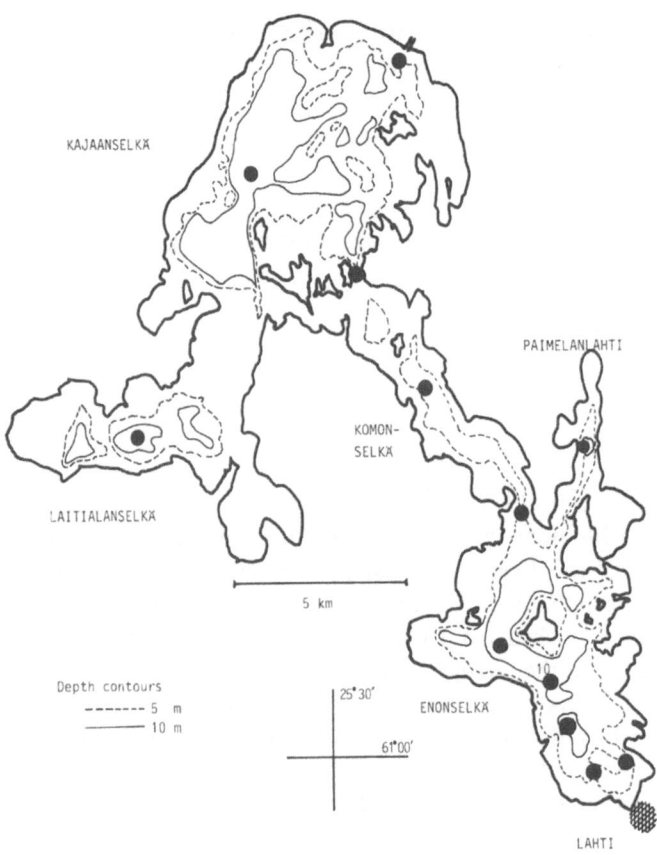

Fig. 1. Map of L. Vesijärvi showing the sampling stations (dots).

by Järnefelt (1929) but the alarming features of pollution was registered in 1960 (Seppänen 1962). During the 1960s L. Vesijärvi became one of the most polluted lakes in Finland and the southern part of the lake lost its value for recreation and fishing.

In 1975–1976 the sewage of Lahti was completely diverted from L. Vesijärvi to the River Porvoonjoki at a cost of 58 million FIM (16 million US$). Hypolimnetic aeration of Enonselkä was started in March 1979 by Hydixor aerator (15 kW) located at station 10 (Fig. 1).

Materials and methods

Existing data on L. Vesijärvi were utilized in order to prepare the follow-up program on the recovery of the lake (Järnefelt 1929; Seppänen 1965; Vaara & Vaara 1971; Keto 1973; Lappalainen 1973; Keto 1976). The data comprised occasional physico-chemical, bacteriological and phytoplankton observations. The first phosphorus analysis was carried out in 1966. At that time the entire hypolimnion of Enonselkä was anaerobic at the end of the winter stagnation period. Due to the primary load and the mobilisation from the sediments, the phosphorus concentrations were high. The average concentration of total phosphorus ranged in March 1966–1973 from 60 to 160 mg m^{-3} and the average concentration of total nitrogen varied between 1 000 and 1 600 mg m^{-3} (Fig. 2). During the late 1960s the influence of sewage load was transferred increasingly to Kajaanselkä where hypolimnetic oxygen depletion became regular. The blue green algal blooms that already predominated in Enonselkä at the beginning of the decade extended all over the lake. The bacterial concentration in Enonselkä were high and the hygiene of the water of Lahti bathing beaches fell short of the health

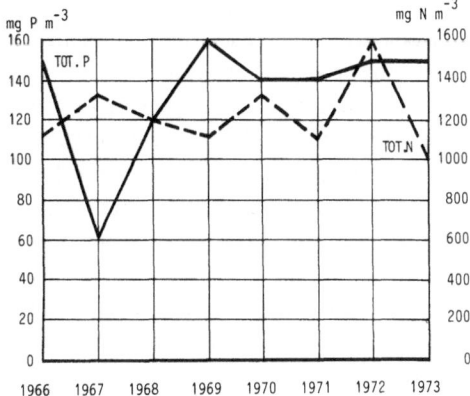

Fig. 2. Changes in the average total phosphorus and total nitrogen concentrations in Enonselkä in March 1966–1973. Values are means of 4–8 measurements.

standards for qualified swimming water (less than one faecal coliform or streptococcus ml^{-1}).

The study was started in the beginning of 1976 when about 70% of the sewage was diverted. The sampling stations are shown in Fig. 1. The vertical samples for physico-chemical analysis were collected in January, March, May, June, August and October. The analyses carried out were dissolved oxygen, pH, electrolytic conductivity, chemical oxygen demand, total P, total N and chloride (Goltermann 1969).

The hygienic quality of the swimming waters of the city was evaluated by samples taken weekly in June–August. For the faecal coliforms the membrane filtration technique (mFc-agar, incubation at 45 °C) was used (APHA *et al.* 1975). The membrane filtration technique was also employed for faecal streptococci analysis. The agar used was n-Streptococcus-agar and plates were incubated at 35 °C for 48 h.

Integrated samples for biological analysis were taken from the euphotic zone every two weeks in June–August. The primary production capacity was measured *in vitro* using the ^{14}C method (Vollenveider 1974). In 1977 and 1979 the composition and freshweight biomass of phytoplankton was determined by the Utermöhl technique (Utermöhl 1958).

Results and discussion

The primary effect of sewage was eliminated in a

couple of months following the total diversion (April 1976). During summer 1976 the concentrations of faecal indicator bacteria decreased below the standard limits. In the following summers the hygienic quality of water did not restrict the recreational use of the lake.

During the high loading period 1960–1975 the quality of water was at its worst at the end of the winter stagnation. In Enonselkä the volume of the oxygen depleted water was about $20 \times 10^6 m^3$. In winter 1976, when 70% of the load was diverted, the respective volume was about $1.1 \times 10^6 m^3$ and in winter 1977 $0.6 \times 10^6 m^3$. No further recovery in oxygen concentrations was observed before the aeration of the lake (Hydixor) since March 1980 (Fig. 3). The volume of the oxygen depleted hypolimnion water during summer stagnation decreased markedly from about $7.6 \times 10^6 m^3$ in 1975 to $2.1 - 3.0 \times 10^6 m^3$ in 1976–1980. During summer 1980, owing to the aeration, the aerobic conditions prevailed almost one month longer than previously.

The average total phosphorus content dropped from 150 mg P m^{-3} in March 1975 to 50 mg P m^{-3} in March 1977. During the following years there was a slight increase in the average phosphorus content (in 1978 61 mg P m^{-3}, in 1979 70 mg P m^{-3}). In March 1980 the aeration decreased the phosphorus concentration (Fig. 4). In May 1977–1980 the average total phosphorus content varied between 36 and 56 mg P m^{-3}. The respective values at the end of summer stagnation ranged from 65 to 79 mg P m^{-3} due to the rapid reduction from sediments. The average reduction rate from sediments was 18 mg P m^{-2}day^{-1} during summer stagnation 1978. This is in accordance with the reduction rates introduced by Kyröläinen (1978) who, in the eutrophic part of L. Saimaa during summer 1975, found under aerobic conditions a respective value of 12 mg P m^{-2}day^{-1}.

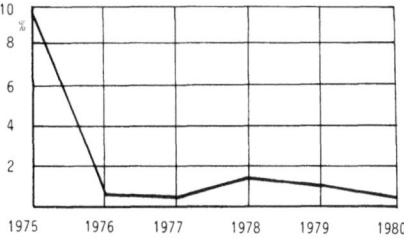

Fig. 3. The percentage of the oxygen depleted water from the whole water body in Enonselkä in March 1975–1980. Values are means of 23–31 measurements.

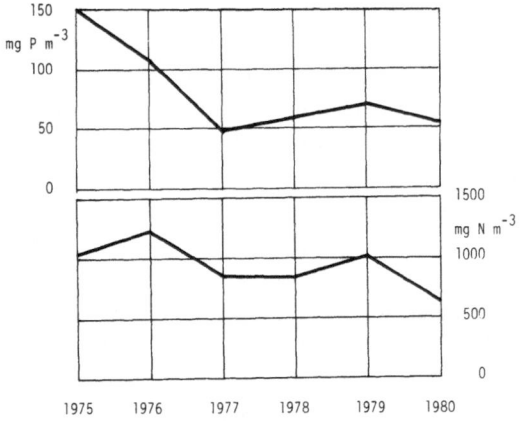

Fig. 4. The average total phosphorus and total nitrogen concentrations in Enonselkä in March 1975-1980. Values are means of 15-17 measurements.

In Kajaanselkä the changes of the phosphorus content were small and the concentrations varied in March 1975-1980 between 15 and 20 mg P m^{-3} (Fig. 5).

The reduction of nitrogen in Enonselkä during 1975-1980 was about 30%. In March 1976 the average total nitrogen content was 1 300 mg N m^{-3}, in March 1977 860 mg N m^{-3}, in March 1978 840 mg N m^{-3}, in March 1979 1 000 mg N m^{-3} and in March 1980 700 mg N m^{-3} (Fig. 4). In Kajaanselkä the changes of nitrogen remained rather small and the average total nitrogen concentrations in March 1975-1980 varied between 450 and 600 mg N m^{-3} (Fig. 5).

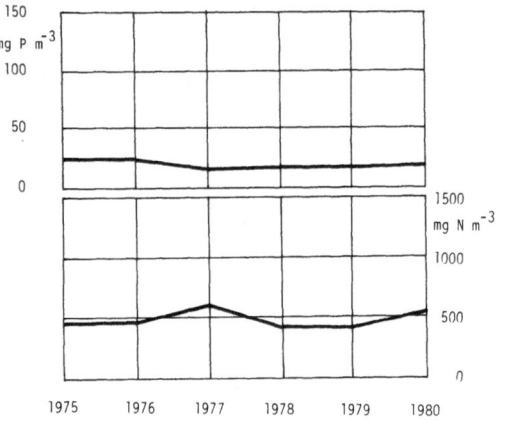

Fig. 5. The average total phosphorus and total nitrogen concentrations in Kajaanselkä in March 1975-1980. Values are means of 6 measurements.

The reduction of chloride in Enonselkä was 17% and the concentration decreased mainly during the first year after the sewage diversion.

Respective phytoplankton analyses in L. Vesijärvi have been done since 1960 but the observations were only occasional (Finni 1979). In the mid 1960s one species, *Oscillatoria agardhii*, reached a major dominancy in Enonselkä and is still prevailing. *Microcystis flos-aquae* was dominant particularly in August. Since 1979 the abundance of heterocystous blue green algae has increased. Blooms of *Anabaena circinalis*, *A. planctonica*, *A. flos-aquae* and *Aphanizomenon flos-aquae* occurred in 1979 in July and August and in 1980 in July. The freshweight biomass values rose to 20 g m^{-3} in 1974 in Enonselkä (Finni 1979) but in 1977 the biomass per growing season was 4.1 and in 1979 9.4 g m^{-3} (Table 1). In Kajaanselkä diatoms dominated *(Asterionella formosa, Fragilaria crotonensis, Tabellaria fenestrata)*. The *Oscillatoria* dominancy occurred in August. The freshweight biomass per growing season in Kajaanselkä in 1977 was 1.1 g m^{-3} and in 1979 1.2 g m^{-3} (Table 1).

The primary productivity *in vitro* indicated significant meiotrophication during 1977-1980 (Table 2). In Enonselkä the reduction of productivity was more than 50%, and the values recently decreased

Table 1. Freshweight biomass of phytoplankton g m^{-3} in Enonselkä and Kajaanselkä. Values are means from samples during open water period.

Year	Enonselkä			Kajaanselkä		
	X	S.D.	n	X	S.D.	n
1977	4.1	2.8	27	1.1	0.9	21
1979	9.4	7.3	18	1.2	0.4	5

Table 2. Primary productivity in vitro mg C_{ass} m^{-3}day^{-1} in Enonselkä and Kajaanselkä, measured during open water period.

Year	Enonselkä			Kajaanselkä		
	X	S.D.	n	X	S.D.	n
1977	1390	770	36	185	71	12
1978	960	715	27	117	84	12
1979	650	410	22	70	55	12
1980	530	438	36	142	87	12

although the physico-chemical data showed no improvement in the water quality.

The restoration of long-polluted lakes is difficult. The unloading of L. Lac d'Annecy carried out in 1955 by complete sewage diversion did not restore the lake in ten years (Chauchoy 1967). Lakes with a shorter loading history may in many cases be rapidly restored (Liepolt 1967).

The results above show that the chemical water quality rapidly improved during the first two years following sewage diversion, but after that there was a standstill. Further recovery of L. Vesijärvi would undoubtedly have taken place slowly without the hypolimnetic aeration. The aeration accelerated the recovery. However, at L. Vesijärvi the aeration capacity was insufficient during summer stagnation.

References

American Public Health Association, American Water Works Association, Water Pollution Control Federation, 1975. Standard Methods for the Examination of Water and Wastewater (14th ed).

Chauchoy, J., 1967. L'assainissement des communes riveraines du Lac d'Annecy. FEG Informationsblatt Nr. 14, 63–66.

Finni, T., 1979. The phytoplankton in L. Vesijärvi during years 1960–1977. Mimeogr. 74 pp. The Municipal Laboratory of Lahti (in Finnish).

Goltermann, H. L., 1969. Methods for Chemical Analysis of Fresh Water. IBP Handbook 8. Blackwell Scientific Publications. Oxford.

Järnefelt, H., 1929. Zur Limnologie einiger Gewässer Finnlands. V. Vesijärvi. Ann. Zool. Soc. Vanamo 8: 8, 1–17.

Keto, J., 1973. The pollution and restoration of L. Vesijärvi. Manuscript. Department of Limnology, University of Helsinki, 68 pp. (in Finnish).

Keto, J., 1976. The present state of L. Vesijärvi and the development leading to that. Ympäristö ja Terveys 3: 299–308 (in Finnish).

Kyröläinen, H., 1978. The exchange of nutrients between sediments and water in L. Saimaa near Mikkeli. National Board of Waters. Report 139: 1–132, 19 app. Helsinki (in Finnish).

Lappalainen, K., 1973. Forecast on the development of L. Vesijärvi after the unloading of the sewage of Lahti City: Mimeogr. 29 pp. National Board of Waters (in Finnish).

Liepolt, R., 1967. Die limnologische Verhältnisse des Zellersees, seine Verunreinigung und Reinigung. FEG Informationsblatt Nr. 14, 59–62.

Seppänen, P., 1962. The pollution of L. Vesijärvi based on the investigations made in 1960: Mimeogr. 91 pp. The Municipal Laboratory of Lahti (in Finnish).

Utermöhl, H., 1958. Zur Vervollkomnung der qualitativen Phytoplanktonmetodik. Mitt. int. Verein. Limnol. 9: 1–38.

Vaara, M. & Vaara, T., 1971. The pollution of L. Vesijärvi based on the investigation made in 1971: Mimeogr. 127 pp. The Municipal Laboratory of Lahti (in Finnish).

Vollenveider, R. A. (ed.), 1974. The Use of Radioactive Carbon (C14) for Measuring Primary Production in Aquatic Environments pp. 1–225. IBP Handbook 12. Blackwell Scientific Publications, Oxford (2nd ed.).

Experimental manipulations of the fish populations in Windermere

T. B. Bagenal
The Freshwater Biological Association, Windermere Laboratory, Ambleside, Cumbria, England

Keywords: perch, pike, predation, Windermere, fishery management

Abstract

A large scale experiment has been carried out on the perch and pike populations of Windermere in the English Lake District for nearly 40 years, and the objectives have changed over this period. Initially the aim was to manage the trout population, but the work has produced data of interest in predator-prey relationships and pike management. Perch were first heavily fished and then monitored for size, and numbers by trapping. Large pike have been culled by gill netting. The effects of pike gill netting have been to increase the growth rate and decrease the average age in the population. These changes in the pike population structure, together with climatic variations, have modified the perch population. The perch did not recover from the heavy fishing for many years and the population numbers remained low due to pike predation supressing recruitment. It was not until a large cohort, mainly controlled by the weather, was produced that the population increased. Cannibalism by these perch later combined with the pike predation to suppress future cohorts. There have also been as yet unexplained changes in the perch mortality rates. The experiments have led to a greater understanding of the relationship between pike and perch populations, but the use of this knowledge for management depends on what the objectives are.

Introduction

The common fish species in Windermere, which is the largest (14.8 km²) natural lake in England, are the char *Salvelinus alpinus*, the trout *Salmo trutta*, the eel *Anguilla anguilla*, the minnow *Phoxinus phoxinus*, the perch *Perca fluviatilis* and the pike *Esox lucius*. Salmon *Salmo salar*, roach *Rutilus rutilus*, rudd *Scardinius erythrophthalmus*, tench *Tinca tinca*, bullhead *Cottus gobio*, stone loach *Noemacheilus barbatulus*, and stickleback *Gasterosteus aculeatus* are also present. Eels and some char are caught commercially.

This paper is concerned with the population changes and interactions of perch and pike.

The Windermere fishery investigations

Allen (1935) showed that the perch can easily be caught in wire netting traps set in 3–9 m from mid April to mid June when males predominate in the catches. During the rest of the year the catches are smaller but contain a larger proportion of females. These observations were made use of during the war years when, from 1941 to 1947, over 90 tonnes of perch were removed by trapping and were canned (Worthington 1950). After the war, large scale trapping ceased in the North Basin of the lake but continued in the South Basin until 1964, but small scale trapping continued in both basins to monitor the populations. The monitoring is done by trapping for a six week period during the spawning season. The catches are counted and each fish is measured for length and weight, and is sexed

Hydrobiologia 86, 201–205 (1982). 0018-8158/82/0862-0201/$01.00.
© Dr W. Junk Publishers, The Hague.

and aged. The fish are aged from the opercular bone. The monitoring has continued until the present day. The data have been fully worked up and papers have been published by Le Cren (1947, 1955, 1958), Le Cren *et al.* (1972, 1977) and Craig *et al.* (1979). The estimated population numbers and biomass of perch in both basins derived from this work are given in Fig. 1. The points to note in this figure are: 1) the initial decline in numbers and biomass in both basins as a result of the 1941–47 fishery; 2) in the North Basin the numbers have never exceeded the 1941 level, and in the South Basin this level has only been exceeded in 1957; 3) the biomass exceeded the 1941 level from 1961 to 1963 in the North Basin and from 1957 to 1975 in the South Basin; 4) there was a catastrophic decline in 1977 in both basins. This last reduction was due to a disease (Bucke *et al.,* 1979) and was not connected with the experimental fishing considered here.

After the initial perch population reduction, it was feared that the Windermere pike would consume more trout and that this would be unpopular with sport fishermen. It was therefore decided to reduce the pike population. After various trials gill nets, first of flax and now terylene, of 64 mm bar, were found to catch pike with a minimum of labour, and allow salmonids of up to 3 lbs (1.36 kg) to escape. Gill nets of this mesh have been used each winter from October to February, from 1944–45 to the present, and they catch pike of about 55 cm and over. Because the females grow faster than the males, the nets catch males of 4 years and older, and females of 3 years and older. From 1944 to 1979–80 12 918 pike weighing 37.7 tonnes have been removed from Windermere. Each fish is measured, weighed, sexed and aged and the stomach contents analysed. After the winter gill netting to remove pike, nets continue to be set to tag pike. From 1949 to 1980 4659 pike have been tagged. A high proportion (up to 80% of some batches) are recaught. The pike data have been worked up and published by Frost (1946, 1954), Frost & Kipling (1967), Kipling & Frost (1970) and Le Cren *et al.* (1972). From the data the population numbers and biomass have been calculated. These are illustrated in Fig. 2. The points to note in this figure are: 1) after an initial reduction the number of pike in Windermere has only fallen below the pre-netting level in 1956; 2) the biomass has only exceeded the pre-netting level in one year, 1962. This suggests that the netting has led to Windermere containing more but smaller pike. This is also shown in Table 1; in the later years there are relatively more younger pike in the catches than in the first year of gill netting. Table 2 indicates that the pike are growing faster, although the differences are not large.

The stomach analyses showed that perch formed an important component of the food of pike from

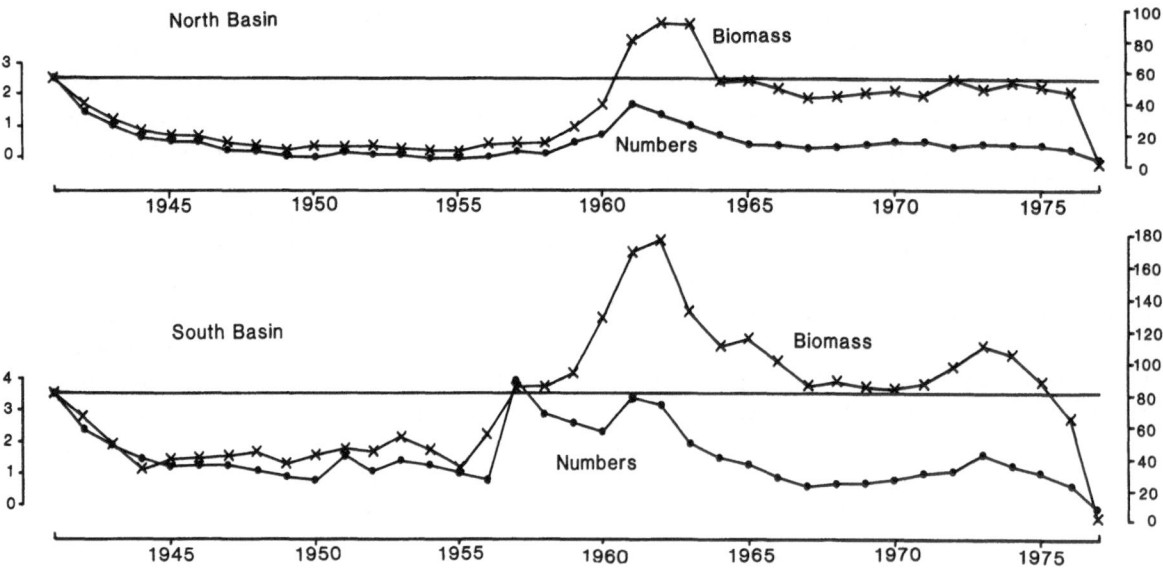

Fig. 1. Estimated numbers and biomass of perch aged 2 and over in Windermere from 1941 to 1977.

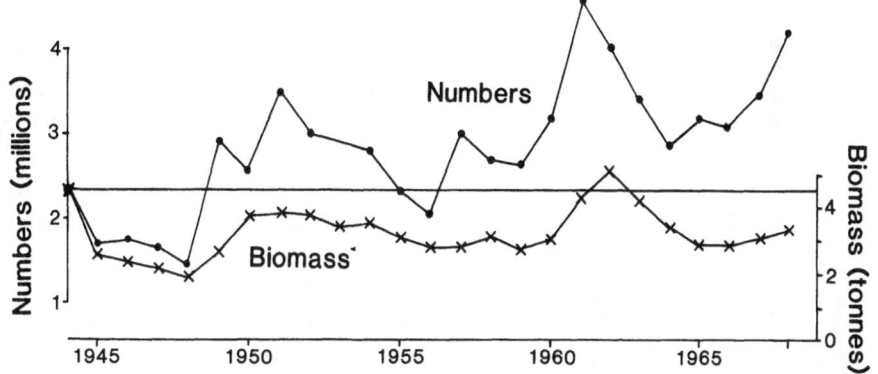

Fig. 2. Estimated numbers of adult pike and their biomass, 1944–68.

Table 1. The percentage age distribution of gill net caught male and female pike for the years 1944–45, 1952–55, 1962–65 and 1972–75 in three age groups, <5 years, 5–9 years and >9 years. (From Bagenal 1977.)

Age (yr)	1944–45		1952–55		1962–65		1972–75	
<5	9.4	26.1	29.8	63.0	23.7	56.9	56.6	79.0
5–9	82.0	62.9	69.7	35.8	75.8	42.7	42.4	20.6
>9	8.6	11.0	0.5	1.2	0.5	0.4	1.0	0.4

Table 2. Mean weight (kg) of gill netted male and female pike aged 4, 7 and 10 years old in the years 1944–45, 1952–55, 1962–65 and 1972–75. The average % increase is the increase in the figures for the last two decades over 1944–45. (From Bagenal 1977.)

Years	Age 4		Age 7		Age 10	
1944–45	1.81	2.65	2.79	5.73	3.49	8.15
1952–55	1.97	2.71	2.95	5.99	–	9.36
1962–65	1.86	2.51	2.64	5.41	4.03	–
1972–75	2.19	3.20	3.20	7.26	3.38	7.10
Average % increase	9.95	5.91	5.02	8.55	6.16	0.98

Table 3. Monthly percentage frequency occurrence of perch in pike stomachs. (From Frost 1954.)

Month	Percentage frequency occurrence	No. examined	Percentage feeding
January	8.2	465	45.2
February	6.6	469	40.3
March	18.0	133	43.0
April	9.2	101	14.8
May	50.6	52	82.7
June	58.7	49	71.4
July	57.3	52	61.5
August	30.5	66	53.0
September	40.2	49	65.3
October	22.8	159	54.0
November	5.9	587	65.7
December	3.5	601	54.9

May to October (Table 3) and that pike tend to eat perch of about one third to a half of their own length (Table 4). If these observations are linked to the changes in the pike population believed to have resulted from the gill netting, we may hypothesize that the effect of the netting has been to increase the consumption of perch, particularly of the small perch. This is believed to have resulted not only from the increased number of pike, and, since young pike grow faster than old ones, from the younger mean age in the population, but also because the actual rate of growth was found to have increased at each age.

Turning back to Fig. 1, it seems that the long period from 1945 to the end of the '50s when the North Basin perch population remained low could reasonably be explained by the increased pike predation preventing successful recruitment. In the South Basin the numbers remained low until 1957. This figure shows the numbers and biomass of perch aged 2–18 years, and the 1957 South Basin increase was due to a massive 1955 cohort (Figure 3). This cohort was not successful in the North

Table 4. The mean lengths of perch eaten by different sized pike. (From Frost 1954.)

Pike length group (cm)	Mean length of perch prey (cm)	Perch length as a percentage of pike length
20–29	6.1	25
30–39	9.3	31
40–49	10.3	26
50–59	12.6	25
60–69	12.8	20
70–79	13.7	20
80–99	13.1	14

204

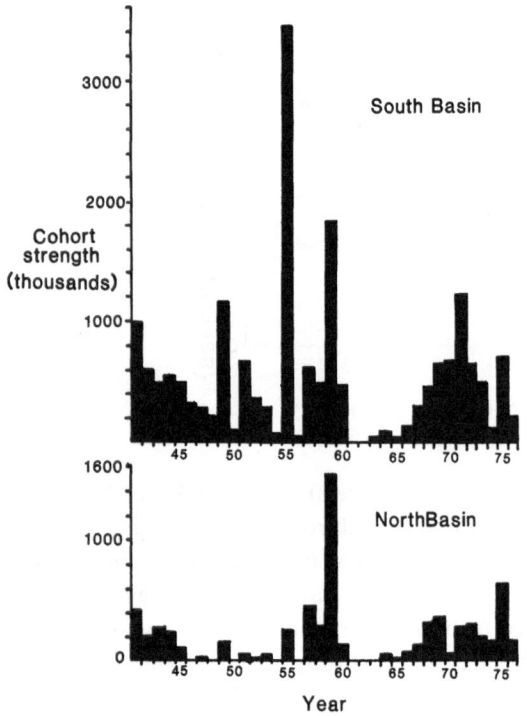

Fig. 3. Cohort strength of perch (at age 3) from 1941 to 1976.

summer is cool, such that the cumulative total of the number of degrees that the surface water temperature is above 14 °C is low, the cohort is always poor. However if the surface temperature is warm the cohort may be successful, but this depends on the perch and pike biomass being low (so reducing predation). Finally, in Windermere, it appears that ice cover in the winter in some way suppresses the subsequent cohort. Unless all the factors are favourable to a cohort, it will not be successful and over the last 40 years there have been only two outstanding cohorts in Windermere, those of 1955 and 1959, and the former was only success-ful in the South Basin. However the successful cohorts grow much more slowly than the less successful and their mortality rate is lower (Craig 1980). This is exemplified by the 1959 and 1968 cohorts (Table 5). The effect of this is that in the South Basin, the 1955 cohort was still being caught in 1976 when it was 21 years old. By this age it is possible that the perch are too big for most of the pike and that mortality is mainly due to physio-logical old age.

Basin and no increase is seen in Fig. 1. In both basins 1959 produced an abundant cohort which (Fig. 1) resulted in the increase in numbers in 1961. The cannabilism by the 1955 cohort led to an increased perch mortality of the 1959 group in the South Basin as compared with the North. The cannabilism by the large perch biomass in the early 1960s, together with the pike predation, prevented a successful cohort emerging through the middle and late '60s, and 1970s.

The predation by the pike was probably effective in keeping the perch numbers low through the 1940s and '50s mainly because the perch population was already low from the war-time fishery. Had there been a balanced perch population, the preda-tion would have been spread over many age groups and the effect on recruitment would have been less severe and the population might have been less affected. However, with the perch heavily fished and the pike predation increased on young perch, a strong perch cohort would not emerge until an exceptional combination of events occurred. Kip-ling (1976) has provided a model to illustrate the determination of cohort strengths. The most im-portant factor appears to be the weather. If the

Fishery management

The results of the experimental manipulations of the pike and perch populations in Windermere have various fishery management implications, but their use depends on what the management objectives are. Many sport fisheries in Britain contain large pike but if the objective is to encourage large numbers of perch, it may be better to keep the large pike in the belief that it is better to have them than many small and fast growing individuals. If on the

Table 5. Population details of the 1955 and 1968 cohorts of male Windermere perch (for the whole lake, or only South Basin where marked S.B.). Data from Craig *et al.* (1979) and Craig (1980).

Cohort	1955	1968
Cohort strength (thousands) (S.B.)	3486	470
Instantaneous mortality rate (Z)	0.37	1.11
L∞ (from Walford plot)	25.43 (S.B.)	26.47
K (from Walford plot)	0.15 (S.B.)	0.58
S.B. population number (all ages >2 yr) (millions)	11	8

other hand the policy is to encourage large perch, this may perhaps be achieved by culling the large pike in the hope of increasing the number of small pike which will control recruitment and so reduce possible overcrowding. The management of freshwater fisheries has a long way to go before it becomes a predictable science, but the Windermere experiments give some indication of how predator-prey interactions may in the future be manipulated to achieve particular management objectives.

Acknowledgements

The results discussed in this paper are based on the work of many people in particular of Mr. K. Shepherd, Miss J. C. McCormack, Miss C. Kipling, Dr. W. E. Frost, Dr. J. F. Craig and Mr. E. D. Le Cren.

References

Allen, K. R., 1935. The food and migration of the perch (Perca fluviatilis) in Windermere. J. Anim. Ecol. 4: 264–273.

Bagenal, T. B., 1977. Effects of fisheries on Eurasian Perch (Perca fluviatilis) in Windermere. J. Fish. Res. Bd Can. 34: 1764–1768.

Buck, D., Cawley, G. D., Graig, J. F., Pickering, A. D. & Willoughby, L. G., 1979. Further studies of an epizootic of uncertain aetiology. J. Fish Dis. 2: 297–311.

Craig, J. F., 1978. A note on ageing in fish with special reference to the perch Perca fluviatilis L. Verh. int. Ver. Limnol. 20: 2060–2064.

Craig, J. F., 1980. Growth and production of the 1955 to 1972 cohorts of perch, Perca fluviatilis L., in Windermere. J. Anim Ecol. 49: 291–315.

Craig, J. F., Kipling, C., Le Cren, E. D. & McCormack, J. C., 1979. Estimates of the numbers, biomass and year-class strengths of perch (Perca fluviatilis L.) in Windermere from 1967 to 1977 and some comparisons with earlier years. J. Anim. Ecol. 48: 315–325.

Frost, W. E., 1946. On the food relationships of fish in Windermere. Biol. Jaarb. 13: 216–231.

Frost, W. E., 1954. The food of pike, Esox lucius L., in Windermere. J. Anim. Ecol. 23: 339–360.

Frost, W. E. & Kipling, C., 1967. A study of reproduction, early life weight-length relationship and growth of pike, Esox lucius L., in Windermere. J. Anim. Ecol. 36: 651–693.

Kipling, C., 1976. Year-class strengths of perch and pike in Windermere. Rep. Freshwater Biol. Ass. 44: 68–75.

Kipling, C. & Frost, W. E., 1970. A study of the mortality, population numbers, year-class strengths and food consumption of pike Esox lucius L. in Windermere from 1944 to 1962. J. Anim. Ecol. 39: 115–155.

Le Cren, E. D., 1947. The determination of the age growth of the perch, Perca fluviatilis from the opercular bone. J. Anim. Ecol. 16: 188–204.

Le Cren, E. D., 1955. Year to year variation in the year-class strength of Perca fluviatilis. Verh. int. Ver. Limnol. 12: 187–192.

Le Cren, E. D., 1958. Observations on the growth of perch (Perca fluviatilis L.) over twenty-two years with special reference to the effects of temperature and changes in population density. J. Anim. Ecol. 27: 287–331.

Le Cren, E. D., Kipling, C. & McCormack, J. C., 1972. Windermere: effects of exploitation and eutrophication on the salmonid community. J. Fish. Res. Bd Can. 29: 819–832.

Le Cren, E. D., Kipling, C. & McCormack, J. C., 1977. A study of the numbers, biomass and year-class strengths of perch (Perca fluviatilis L.) in Windermere from 1941 to 1966. J. Anim. Ecol. 46: 281–307.

Worthington, E. B., 1950. An experiment with populations of fish in Windermere 1939–48. Proc. Zool. Soc. Lond. 120: 113–149.

The management of fisheries in Lake Vörtsjärv

Ervin Pihu & Aare Mäemets
Institute of Zoology and Botany of the Academy of Sciences of the Estonian S.S.R., Tallinn, U.S.S.R.

Keywords: L. Vörtsjärv, fisheries management, eels, pike-perch

Abstract

L. Vörtsjärv is a eutrophic lake in Central Estonia. The area of the lake is 270 km², the average depth only 2.8 m. The biological productivity of the lake is rather low. Twelve to fifteen years ago the main fish were ruff and perch (80–90% of catch), while the numbers of valuable fish were small. At the same time the total catch of fish was relatively high – about 300 tons a year. Attempts were made to reduce the numbers of undesirable fish by intense trawl-catch but no marked results were achieved. Subsequently trawl-catch was ended, elvers were regularly introduced into the lake and protection of valuable fish was improved. At the present time the total catch of fish has decreased (184 tons in 1979), but pike-perch and eels are now the main game fish (50–60% of catch) and the value of the catch of fish has risen 2–3 times.

Lake Vörtsjärv

Lake Vörtsjärv is situated in the Central Estonian depression of preglacial origin. It is connected with Lake Peipus by the Suur-Emajögi River. The area of Lake Vörtsjärv is 270 km², the maximum depth is 6 m, and the average depth only 2.8 m. The lake is of glacial origin and has passed several stages of development. Due to the postglacial rise of the surface of the earth and formation the new outflow (Suur-Emajögi River) the area of the lake has constantly decreased. The water-level of the lake has also fallen. In the 19th century it was 35 m above sea level, at the beginning of the 20th century 34.4 m and at the present time is about 33.7 m above sea level. The drainage area of the lake is over 12 times larger than the area of the lake. The mean retention time is one year.

Lake Vörtsjärv is eutrophic but its biological productivity is low because of its shallowness and the great fluctuations in water level (in 1979, for instance, over 2 m). Due to the small depth, large areas of the lake are under the influence of turbulence which causes the rise of detritus from the bottom, especially many particles of sand and clay. Therefore, during storms the water of Lake Vörtsjärv, which is usually greenish-yellow, becomes red and muddy. The detritus clogs the filter-apparatus of zooplankters and obstructs the normal life of the zoobenthos. This is the reason for the low productivity of the zooplankton and benthos of Lake Vörtsjärv. The lake may be considered eutrophic with argillotrophic features. The summer productivity of its phytoplankton is high: the average biomass during the vegetation period is 28.6–55.6 g m^{-3} and maximum 189 g m^{-3} (Laugaste 1975, 1978), while water-blooms commonly last the whole summer. The average annual biomass of zooplankton in the lake is only 0.50–1.50 g m^{-3} (J. Haberman, pers. comm.) and the average summer biomass of zoobenthos 2–4 g m^{-2} (Timm 1975). Macroflora covers about 15% of Lake Vörtsjärv (Mäemets 1973). Therefore, food reserves for fish in the shallow lake are scanty. The trophic state and hydrobiological character of Lake Vörtsjärv have not changed essentially during the

Hydrobiologia 86, 207–210 (1982).   0018-8158/82/0862-0207/$00.80.
© Dr W. Junk Publishers, The Hague.

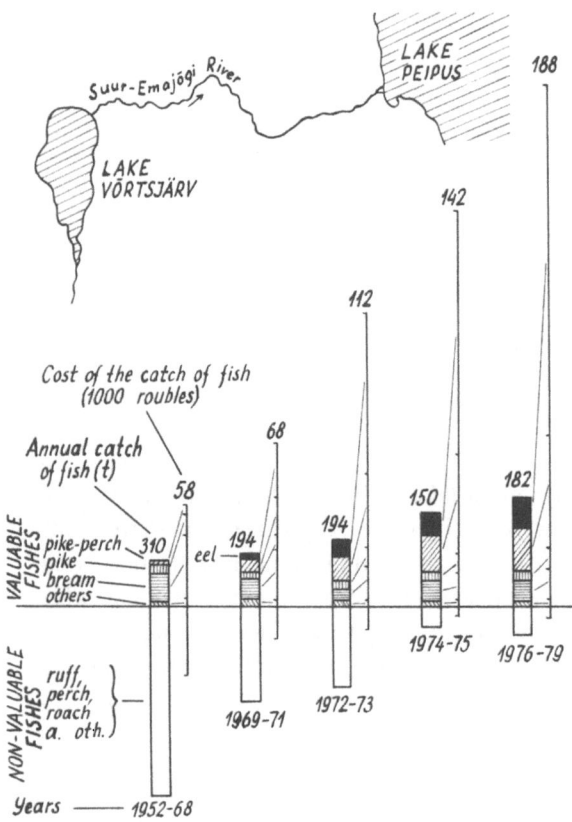

Fig. 1. Catch of fish and its cost in Lake Vörtsjärv.

over 1 m in summer. It is known that pike-perch catch their prey more easily in turbid water. It should be noted that the name of this lake means dung-water in English. Nevertheless, oxygen conditions in summer are good in Lake Vörtsjärv (O_2 saturation 100–130%). In winter in the southern part where the bottom is muddy there is oxygen deficit and an abundance of CO_2 (up to 92 mg l^{-1} in March 1979). Therefore many fish leave the southern part in winter. From time to time (since 1939) there have occurred considerable winter-kills of fish. The abundance of vegetation is scanty. In the northern part of the lake there are many good spawning sites with sandy and stony bottoms for pike-perch. In Lake Vörtsjärv the pike-perch also has very favourable feeding conditions because of the high abundance of dwarf smelt and fry of other fishes.

last 30 years. On the other hand, the activity of man has caused noticeable changes in the composition of the fish fauna.

Fishes

Over 30 species of fish have been identified in Lake Vörtsjärv and the mouths of its tributaries. The main game fish are pike-perch and, in recent years, also eel and bream, pike, roach, ruff, perch and burbot. Vendace, smelt, ide, tench, crucian carp, bleak and some other species are less abundant. The catch has been comparatively low and rarely has it exceeded 400 tons or 15 kg ha^{-1} yr^{-1}.

According to fishery classification, Lake Vörtsjärv belongs to the pike-perch lakes. It has all the qualities which characterize such lakes. The water in the lake is shallow, relatively warm and turbid because of seston, having transparency not

Fish catch and fishery

Until recently Lake Vörtsjärv was not considered a pike-perch lake. It was disparagingly referred to as a ruff lake. The bulk of the catch (80–90%) consisted of ruff, young perch, roach and other worthless small fish, which were caught mainly with fine-meshed trawls. Attempts were made to decrease the number of undesirable fish (first of all ruff) by intense trawl-catch, but were unsuccessful (Kangur 1969). The stocks of ruff, perch, roach and other worthless fish did not decrease owing to the high rates of reproduction of these fish. Trawl catches killed immature individuals of valuable fish (mostly pike-perch) in large numbers, so that their stocks (except for bream) became scarce. The annual catch of pike-perch fell to under 1 ton. The bream was the most important among the valuable fish, but because of lack of food its meat is lean and not tasty.

Consequently, the total catch of fish in Lake Vörtsjärv (taking into consideration its small productivity) was relatively large – about 300 tons a year, but the commercial value of the catch was low. The bulk of the fish taken here was used as feed for pigs and fur-bearing animals.

In order to increase the stocks of valuable fish in Lake Vörtsjärv, a number of measures advised by research workers at Vörtsjärv Limnological Station have been put into practice within the last quarter of a century. Since 1956 elvers (i.e. glass eels) have been regularly released into the lake, 15 million individuals to date. In 1958, quotas and a spring close season were established to preserve valuable fish. Trawling for small fish was restricted in 1966–1970 and subsequently stopped altogether.

As a result of these measures the total catch decreased, but the number and catches of valuable species as well as the composition of catches have greatly changed in the lake during the last decade.

In former times eels had no commercial significance in Lake Vörtsjärv. Thanks to regular introduction of elvers from France the abundance and catches of eels are increasing from year to year. In recent years the total catch of this valuable fish has risen to 50 tons a year (1977). Two thirds of the income from fish catches is from eels. The return catch of the eel is 10–15%. The income from eel catches exceeds the costs of introducing elvers by 40 times. Eels feed mainly on benthic organisms (especially larvae of *Chironomus plumosus*) and are to some extent competitors of bream. But eels are also predatory fish, taking inferior small fish, especially ruff, and are therefore good biological ameliorators.

In addition to the prohibition of fine-meshed trawls several other measures have been taken to protect fish. In the spawning sites of fish in the southern part of the lake, catching is forbidden throughout the year. From April 1 to June 1, the catching of pike-perch, pike and some other game fish is forbidden in the whole lake. Limits have been introduced for the size of fish and catches. Due to these measures the mortality of the fish, especially that of pike-perch, has become lower and their number and catch higher. The catch of pike-perch reached its maximum in 1979 (73 tons). At present pike-perch and eels are the main game fish in the lake.

The numbers of pike have also increased in recent years. At the beginning of the 1970s their numbers decreased due to the low water level, drying out of flood plains and partial amelioration in the vicinity of Lake Vörtsjärv.

The increasing pressure of predatory fish has led to a sudden fall in the abundance of non-commercial fish, their stocks having decreased about 5–7 times, and the total catch has inevitably diminished (184 tons in 1979). In consequence, the income of fishermen has not decreased, but increased (largely from eels and pike-perch) 2.5

times in comparison with previous decades (1952–1968).

Pike, pike-perch and burbot earlier fed mainly on perch, ruff and roach as these were abundant and small in size. Due to the actions of the predatory fish the number of the small fishes became low and now the main food of pike, pike-perch and burbot is bream fry. Bream are usually valuable fish, but Lake Vörtsjärv is an exception. Its zoobenthos is scarce, and bream compete for food with the still more valuable eels. Good spawning conditions mean the numbers of bream in the lake are high, so all catch limitations on bream have been cancelled and they are not considered valuable fish. No measures have been taken to hinder the use of bream fry as food by predatory fish.

Discussion

In the present situation in Lake Vörtsjärv, the stocks of valuable fish have reached an optimum level. If more fish are wanted from this lake, the fishery must intensify. This intensive catch would inevitably reduce stocks of valuable fish and return the lake to its previous state, i.e. to a ruff lake. Introduction of other valuable species (such as carp, peled and others) is impossible with scarce zooplankton and benthos and large number of predators in the lake.

To raise the biological productivity of Lake Vörtsjärv and the number of fish we propose to restore the water level which existed at the beginning of the century, i.e. to raise the level by 0.7 m. This would end the low level in summer and winter. The plan involves a lock-regulator on the outflow towards the Suur-Emajögi. This is intended to create better conditions for the development of zooplankton and benthos and double the biological productivity of the lake. To raise the numbers of zoobenthos, *Paramysis lacustris* and *Limnomysis benedeni* of Ponto-Caspian origin have been introduced to the lake (Timm 1977). They also serve as feed for the fish. Regulation of the water level will create more favourable conditions for spawning and for fry for valuable fish (mainly pike). With regulation of the water level, the total catch of fish from Lake Vörtsjärv will rise to 200–250 tons a year, the catch of eels rising to 100 tons. In order to ensure such a catch of eel it is necessary to release at least 1 million elvers into the lake each year.

210

References

Kangur, M., 1969. Fishery in Lake Vörtsjärv. Eesti Loodus. 9: 557–560 (In Estonian).

Laugaste, R., 1975. Productivity of phytoplankton in Lake Vörtsjärv in 1970–1972. Estonian Contrib. int. Biol. Programme 6: 161–164.

Laugaste, R., 1978. The winter phytoplankton in L. Vörtsjärv. Hydrobiol. Res. 7: 7–19.

Mäemets, A., 1973. Macrophytes. In: Timm (ed.), Vörtsjärv: 77–82 (In Estonian).

Timm, T., 1975. Zoobenthos of Lake Vörtsjärv in 1964–1972. Estonian Contrib. int. Biol. Programme 6: 165–200.

Timm, T., 1977. Bottom fauna introduced to Lake Vörtsjärv. Eesti Loodus 9: 572–576 (In Estonian).

A plan for fisheries management in the lakes drained by the Oulujoki river

K. Salojärvi, H. Auvinen & E. Ikonen
Finnish Game and Fisheries Research Institute, Fisheries Division, Po. Box 193, SF-00131 Helsinki 13, Finland

Keywords: fishery management, stock assessment, multispecies fishery, stocking results

Abstract

Freshwater fishery management is treated as a dynamic system comprising environment protection and improvement, fishing, fishery resources allocation, fish stocking programs and marketing policy. The aims of the plan are to increase the economic value of the catch and to protect the professional fishery. Fisheries statistics, catch per unit effort data and other material were collected during 1972–1976. The total allowable catch (TAC) for the most important fish species was estimated with MSY and population analysis models. The results of fish stockings were studied by tagging and population analysis calculations. Fishing of vendace and non-valuable species (perch, roach, smelt) can be increased, but the fishing pressure on other species should not be raised above the present level. Restrictions on whitefish fishing are needed in some areas. A balanced multispecies fishery is desirable, and suggestions are given for the composition of the fishing gear. Fish stocking can make possible a larger and more valuable catch, but at present its profitability is rather low. The stocking results are strongly affected by the fishery and the gear composition.

Introduction

The present paper discusses the plan for the fisheries management of the water system drained by the Oulujoki (Salojärvi *et al.* 1978) and its implementation. This plan treated the freshwater fisheries as a dynamic system comprising environmental protection and improvement, fishing, fishery resources allocation, fish stocking programs, fish culture and marketing policy. The aim of the plan was to compensate for the damage caused by man to the natural fish stocks and fishery, to increase the economic value of the catch and to protect the commercial fishery.

Study area

The Oulujoki water system (Fig. 1), draining to the Gulf of Bothnia, has a drainage basin measuring 22 572 km². The total number of lakes in the basin is 3 303, and they have a combined area of 2 558 km² (Anon. 1977). The Oulujoki water system is strongly influenced by man. About 60% of the water area is regulated for hydro-electric power purposes. Due to the construction of 17 hydro-electric power plants and dredging of the rivers (77%), most of the rapids have disappeared. In addition, some parts of Oulujärvi, the central lake of the system, have been polluted by pulp and paper mill effluents.

Materials and methods

In 1973 and 1976 catch statistics were collected from commercial and non-commercial fishermen. During the period 1973–79 data on the catch per unit effort were received from 50 local fishermen using different types of fishing gear. In the period 1973–76 test fishing with gill net series (mesh sizes

Hydrobiologia 86, 211–217 (1982). 0018-8158/82/0862-0211/$01.40.
© Dr W. Junk Publishers, The Hague.

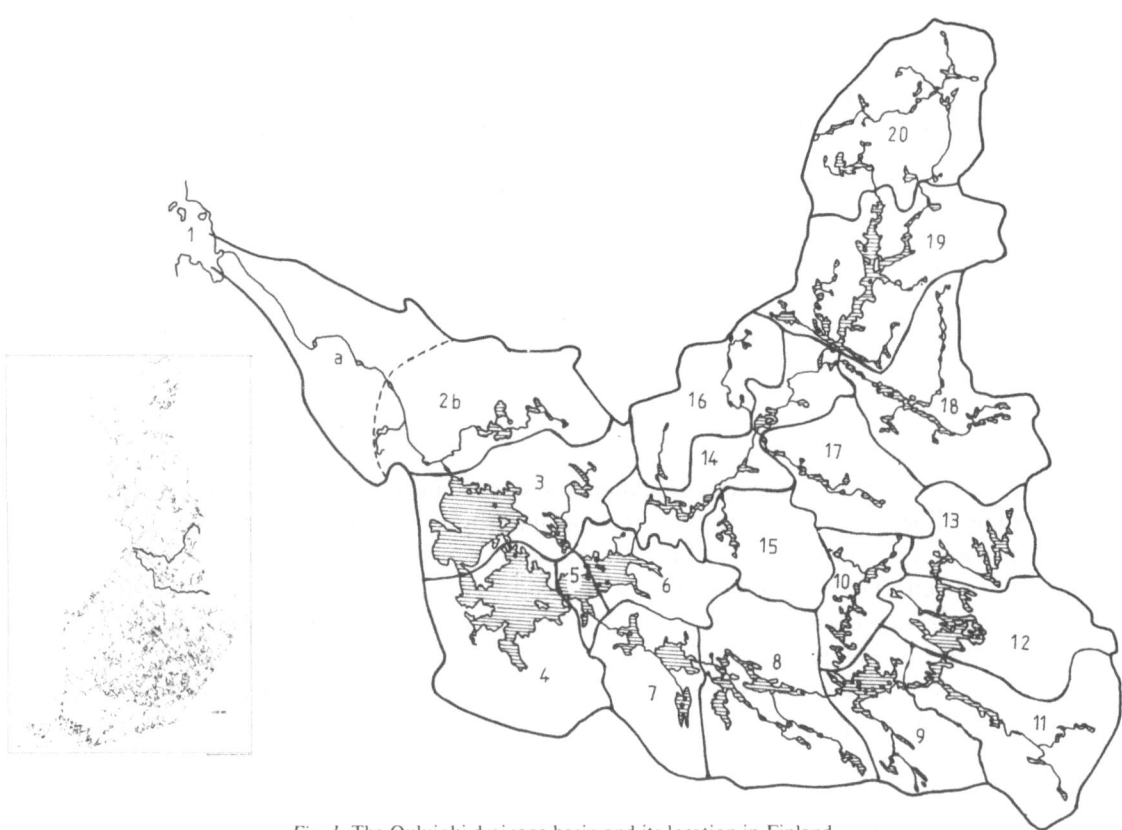

Fig. 1. The Oulujoki drainage basin and its location in Finland.

knot to knot 12, 15, 20, 25, 35, 45, 60 and 75 mm) was carried out by local fishermen in the littoral, pelagic and profundal zones of the ten largest lakes The total number of gill net series nights was 4 000. Electric fishing was carried out in 19 test areas in 9 tributaries. Since 1959 162 600 fish have been tagged (75 300 salmon, 22 300 sea trout, 17 000 brown trout, 15 400 land-locked salmon, 14 800 splake and 14 other species). The number of fish aged in catch samples was 17 200 (6 000 whitefish, 5 700 vendace, 2 600 perch, 1 700 roach and 5 other species). Other material includes replies to enquiries about spawning places of whitefish and vendace, samples of whitefish and vendace fry and their food, and stomach samples of adult whitefish and vendace. Sympatric whitefish species were separated in catch samples by the CORSPEC-1 computer program (Salojärvi & Auvinen 1980). The method of Beverton & Holt (1957) modified by Jones (1957) and modified population analyses were used in stock assessment and in the calculation of stocking results.

Results

The number of professional fishermen in the water system drained by the Oulujoki in 1973 was 245 and in 1976 135. In 1976 there were about 16 000 persons fishing for their domestic needs and 14 000 fishing for recreation. In 1977 the fishing permits sold in the special sport fishing areas numbered 7 000. The water area per fisherman was approximately 9 ha.

The most important gear used by professional fishermen is the seine (summer and winter seining), though trap nets, gill nets, etc. are used as well. Various gill nets are the most popular types of gear used by subsistence and recreational fishermen (Table 1).

The catch per unit effort (cpue) with the traditional gear is low from the point of view of professional fishing (Table 2). The gear is highly selective, fishing pressure being highest on the predator species.

In 1973 the total catch was 1 400 tons and in 1976

Table 1. Percentage distribution of catches by types of fishing gear, and numbers of implements used in 1976.

Fishing gear	Number used	Perch Perca fluviatilis L.	Pike Esox lucius L.	Burbot Lota lota L.	Vendace Coregonus albula L.	Whitefish Coregonus lavaretus L.	Roach Rutilus rutilus L.
Vendace gill nets	19 110	4.8			95.3		3.6
Gill nets, mesh 27–33 mm	21 870	6.4	2.9	2.2		60.9	54.3
Gill nets, mesh over 33 mm	40 630	3.9	68.6	30.9		38.3	6.8
Seines	89	0.4			4.7	0.8	0.7
Wire traps	23 800	81.6	8.5	48.0			33.1
Trap nets	2 710		3.0	11.9			1.5
Hooks and longlines	87 820	2.9	17.0	7.0			
	196 029						

Table 2. Cpue (kg/fishing effort/gear) of the most important types of gear used (1974–79) in the water system drained by the Oulujoki river.

Fishing gear	Whitefish Coregonus lavaretus L.	Vendace Coregonus albula L.	Perch Perca fluviatilis L.	Pike Esox lucius L.	Burbot Lota lota L.	Roach Rutilus rutilus L.	Total
Seine	0.63	11.27	0.87	0.12	0.00	3.49	16.38
Wire trap	–	0.00	0.41	0.05	0.06	0.05	0.57
Trap net	0.02	0.01	0.07	0.55	0.24	0.05	0.92
Vendace gillnet	0.00	1.46	0.02	0.01	0.00	0.02	1.51
Gillnet, mesh 27–33 mm	0.27	0.001	0.05	0.03	0.02	0.19	0.56
Gillnet, mesh 34–40 mm	0.14	–	0.02	0.11	0.08	0.04	0.38
Gillnet, mesh over 40 mm	0.03	–	0.01	0.21	0.11	0.00	0.36
Different kinds of hooks	–	–	0.01	0.06	0.07	0.00	0.13

nearly 1 000 tons (Table 3). In 1973 the professional catch was 30% of the total catch and in 1976 it was 9%. In the latter year the value of the professional catch was only 500 000 FIM (about 1 250 000 US $). The expenditure of subsistence and recreational fishermen on fishing in 1976 was approximately 5 million FIM (125 000 US $). The value of the fishing gear and boats used in that year was about 24 million FIM (6 million US $).

Selective fishing together with water level regulation, construction of hydro-electric power plants, dredging and pollution have caused changes in the species composition of the fish. The lakes that are still in a natural state have large stocks of perch, vendace and whitefish, but in the polluted and regulated lakes the stocks of cyprinids and pike have become stronger. The salmonid stocks have collapsed.

According to MSY calculations, fishing of river-spawning whitefish, pike, pike-perch and bream (*Abramis brama* L.) cannot be increased, whereas fishing of lake-spawning dwarfed whitefish, perch, roach, smelt and vendace should be intensified.

The management of the fisheries in the area has mostly included fish stocking. In the 1970s, the numbers stocked annually were over 60 million whitefish larvae, 97 000 whitefish fingerlings, 90 000 salmon smolts, 102 000 sea trout smolts, 48 000 brown trout smolts and small numbers of other species. The annual value of the stocking was 708 000 FIM (177 000 US $). The results of the stocking are presented in Table 4.

In the early 1970s there was no organized fish-marketing system in the area and the fishermen had to arrange the sale of their own catch. No fish-processing factories existed locally.

Table 3. The catch (tons) of professional and non-professional fishermen in 1979 and 1976 and its distribution by species in the water system drained by the Oulujoki.

	1973					1976				
	Professional fishermen	%	Non-professional fishermen	%	Total	Professional fishermen	%	Non-professional fishermen	%	Total
Fishermen	247		10 542			135		12 323		
Whitefish										
Coregonus lavaretus (L.)	10.4	2.6	54.7	5.7	65.1	6.3	6.7	63.8	7.5	70.1
Vendace										
Coregonus albula (L.)	247.8	61.7	198.3	20.7	446.1	33.6	40.8	103.0	12.1	136.6
Brown trout										
Salmo trutta L.	0.6	0.1	7.4	0.8	8.0	0.1	0.1	7.3	0.9	7.4
Smelt										
Osmerus eperlanus (L.)	36.9	9.2	7.5	0.8	44.4	7.1	8.6	15.2	1.8	22.3
Pike										
Esox lucius L.	31.9	7.9	190.5	19.9	222.4	15.1	18.3	195.5	22.9	210.6
Bream										
Abramis brama (L.)	2.8	0.7	11.7	1.2	14.5	0.6	0.7	11.2	1.3	11.8
Burbot										
Lota lota (L.)	24.5	6.1	83.7	8.7	108.2	9.0	10.9	77.2	9.0	86.2
Pike-perch										
Stizostedion lucioperca (L.)	0.5	0.1	2.2	0.2	2.7	0.2	0.2	2.4	0.3	2.6
Perch										
Perca fluviatilis L.	21.6	5.4	252.2	26.3	273.8	4.8	5.8	225.5	26.3	230.3
Roach										
Rutilus rutilus L.	16.9	4.2	135.0	14.1	151.9	5.3	6.4	137.8	16.1	143.1
Others	7.7	1.9	16.4	1.7	24.1	0.3	0.4	15.6	1.8	15.9
Total	401.6	100.0	959.6	100.0	1361.2	82.4	100.0	854.5	100.0	936.9

Table 4. The results of the stocking in the 1970s in the water system drained by the Oulujoki

Species	Average annual catch (kg) from stocking	Value of the catch from stocking FIM
Salmon	28 500	600 000
Sea trout	18 500	300 000
Whitefish	7 500	50 000
Brown trout	500	10 000
	55 000	960 000

Plan for the fisheries management

The following general objectives were chosen:
- to decrease the damage caused by man to the fish stocks (environmental protection and improvement, fish stocking programs);
- to develop and protect professional fishing (reconciliation of rights of land owners with needs of professional fishermen, development of fishing technology, marketing policy, fishprocessing industry, direction of professional fishing towards multispecies fishing, etc.);
- to develop subsistence and recreational fishing (fish stocking, free ice-fishing, sport fishing areas);
- to change the species composition from non-valuable species to highly valued species (fishery regulations and recommendations, environmental improvement, fish stocking programs).

In the plan the present state of the environment was considered to be poor, but the fisheries can be improved if its deterioration can be stopped. To improve the environment from the fisheries point of view, it is important to decrease the amplitude of water level regulation. The damaging effects of regulation could also be reduced by building refuges for crayfish and spawning places for pike and bream in the lakes. The harmful effects of lumber floating and traffic can be reduced by restoration of rapids, elimination of barriers harmful to migratory species and using ferrys rather than short bridges with long embankments. More effective waste water treatment is needed in factories and municipial waste water plants.

It was recommended that the area should be divided into professional fishing areas, subsistence and recreational fishing areas, sport fishing areas, and spawning and fingerling areas (Fig. 2). In the professional fishing area, fishing is carried out with efficient gear like seines, and is mainly directed to pelagic fish stocks, though fishing of coarse fish is also recommended. In the subsistence and recreational fishing areas the most important gear is gill nets. Predators of pelagic fish stocks (like brown trout) should be stocked within these areas. In the sport-fishing areas fishing will be mostly based on the put and take method. In the spawning and

fingerling areas fishing is forbidden. In the first two areas the fishing regulations and stocking are aimed at creating a balanced multi-species fishery and increasing the use of seines, wire traps and ice fishing rods.

To offset damage to the reproduction of valuable fish species, the following stocking program was recommended: 6.3 mill. whitefish fingerlings, 0.7 mill. brown trout larvae, 30 000 3-yr-old (over 25 cm) brown trout and 1.8 mill. grayling larvae. To produce the stocking material, it was planned to construct a new fish hatchery, to use private fish farms and to rear whitefish fingerlings in ponds with a natural food supply.

For the transportation, processing and marketing of the fish the following facilities were recommended (Fig. 3).
- 4 fish landing harbours,
- 1 central fish harbour,
- transportation routes from the fish landing harbours to the central harbour and to processing factories outside the area,

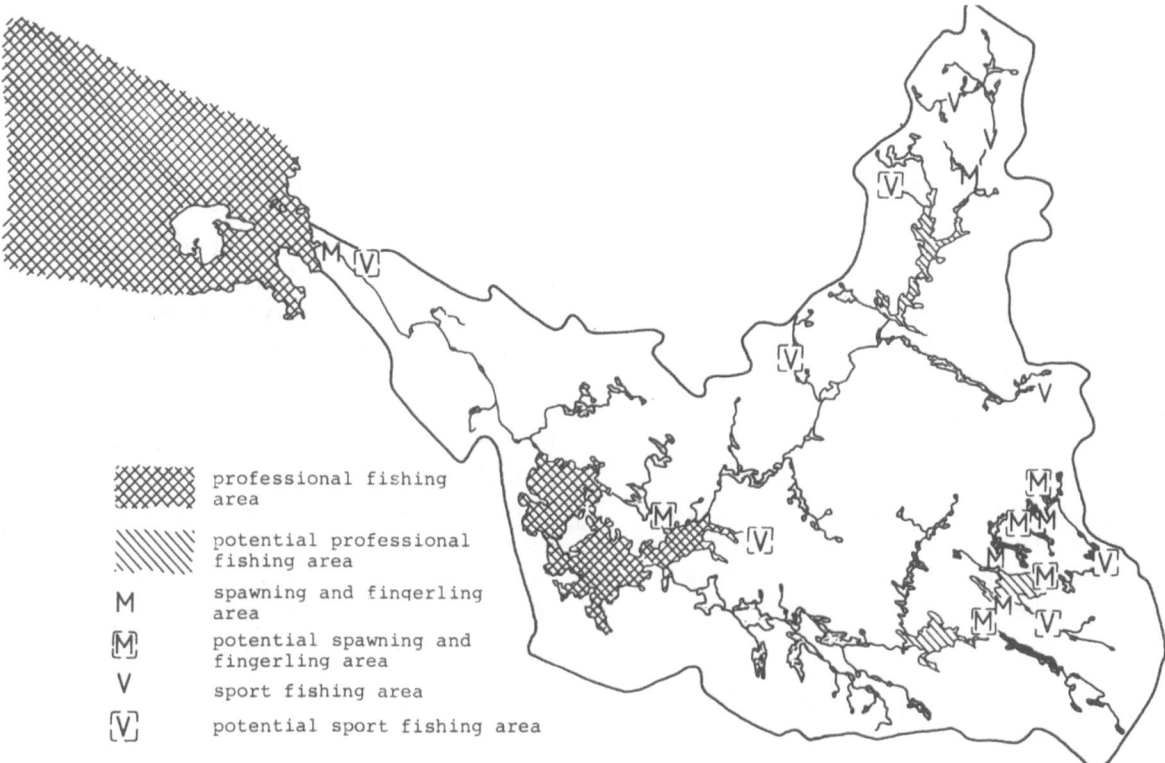

Fig. 2. Recommended professional fishing, sport fishing and spawning and fingerling areas. All the other water areas are for subsistence and recreational fishing.

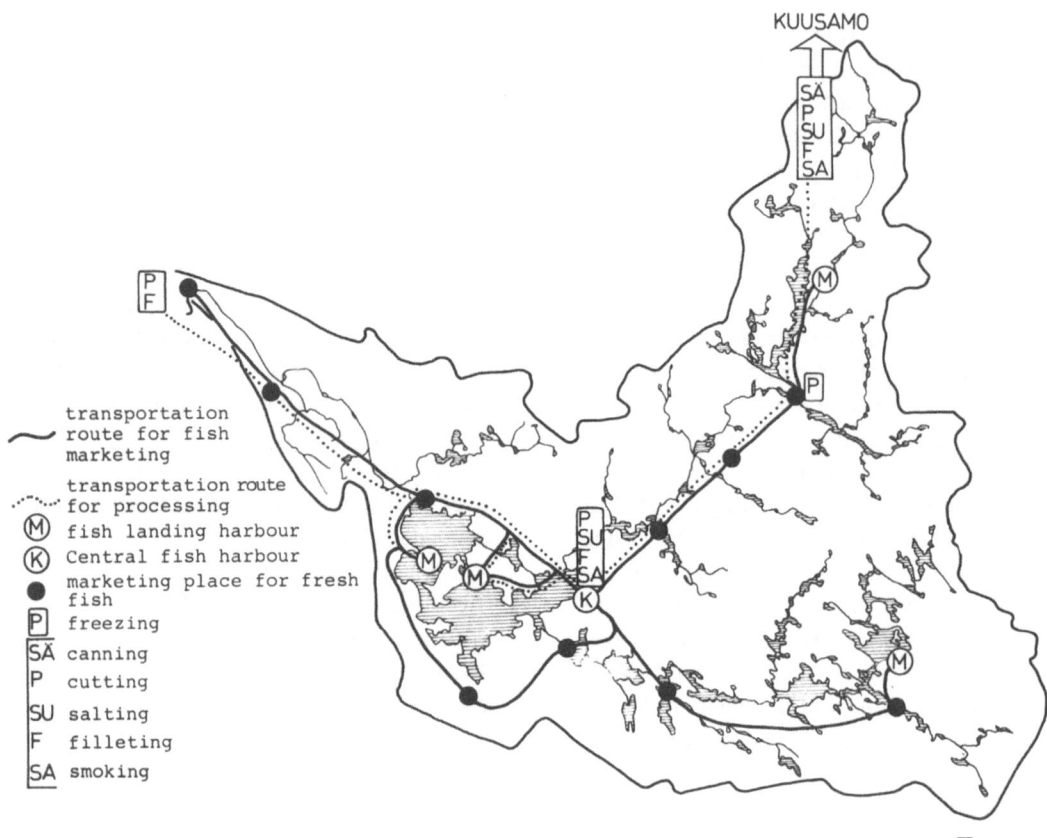

Fig. 3. Recommended transportation routes for fresh fish marketing and fish processing.

- fish-processing facilities near the central harbour,
- marketing places for fresh fish on the transportation routes.

The arrangements for transportation, processing and marketing should include both the catch by commercial fishermen and farmed rainbow trout.

Fisheries management is a complex system comprising biological, economic and social elements. Their dynamic interaction was recognized, but accurate long-term forecasting or modelling of all the important factors is difficult. Great importance thus attaches to an efficient system for following up the results of the fisheries management.

Implementation of the plan up to 1980

The following measures have been taken since the plan was presented:
- Kainuun Kalatoimisto (The Fisheries bureau of Kainuu) has been established to implement and update the plan.
- A field station for development of fishing technology has been established.
- The Central Fish Hatchery for Kainuu is being designed and will be constructed as soon as possible.
- Ponds with a total area of about 300 ha have been built for rearing whitefish fingerlings with natural food supply.
- Two small-scale private fish-processing plants have been built.
- Transportation routes have been arranged and a landing harbour has been built.
- Private fish farming is increasing rapidly.
- Restoration of the dredged rivers is in progress. About 10 rivers have been restored.
- The National Board of Forestry and private water owners associations have followed the main lines of the recommendations concerning fishery regulations.

- Stocking experiments with brown trout smolts over 25 cm have given 20–30 times as high catches as the traditional stocking with smaller smolts.
- Planning and construction of the recommended artificial spawning areas for pike and bream have been started by the National Board of Waters, Kainuu Water District.

References

Anon., 1977. A regional water resources development plan for the Oulujoki river basin. Part I. The planning region and its water resources. Rep. Nat. Bd Wat. Finl. 125: 1–101 (Finnish).

Bergman, G., Aro, M., Mattila, S., Sipilä, P., Sjöblom, V., Särkkä, M., Tuunainen, P. & Ervola, S., 1977. The report of a working group planning the use of water resources for fishery. Min. Agr. and Forestry, Helsinki, 25 pp. (mimeo). (Finnish).

Beverton, R. J. H. & Holt, S. J., 1957. On the dynamics of exploited fish populations. Great Britain Min. Agr. Fish. Invest. Ser. 2, 19, 533 pp.

Jones, R., 1957. A much simplifield version of the fish yield equation. Doc. P21 presented at the Lisbon joint meeting of ICNAF/ICES/FAO Lisbon, 27 May – 3 June 1957.

Salojärvi, K. & Auvinen, H., 1980. A computer program for classifying sympatric whitefish (Coregonus lavaretus L. s.l.) stocks. Finnish Fish. Res. 3: 23–28.

Salojärvi, K., Auvinen, H. & Ikonen, E., 1978. A plan for fisheries management in the lakes drained by the Oulujoki river. Finnish Game and Fisheries Research Institute, Fisheries Division, Helsinki, 281 pp. (mimeo). (Finnish).

Fish stock assessments in Lake Konnevesi

J. Toivonen, H. Auvinen & P. Valkeajärvi
Finnish Game and Fisheries Research Institute, P.O. Box 193, SF-00131 Helsinki 13, Finland

Keywords: fish stock assessment, management

Abstract

In 1977, the average fish yield in the oligotrophic and oligohumic lake Konnevesi was 6.4 kg ha^{-1} (2.6 kg vendace, 1.3 kg perch, 0.7 kg whitefish, 0.5 kg roach, 0.3 kg pike and 0.3 kg burbot). In that year the stock of vendace was exceptionally poor. A study was made to determine the optimum strategy for exploiting the stocks. The field work included test fishing (1970–78), sampling of catches (1977–80) and collection of catch per unit effort data (1978–80), and catch statistics (1969, 1970, 1977). The stocks were assessed by the Beverton and Holt method. Whitefish is exploited at a proper level. Fishing of vendace and roach can be increased, and the yield of pike would be greater, if the recruiting age were higher than at present.

Introduction

Lake Konnevesi, which belongs to the Kymijoki watercourse, has a total area of 188 km², the southern part measuring 119.5 km² and the northern part 68.5 km². The mean depth in southern Konnevesi is 12.5 m and in northern Konnevesi 7.5 m. The greatest depth is 56 m. The properties of the water are typical of an oligòtrophic and oligo-humic lake without any water level control (Tuunainen 1972). According to the measurements of primary production (Granberg 1972) and zooplankton studies (Hakkari 1972), the northern part is slightly more eutrophic than the southern part.

The purpose of this study was to determine the optimum strategy for exploiting the fish stocks.

Material and methods

Catch statistics have been collected from Lake Konnevesi in 1969, 1970 and 1977. In 1977, the statistics were collected from commercial fishermen,

local non-commercial fishermen and sport fisher-men. Data on catch, effort and gear composition were requested. Data on catch per unit effort for different types of gear have been collected since 1978 from 25 local fishermen. Test fishing with a set of gill nets has been carried out since 1970 at permanent stations (Toivonen 1972). Fish sampled from the catches of commercial fisheries and test fishing have been measured, weighed and aged by reading their scales. The mortality has then been estimated from the age composition of the catch sample. For vendace the catch per unit effort data has also been utilized. In the stock assessment the method used is that of Beverton and Holt (1957) modified by Jones (1957). The growth parameters for the calculations have been obtained from von Bertalanffy's (1938) growth curve and a weight-length analysis (Abramson 1971) of the catch samples.

Hydrobiologia 86, 219–222 (1982).   0018-8158/82/0862-0219/$00.80.

Table 1. The catches on Lake Konnevesi (kg/ha).

Fish species	1969			1970			1977		
	South	North	Whole lake	South	North	Whole lake	South	North	Whole lake
Vendace *(Coregonus albula)*	6.0	3.9	5.3	6.6	3.2	5.3	3.2	1.5	2.6
Perch *(Perca fluviatilis)*	0.4	0.9	0.6	0.5	0.9	0.7	0.8	2.3	1.3
Whitefish *(Coregonus lavaretus)*	0.1	0.3	0.2	0.2	0.4	0.3	0.5	1.0	0.7
Roach *(Rutilus rutilus)*	0.1	0.2	0.2	0.1	0.3	0.2	0.3	0.9	0.5
Burbot *(Lota lota)*	0.3	0.2	0.2	0.3	0.2	0.3	0.4	0.3	0.4
Pike *(Esox lucius)*	0.3	0.4	0.4	0.3	0.4	0.4	0.3	0.4	0.4
Smelt *(Osmerus eperlanus)*	–	0.1	0.0	0.0	0.1	0.0	0.2	0.3	0.2
Bream *(Abramis brama)*	0.1	0.2	0.1	0.1	0.2	0.1	0.0	0.2	0.1
Brown trout *(Salmo trutta m. lacustris)*	0.1	0.0	0.0	0.1	0.0	0.0	0.1	0.1	0.1
Others	0.0	0.2	0.1	0.0	0.0	0.0	0.0	0.2	0.1
Total	7.4	6.4	7.1	8.1	5.9	7.3	5.8	7.2	6.4
Total catches (tn)	88.9	43.9	132.9	98.2	39.4	137.6	71.6	48.1	119.7

Table 2. Composition of catch (weight %) in 1969, 1970 and 1977 and in test fishing (1970–78) in L. Konnevesi.

	Southern Konnevesi				Northern Konnevesi				Whole of Konnevesi			
	Catch statistics			Test fishing	Catch statistics			Test fishing	Catch statistics			Test fishing
	1969	1970	1977	1970–78	1969	1970	1977	1970–78	1969	1970	1977	1970–78
Vendace	81.2	79.8	54.1	33.2	61.2	55.9	20.9	29.0	74.6	72.9	40.8	31.1
Perch	5.8	6.7	13.3	31.5	14.2	16.1	32.8	33.3	8.6	9.4	21.2	32.3
Whitefish	1.6	2.4	9.0	12.1	4.2	6.2	14.5	9.7	2.6	3.4	11.2	10.9
Roach	1.5	1.5	4.4	3.7	3.5	4.5	12.3	16.5	2.2	2.4	7.6	10.4
Burbot	3.6	3.7	7.0	15.6	3.0	3.9	4.1	4.4	3.5	3.7	5.8	9.7
Pike	4.5	4.4	5.1	1.8	6.1	7.5	6.4	2.8	5.0	5.3	5.6	2.3
Smelt	0.0	0.1	3.9	1.3	1.2	1.1	3.6	3.8	0.3	0.4	3.8	2.6
Bream	0.8	0.7	0.7	0.0	2.8	3.0	2.4	0.0	1.5	1.3	1.4	0.0
Brown trout	0.7	0.8	1.8	0.0	0.3	0.1	0.8	0.0	0.6	0.6	1.4	0.0
Others	0.3	0.0	0.6	0.8	3.5	1.7	2.3	0.5	1.2	0.6	1.3	0.6
Total catches (tn)	88.9	98.2	71.6	0.374	43.9	39.4	48.1	0.357	132.9	137.6	119.7	0.731

Results

Catch statistics

The catches are presented in Table 1. In 1977, the stock of vendace was very poor. The catches of perch, roach and whitefish have increased from the beginning of the decade. In 1977, the lake was fished by 45 commercial fishermen, 700 local subsistence fishermen, and 90 sport fishermen.

Commercial fishermen mainly fish vendace with seines and gill nets. Local subsistence fishermen use many kinds of gear. In 1977, the commercial fishermen caught 40% of the total catch.

The changes in fish stocks have affected the amount of fishing gear. In 1969, there were 33 seines but in 1977 only 25. In the last few years the number of seines have increased again because of better vendace stocks. In 1969, the gill nets for vendace numbered 742 and in 1977 1884. This is due to the decreasing vendace populations and increasing average size of vendace.

According to the test fishing results the dominant species in the northern part of the lake are perch, vendace, roach and whitefish, whereas the species predominating in the southern part are vendace, perch, burbot and whitefish (Toivonen 1972). In northern Konnevesi the vendace stock fluctuates,

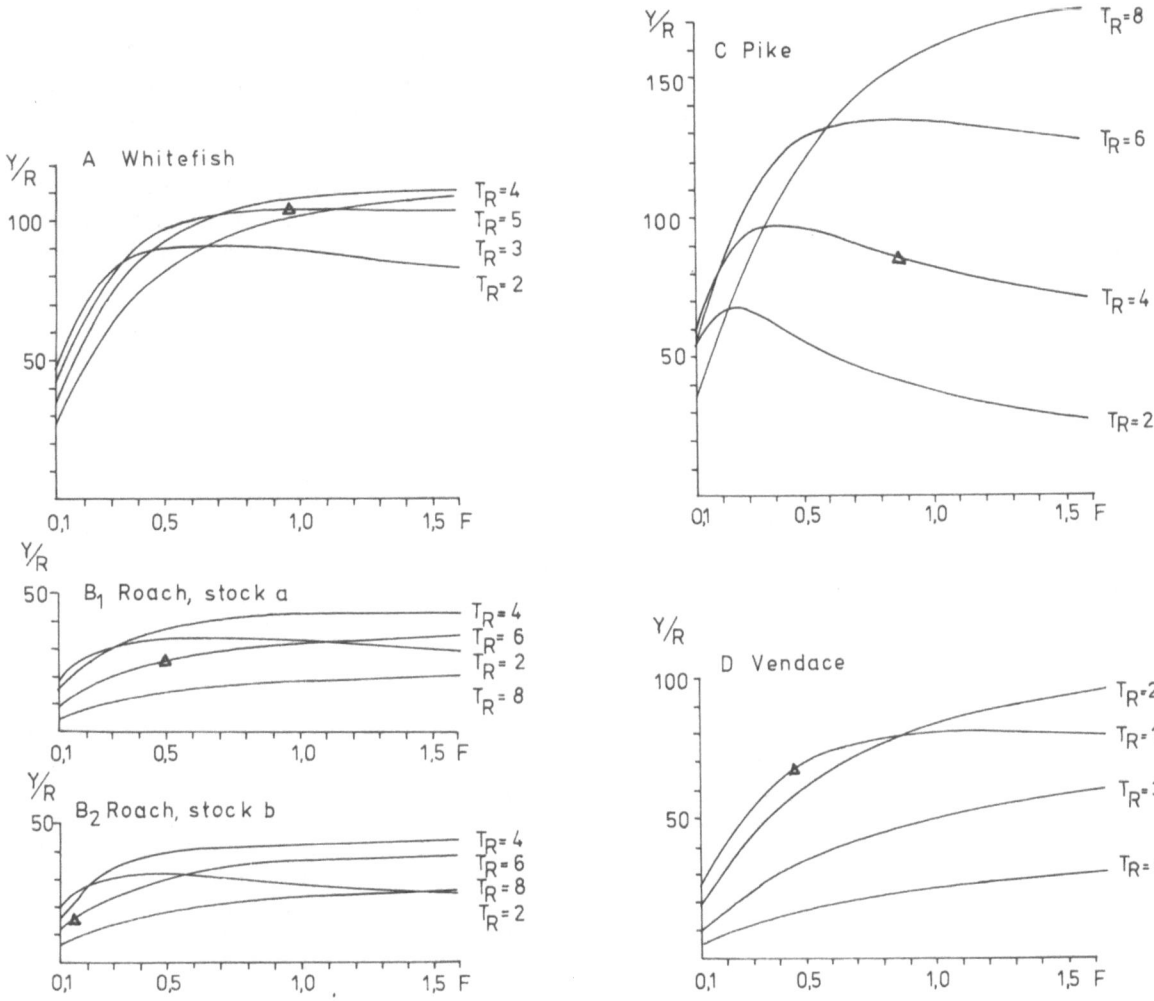

Fig. 1. Effects of fishing mortality (F) and age of recruitment (T_R) on the catch (Y/R) in Konnevesi. The present situation is marked with a triangle. The parameters used in the calculations were in A: N = 318, M = 0.2, K = 0.3082, T_0 = 0.6785, T_λ = 8, T_R = 3, W_∞ = 289.2, and b = 2.8769; in B_1 (stock under constant fishing pressure): N = 206, M = 0.5, K = 0.1470, T_0 = 0.077, T_λ = 13, T_R = 6, W_∞ = 212,6, and b = 3.236, in B_2 (stock under minimum fishing pressure): N = 419, M = 0.5, K = 0.1223, T_0 = 0.095, T_λ = 13, T_R = 6, W_∞ = 298.2, and b = 3.321; in C: N = 328, M = 0.15, K = 0.0223, T_0 = 1.1257, T_λ = 10, T_R = 4, W_∞ = 68261, and b = 3.198; and in D: N = 1242, M = 0.8, K = 0.9436, T_0 = 0.4424, T_λ = 7, T_R = 1, W_∞ = 53,9 and b = 2.987.

showing a 2-year periodicity. In the southern part the fluctuations show no clear periodicity. The percentages of vendace, pike and whitefish are higher in the catch statistics than in the test fishing catch (Table 2).

Stock assessments

The yield per recruit calculations were made for whitefish, vendace, roach and pike (Fig. 1). The parameters used in the calculations are given in figure caption.

The present fishing pressure on whitefish is at a suitable level, and only a small increase in the catch could be achieved by a change in fishing (Fig. 1A).

The fishing pressure on the roach populations in Konnevesi differs between the stocks. Most of them are caught only as a by-catch in gill net fishing for whitefish. The total mortality (Z) in the two stocks studied varied from 1.0 (Fig. $1B_1$) to 0.6 (Fig. $1B_2$). The catch of roach could be raised by lowering the recruiting age and increasing the effort, especially by the former measure. This would need specialized roach fishing with pound nets or seines. At present,

222

the catch of the utilized stock (stock a) is $2.2 \, \text{kg ha}^{-1}$.

The pike taken at present are too young (Fig. 1C). The catch could be increased by raising the recruiting age.

Vendace are recruited to the fishery in Konnevesi in the first autumn of their life. This situation seems to be in good agreement with the maximum sustainable yield (MSY) calculations (Fig. 1D). Fishing of vendace should be started in their first autumn (age 0+), and an increase of effort is also acceptable. However, this recommendation may not be supported by the market demand.

Discussion

The material available at present on the fish stocks of Konnevesi allows us to make recommendations regarding the fishing strategy for four species. The MSY calculations should be made separately for different stocks, but the knowledge of the stocks in Konnevesi is not yet sufficient to allow this. The Beverton & Holt (1957) method is a single species method and thus does not predict the effects that changes in the fishing of one species may have on other fish. The present method also ignores reproduction. This has to be taken into consideration if the recruiting age that gives the maximum yield is much below the age of maturity, as is often the case, when the natural mortality is high. For such stocks consideration of the yield per recruit alone could favour dangerously high levels of exploitation and in these cases, as with vendace, it is particularly important to understand the form of the stock recruitment relationship. A dome-shaped

relationship has been suggested for vendace (Auvinen 1978).

The method used gives us information on the direction in which the fishing of a certain species should be developed. After the recommendations have been given and put into effect, it is most important that the development of the stocks is followed carefully and that new recommendations are made after a certain interval, if this seems necessary.

References

Abramson, N. J., 1971., 1971. Computer programs for fish stock assessment. FAO Fish. Techn. Pap. 101: 1–154.

Auvinen, H., 1978. On the factors affecting the size and fluctuations of vendace (Coregonus albula) stocks. (In Finnish). Kalamies 1978 (9): 3.

Bertalanffy, L. von, 1938. A quantitative theory of organic growth. Human Biol. 10: 181–213.

Beverton, R. J. H. & Holt, S. J., 1957. On the dynamics of exploited fish populations. Great Britain Min. Agric. and Fish., Fish. Invest. Ser. 2, 19. 533 pp.

Granberg, K., 1972. Phytoplankton and primary production of Lake Konnevesi, Central Finland, in 1970. (In Finnish). Suomen kalatalous 46: 11–19.

Hakkari, L., 1972. The rôle of zooplankton as food for vendace (Coregonus albula) in Lake Konnevesi, Central Finland. (In Finnish). Suomen kalatalous 46: 21–28.

Jones, R., 1957. A much simplified version of the fish yield equation. Doc. No. P21 presented at the Lisbon joint meeting of ICNAF/ICES/FAO Lisbon, 27 May – 3 June, 1957. (Mimeo).

Toivonen, J., 1972. The fish fauna of Lake Konnevesi (In Finnish). Suomen kalatalous 46: 39–44.

Tuunainen, P., 1972. On the hydrological, physical and chemical conditions of Lake Konnevesi, Central Finland. (In Finnish). Suomen kalatalous 46: 3–10.